College Algebra Demystified

Demystified Series

Advanced Statistics Demystified
Algebra Demystified
Anatomy Demystified
Astronomy Demystified
Biology Demystified
Business Statistics Demystified
Calculus Demystified
Chemistry Demystified
College Algebra Demystified
Earth Science Demystified
Everyday Math Demystified
Geometry Demystified
Physics Demystified
Physiology Demystified
Pre-Algebra Demystified
Project Management Demystified
Statistics Demystified
Trigonometry Demystified

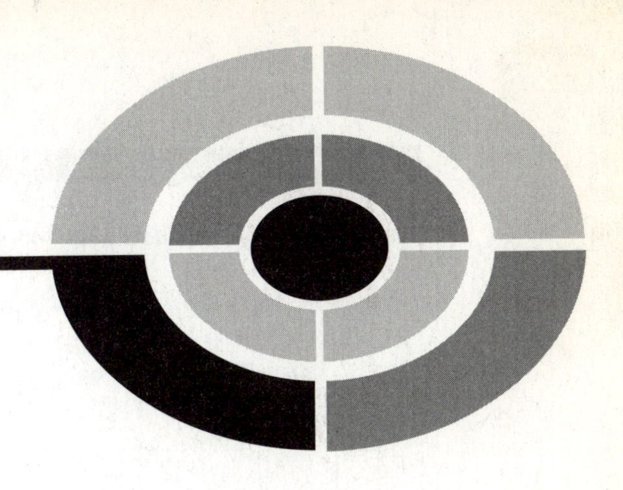

COLLEGE ALGEBRA DEMYSTIFIED

Rhonda Huettenmueller

McGRAW-HILL

New York Chicago San Francisco Lisbon London
Madrid Mexico City Milan New Delhi San Juan
Seoul Singapore Sydney Toronto

The McGraw-Hill Companies

Cataloging-in-Publication Data is on file with the Library of Congress

14 15 16 17 18 DOC/DOC 1 5 4 3 2

ISBN 0-07-143928-5

The sponsoring editor for this book was Judy Bass and the production supervisor was Pamela Pelton. The art director for the cover was Margaret Webster-Shapiro. It was set in Times Roman by Keyword Publishing Services Ltd.

Printed and bound by RR Donnelley.

CONTENTS

PREFACE

Early in my teaching career, I realized two seemingly contradictory facts—that students are fully capable of understanding mathematical concepts but that many have had little success with mathematics. There are several reasons people struggle with mathematics. One is a weak background in basic mathematics. Most topics in mathematics are sequential. Weaknesses in any area will likely cause problems later. Another is that textbooks tend to present too many concepts at once, keeping students from being able to absorb them. I wrote this book (as well as my previous book, *Algebra Demystified*) with these issues in mind. Each section is short, containing *exactly* one new concept. This gives you a chance to absorb the material. Also, I have included detailed examples and solutions so that you can concentrate on the new lesson without being distracted by missing steps. The extra detail will also help you to review important skills.

You will get the most out of this book if you work on it several times a week, a little at a time. Before working on a new section, review the previous sections. Most sections expand on the ideas in previous sections. Study for the end-of-chapter reviews and final exam as you would a regular test. This will help you to see the big picture. Finally, study the graphs and their equations. Even with graphing calculators to plot graphs, it is important in college algebra and more advanced courses to understand why graphs behave the way they do. Because testing has become so important, I would like to leave you with a few tips on how to study for and to take a mathematics test.

- Study at regular, frequent intervals. Do not cram.
- Prepare one sheet of notes as if you were allowed to bring it into the test. This exercise will force you to summarize the concepts and to focus on what is important.

- Imagine explaining the material to someone else. You will have mastered the material only when you can explain it in your own words.
- When taking a test, read it over before answering any questions. Answer the easy questions first. By the time you get to the more difficult problems, your mind will already be thinking mathematically. Also, this can keep you from spending too much valuable test time on harder problems.
- Be patient with yourself while you are learning. Understanding will not come all at once. But it will come.

Acknowledgments

I am very grateful to my family for tolerating my neglect while finishing this book. I also want to express my appreciation to my friends at the University of North Texas for their encouragement. In particular, I want to thank my colleague Mary Ann Teel for her suggestions. Finally, I want to thank my editor Judy Bass for her enthusiasm and support.

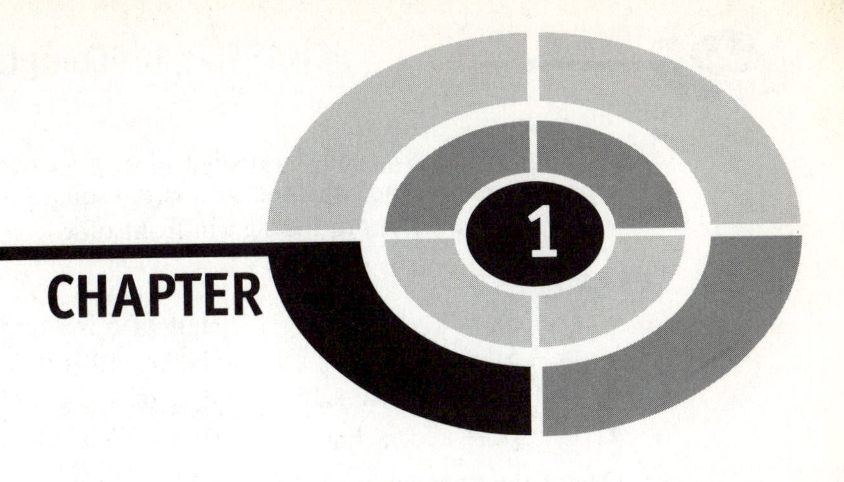

Completing the Square

Quadratic equations (those of the form $ax^2 + bx + c = 0$, where $a \neq 0$) are usually solved by factoring and setting each factor equal to zero or by using the quadratic formula

$$x = \frac{-b \pm \sqrt{b^2 - 4ac}}{2a}.$$

Another method used to solve quadratic equations is called *completing the square*. This method is also useful in graphing circles and parabolas. The goal is to rewrite the quadratic equation in the form "$(x + a)^2$ = number" or "$(x - a)^2$ = number."

To see how we can begin, we will use the FOIL method (**F**irst × first + **O**uter × outer + **I**nner × inner + **L**ast × last) on two perfect squares.

$$(x + a)^2 = (x + a)(x + a) \qquad (x - a)^2 = (x - a)(x - a)$$
$$= x^2 + 2ax + a^2 \qquad = x^2 - 2ax + a^2$$

The constant term is a^2 and the coefficient of x is $2a$ or $-2a$. This means that, in a perfect square, the constant term is the square of half of the coefficient of x: $(2a/2)^2 = a^2$. (Ignore the sign in front of x.)

EXAMPLES

- $(x+3)^2 = x^2 + 6x + 9$ Half of 6 is 3 and 3^2 is 9.
- $(x-5)^2 = x^2 - 10x + 25$ Half of 10 is 5 and 5^2 is 25.
- $(x+4)^2 = x^2 + 8x + 16$ Half of 8 is 4 and 4^2 is 16.
- $(x-\frac{1}{2})^2 = x^2 - x + \frac{1}{4}$ Half of 1 is $\frac{1}{2}$ and $(\frac{1}{2})^2$ is $\frac{1}{4}$.

One of the steps on any completing the square problem is to decide what constant term should be added to the x^2 and x terms to "complete the square." Divide the coefficient of x by 2, then square that number.

EXAMPLES

Fill in the blank with the number that completes the square.

- $x^2 + 12x + \_$ $\frac{12}{2} = 6$ and $6^2 = 36$ $x^2 + 12x + \underline{36}$ is a perfect square.
- $x^2 - 4x + \_$ $\frac{4}{2} = 2$ and $2^2 = 4$ $x^2 - 4x + \underline{4}$ is a perfect square.
- $x^2 + 16x + \_$ $\frac{16}{2} = 8$ and $8^2 = 64$ $x^2 + 16x + \underline{64}$ is a perfect square.
- $x^2 + 2x + \_$ $\frac{2}{2} = 1$ and $1^2 = 1$ $x^2 + 2x + \underline{1}$ is a perfect square.
- $x^2 + \frac{1}{3}x + \_$ $\frac{1}{2}\cdot\frac{1}{3} = \frac{1}{6}$ and $\left(\frac{1}{6}\right)^2 = \frac{1}{36}$ $x^2 + \frac{1}{3}x + \underline{\frac{1}{36}}$ is a perfect square.
- $x^2 - \frac{2}{5}x + \_$ $\frac{1}{2}\cdot\frac{2}{5} = \frac{1}{5}$ and $\left(\frac{1}{5}\right)^2 = \frac{1}{25}$ $x^2 - \frac{2}{5}x + \underline{\frac{1}{25}}$ is a perfect square.

PRACTICE

Fill in the blank with the number that completes the square.

1. $x^2 + 18x + \_$
2. $x^2 + 14x + \_$
3. $x^2 - 22x + \_$
4. $x^2 + 30x + \_$
5. $x^2 - 7x + \_$
6. $x^2 + \frac{1}{4}x + \_$
7. $x^2 + \frac{4}{3}x + \_$

SOLUTIONS

1. $x^2 + 18x + \underline{81}$
2. $x^2 + 14x + \underline{49}$
3. $x^2 - 22x + \underline{121}$
4. $x^2 + 30x + \underline{225}$
5. $x^2 - 7x + \underline{\frac{49}{4}}$
6. $x^2 + \frac{1}{4}x + \underline{\frac{1}{64}}$
7. $x^2 + \frac{4}{3}x + \underline{\frac{4}{9}}$

Another step in completing the square is to rewrite the expression as a perfect square. First we write $(x + \_\_)^2$ if the first sign is a plus sign, and write $(x - \_\_)^2$ if the first sign is a minus sign. Then we can fill in the blank in one of two ways. Divide the coefficient of x by 2 (multiplying by $\frac{1}{2}$ is the same thing) or take the square root of the constant term.

EXAMPLES

- $x^2 + 12x + 36 = (x + \_\_)^2 = (x + 6)^2$ Use 6 in the blank because $6 = \frac{12}{2} = \sqrt{36}$.
- $x^2 - 4x + 4 = (x - \_\_)^2 = (x - 2)^2$ Use 2 in the blank because $2 = \frac{4}{2} = \sqrt{4}$.
- $x^2 + 16x + 64 = (x + \_\_)^2 = (x + 8)^2$ Use 8 in the blank because $8 = \frac{16}{2} = \sqrt{64}$.
- $x^2 + 2x + 1 = (x + \_\_)^2 = (x + 1)^2$ Use 1 in the blank because $1 = \frac{2}{2} = \sqrt{1}$.
- $x^2 + \frac{1}{3}x + \frac{1}{36} = (x + \_\_)^2 = \left(x + \frac{1}{6}\right)^2$ Use $\frac{1}{6}$ in the blank because $\frac{1}{6} = \frac{1}{2} \cdot \frac{1}{3} = \sqrt{\frac{1}{36}}$.
- $x^2 - \frac{2}{5}x + \frac{1}{25} = (x - \_\_)^2 = \left(x - \frac{1}{5}\right)^2$ Use $\frac{1}{5}$ in the blank because $\frac{1}{5} = \frac{1}{2} \cdot \frac{2}{5} = \sqrt{\frac{1}{25}}$.

PRACTICE

Write the quadratic expression as a perfect square. These are the same problems as used in the previous practice problems.

1. $x^2 + 18x + 81 =$
2. $x^2 + 14x + 49 =$
3. $x^2 - 22x + 121 =$
4. $x^2 + 30x + 225 =$

5. $x^2 - 7x + \frac{49}{4} =$
6. $x^2 + \frac{1}{4}x + \frac{1}{64} =$
7. $x^2 + \frac{4}{3}x + \frac{4}{9} =$

SOLUTIONS

1. $x^2 + 18x + 81 = (x + 9)^2$
2. $x^2 + 14x + 49 = (x + 7)^2$
3. $x^2 - 22x + 121 = (x - 11)^2$
4. $x^2 + 30x + 225 = (x + 15)^2$
5. $x^2 - 7x + \frac{49}{4} = (x - \frac{7}{2})^2$
6. $x^2 + \frac{1}{4}x + \frac{1}{64} = (x + \frac{1}{8})^2$ (since $\frac{1}{2} \cdot \frac{1}{4} = \frac{1}{8}$)
7. $x^2 + \frac{4}{3}x + \frac{4}{9} = (x + \frac{2}{3})^2$ (since $\frac{1}{2} \cdot \frac{4}{3} = \frac{2}{3}$)

To solve an equation of the form $(x + a)^2 = \text{number}$ or $(x - a)^2 = \text{number}$, we will take the square root of each side of the equation, then solve for x.

$$\begin{aligned} (x - a)^2 &= \text{number} \\ x - a &= \pm\sqrt{\text{number}} \\ x &= a \pm \sqrt{\text{number}} \end{aligned} \qquad \begin{aligned} (x + a)^2 &= \text{number} \\ x + a &= \pm\sqrt{\text{number}} \\ x &= -a \pm \sqrt{\text{number}} \end{aligned}$$

We need to use the "±" symbol in the second and third steps to get *both* solutions (most quadratic equations have two solutions).

EXAMPLES

- $(x - 1)^2 = 9$

$$\begin{aligned} x - 1 &= \pm\sqrt{9} = \pm 3 \\ x &= 1 \pm 3 = 1 + 3, 1 - 3 \\ x &= 4, -2 \end{aligned}$$

- $(x + \frac{1}{2})^2 = 5$

$$\begin{aligned} x + \frac{1}{2} &= \pm\sqrt{5} \\ x &= -\frac{1}{2} \pm \sqrt{5} \end{aligned}$$

- $(x-6)^2 = 0$

$$x - 6 = \pm\sqrt{0} = 0$$
$$x = 6$$

PRACTICE

Solve for x.

1. $(x-2)^2 = 4$
2. $(x+1)^2 = 25$
3. $(x-4)^2 = 9$
4. $(x+5)^2 = 10$
5. $(x+\frac{1}{3})^2 = 1$
6. $(x-\frac{2}{5})^2 = 0$

SOLUTIONS

1.

$$(x-2)^2 = 4$$
$$x - 2 = \pm 2$$
$$x = 2 \pm 2 = 2+2, 2-2$$
$$x = 4, 0$$

2.

$$(x+1)^2 = 25$$
$$x + 1 = \pm 5$$
$$x = -1 \pm 5 = -1+5, \ -1-5$$
$$x = 4, \ -6$$

3.

$$(x-4)^2 = 9$$
$$x - 4 = \pm 3$$
$$x = 4 \pm 3 = 4+3, 4-3$$
$$x = 7, 1$$

4.

$$(x+5)^2 = 10$$
$$x+5 = \pm\sqrt{10}$$
$$x = -5 \pm \sqrt{10}$$

5.

$$\left(x+\frac{1}{3}\right)^2 = 1$$
$$x+\frac{1}{3} = \pm\sqrt{1} = \pm 1$$
$$x = -\frac{1}{3} \pm 1 = -\frac{1}{3} \pm \frac{3}{3}$$
$$x = \frac{2}{3}, -\frac{4}{3}$$

6.

$$\left(x-\frac{2}{5}\right)^2 = 0$$
$$x-\frac{2}{5} = \pm 0 = 0$$
$$x = \frac{2}{5}$$

Completing the Square To Solve a Quadratic Equation

We can solve a quadratic equation in the form $ax^2 + bx + c = 0$, with $a \neq 0$, by completing the square if we follow the steps below.

1. Move the constant term to the other side of the equation. (Sometimes this step is not necessary.)
2. Divide both sides of the equation by a. (Sometimes this step is not necessary.)
3. Find the constant that would make the left-hand side of the equation a perfect square. (This is what we did in earlier practice problems.) Add this number to both sides of the equation.
4. Rewrite the left-hand side as a perfect square.

5. Take the square root of both sides of the equation. Remember to use a "±" symbol on the right-hand side of the equation.
6. Move the constant to the right-hand side of the equation.
7. Simplify the right-hand side. (Sometimes this step is not necessary.)

EXAMPLES

- $x^2 + 6x - 7 = 0$

$x^2 + 6x = 7$	Step 1
$x^2 + 6x + 9 = 7 + 9$	Step 3
$(x + 3)^2 = 16$	Step 4
$x + 3 = \pm\sqrt{16} = \pm 4$	Step 5
$x = -3 \pm 4 = -3 + 4, \; -3 - 4$	Step 6
$x = 1, \; -7$	Step 7

- $x^2 + 4x = -1$

$x^2 + 4x + 4 = -1 + 4$	Step 3
$(x + 2)^2 = 3$	Step 4
$x + 2 = \pm\sqrt{3}$	Step 5
$x = -2 \pm \sqrt{3}$	Step 6

- $2x^2 - 2x - 24 = 0$

$2x^2 - 2x = 24$	Step 1
$\frac{2}{2}x^2 - \frac{2}{2}x = \frac{24}{2}$	Step 2
$x^2 - x = 12$	
$x^2 - x + \frac{1}{4} = 12 + \frac{1}{4}$	Step 3
$= \frac{48}{4} + \frac{1}{4} = \frac{49}{4}$	
$\left(x - \frac{1}{2}\right)^2 = \frac{49}{4}$	Step 4

$$x - \frac{1}{2} = \pm\sqrt{\frac{49}{4}} \qquad \text{Step 5}$$

$$x - \frac{1}{2} = \pm\frac{7}{2}$$

$$x = \frac{1}{2} \pm \frac{7}{2} \qquad \text{Step 6}$$

$$x = \frac{1}{2} + \frac{7}{2}, \frac{1}{2} - \frac{7}{2}$$

$$x = 4, -3 \qquad \text{Step 7}$$

- $3x^2 + 15x = -4$

$$\frac{3}{3}x^2 + \frac{15}{3}x = -\frac{4}{3} \qquad \text{Step 2}$$

$$x^2 + 5x = \frac{-4}{3}$$

$$x^2 + 5x + \frac{25}{4} = -\frac{4}{3} + \frac{25}{4} \qquad \text{Step 3}$$

$$= \frac{-4}{3} + \frac{25}{4} = -\frac{16}{12} + \frac{75}{12} = \frac{59}{12}$$

$$\left(x + \frac{5}{2}\right)^2 = \frac{59}{12} \qquad \text{Step 4}$$

$$x + \frac{5}{2} = \pm\sqrt{\frac{59}{12}} = \pm\frac{\sqrt{59}}{\sqrt{12}} \qquad \text{Step 5}$$

$$= \pm\frac{\sqrt{59}}{2\sqrt{3}} = \frac{\sqrt{59}\cdot\sqrt{3}}{2\sqrt{3}\cdot\sqrt{3}}$$

$$x + \frac{5}{2} = \frac{\sqrt{177}}{2\cdot 3} = \frac{\sqrt{177}}{6}$$

$$x = -\frac{5}{2} \pm \frac{\sqrt{177}}{6} \qquad \text{Step 6}$$

PRACTICE

Solve for x by completing the square.

1. $x^2 - 10x + 24 = 0$
2. $x^2 + 6x + 5 = 0$
3. $2x^2 - 8x - 24 = 0$

4. $x^2 + 5x + 6 = 0$
5. $x^2 - 3x = 4$
6. $4x^2 + 11x = -6$
7. $x^2 + 7x + 2 = 0$
8. $3x^2 + 9x - 2 = 0$

SOLUTIONS

1.

$$\begin{aligned} x^2 - 10x + 24 &= 0 \\ x^2 - 10x &= -24 \\ x^2 - 10x + 25 &= -24 + 25 \\ (x - 5)^2 &= 1 \\ x - 5 &= \pm 1 \\ x &= 5 \pm 1 = 5 + 1, 5 - 1 \\ x &= 6, 4 \end{aligned}$$

2.

$$\begin{aligned} x^2 + 6x + 5 &= 0 \\ x^2 + 6x &= -5 \\ x^2 + 6x + 9 &= -5 + 9 \\ (x + 3)^2 &= 4 \\ x + 3 &= \pm 2 \\ x &= -3 \pm 2 = -3 + 2, \; -3 - 2 \\ x &= -1, \; -5 \end{aligned}$$

3.

$$\begin{aligned} 2x^2 - 8x - 24 &= 0 \\ 2x^2 - 8x &= 24 \\ \frac{2}{2}x^2 - \frac{8}{2}x &= \frac{24}{2} \\ x^2 - 4x &= 12 \\ x^2 - 4x + 4 &= 12 + 4 \\ (x - 2)^2 &= 16 \\ x - 2 &= \pm 4 \\ x &= 2 \pm 4 = 2 + 4, 2 - 4 \\ x &= 6, \; -2 \end{aligned}$$

4.

$$
\begin{aligned}
x^2 + 5x + 6 &= 0 \\
x^2 + 5x &= -6 \\
x^2 + 5x + \frac{25}{4} &= -6 + \frac{25}{4} = -\frac{24}{4} + \frac{25}{4} \\
\left(x + \frac{5}{2}\right)^2 &= \frac{1}{4} \\
x + \frac{5}{2} &= \pm\sqrt{\frac{1}{4}} = \pm\frac{1}{2} \\
x &= -\frac{5}{2} \pm \frac{1}{2} = -\frac{5}{2} + \frac{1}{2}, \ -\frac{5}{2} - \frac{1}{2} \\
x &= -\frac{4}{2}, \ -\frac{6}{2} = -2, \ -3
\end{aligned}
$$

5.

$$
\begin{aligned}
x^2 - 3x &= 4 \\
x^2 - 3x + \frac{9}{4} &= 4 + \frac{9}{4} = \frac{16}{4} + \frac{9}{4} \\
\left(x - \frac{3}{2}\right)^2 &= \frac{25}{4} \\
x - \frac{3}{2} &= \pm\sqrt{\frac{25}{4}} = \pm\frac{5}{2} \\
x &= \frac{3}{2} \pm \frac{5}{2} = \frac{3}{2} + \frac{5}{2}, \frac{3}{2} - \frac{5}{2} \\
x &= \frac{8}{2}, \ -\frac{2}{2} = 4, \ -1
\end{aligned}
$$

6.

$$
\begin{aligned}
4x^2 + 11x &= -6 \\
\frac{4}{4}x^2 + \frac{11}{4}x &= -\frac{6}{4} = -\frac{3}{2} \\
x^2 + \frac{11}{4}x + \frac{121}{64} &= -\frac{3}{2} + \frac{121}{64} \quad \left(\text{since } \left(\frac{1}{2} \cdot \frac{11}{4}\right)^2 = \frac{121}{64}\right)
\end{aligned}
$$

$$\left(x+\frac{11}{8}\right)^2 = -\frac{96}{64}+\frac{121}{64}$$

$$= \frac{25}{64}$$

$$x+\frac{11}{8} = \pm\sqrt{\frac{25}{64}} = \pm\frac{5}{8}$$

$$x = -\frac{11}{8}\pm\frac{5}{8} = -\frac{11}{8}+\frac{5}{8}, -\frac{11}{8}-\frac{5}{8}$$

$$x = -\frac{6}{8}, -\frac{16}{8} = -\frac{3}{4}, -2$$

7.

$$x^2+7x+2=0$$

$$x^2+7x=-2$$

$$x^2+7x+\frac{49}{4} = -2+\frac{49}{4} = -\frac{8}{4}+\frac{49}{4}$$

$$\left(x+\frac{7}{2}\right)^2 = \frac{41}{4}$$

$$x+\frac{7}{2} = \pm\sqrt{\frac{41}{4}} = \pm\frac{\sqrt{41}}{\sqrt{4}} = \pm\frac{\sqrt{41}}{2}$$

$$x = -\frac{7}{2}\pm\frac{\sqrt{41}}{2} \quad \text{or} \quad \frac{-7\pm\sqrt{41}}{2}$$

8.

$$3x^2+9x-2=0$$

$$3x^2+9x=2$$

$$\frac{3}{3}x^2+\frac{9}{3}x=\frac{2}{3}$$

$$x^2+3x=\frac{2}{3}$$

$$x^2+3x+\frac{9}{4} = \frac{2}{3}+\frac{9}{4} = \frac{8}{12}+\frac{27}{12}$$

$$\left(x+\frac{3}{2}\right)^2=\frac{35}{12}$$

$$x+\frac{3}{2}=\pm\sqrt{\frac{35}{12}}=\pm\frac{\sqrt{35}}{\sqrt{12}}=\pm\frac{\sqrt{35}}{\sqrt{4}\sqrt{3}}$$

$$x+\frac{3}{2}=\pm\frac{\sqrt{35}}{2\sqrt{3}}\cdot\frac{\sqrt{3}}{\sqrt{3}}=\pm\frac{\sqrt{105}}{2\cdot 3}=\pm\frac{\sqrt{105}}{6}$$

$$x=-\frac{3}{2}\pm\frac{\sqrt{105}}{6}$$

Not every quadratic equation has real number solutions. For example, $(x-1)^2=-10$ has no real number solutions. This is because $\sqrt{-10}$ is not a real number. The equation does have two complex number solutions, though.

Now that we are experienced at solving quadratic equations by completing the square, we can see why the quadratic formula works. The quadratic formula comes from solving $ax^2+bx+c=0$ for x by completing the square.

$$ax^2+bx+c=0$$

$$ax^2+bx=-c \qquad \text{Step 1}$$

$$\frac{a}{a}x^2+\frac{b}{a}x=-\frac{c}{a} \qquad \text{Step 2}$$

$$x^2+\frac{b}{a}x+\frac{b^2}{4a^2}=-\frac{c}{a}+\frac{b^2}{4a^2} \qquad \text{Step 3}$$

$$x^2+\frac{b}{a}x+\frac{b^2}{4a^2}=-\frac{c}{a}\cdot\frac{4a}{4a}+\frac{b^2}{4a^2} \qquad \text{Simplify}$$

$$x^2+\frac{b}{a}x+\frac{b^2}{4a^2}=\frac{-4ac+b^2}{4a^2} \qquad \text{Simplify}$$

$$\left(x+\frac{b}{2a}\right)^2=\frac{b^2-4ac}{4a^2} \qquad \text{Step 4}$$

$$x+\frac{b}{2a}=\pm\sqrt{\frac{b^2-4ac}{4a^2}} \qquad \text{Step 5}$$

$$x+\frac{b}{2a}=\pm\frac{\sqrt{b^2-4ac}}{2a} \qquad \text{Simplify}$$

$$x = -\frac{b}{2a} \pm \frac{\sqrt{b^2 - 4ac}}{2a} \qquad \text{Step 6}$$

$$x = \frac{-b \pm \sqrt{b^2 - 4ac}}{2a} \qquad \text{Step 7}$$

Chapter 1 Review

1. What number completes the square for $x^2 - 8x$?
   a) 4 b) -4 c) 16 d) -16
2. $x^2 + 5x + \frac{25}{4} =$
   a) $(x + \frac{5}{2})^2$ b) $(x + \frac{5}{4})^2$ c) $(x + \frac{25}{2})^2$ d) $(x + \frac{25}{4})^2$
3. What are the solutions for $(x + 1)^2 = 9$?
   a) $x = 2$ and $x = -4$ b) $x = 2$ and $x = 4$ c) $x = 8$ and $x = 10$
   d) $x = -8$ and $x = -10$
4. What number completes the square for $x^2 + \frac{2}{3}x$?
   a) $\frac{1}{9}$ b) $\frac{4}{9}$ c) $\frac{1}{3}$ d) $\frac{1}{6}$
5. $x^2 + \frac{1}{4}x - 2 = 0$ is equivalent to
   a) $(x + \frac{1}{16})^2 = \frac{17}{8}$ b) $(x + \frac{1}{16})^2 = \frac{33}{16}$ c) $(x + \frac{1}{8})^2 = \frac{129}{64}$
   d) $(x + \frac{1}{8})^2 = \frac{17}{8}$
6. What are the solutions for $(x - 3)^2 = 12$?
   a) $x = -3 \pm 2\sqrt{3}$ b) $x = 3 \pm 2\sqrt{3}$ c) $x = -3 \pm 3\sqrt{2}$
   d) $x = 3 \pm 3\sqrt{2}$
7. $3x^2 - 6x - 2 = 0$ is equivalent to
   a) $(x - 3)^2 = 11$ b) $(x - 3)^2 = 7$ c) $(x - 1)^2 = 3$
   d) $(x - 1)^2 = \frac{5}{3}$

SOLUTIONS

1. c) 2. a) 3. a) 4. a) 5. c) 6. b) 7. d)

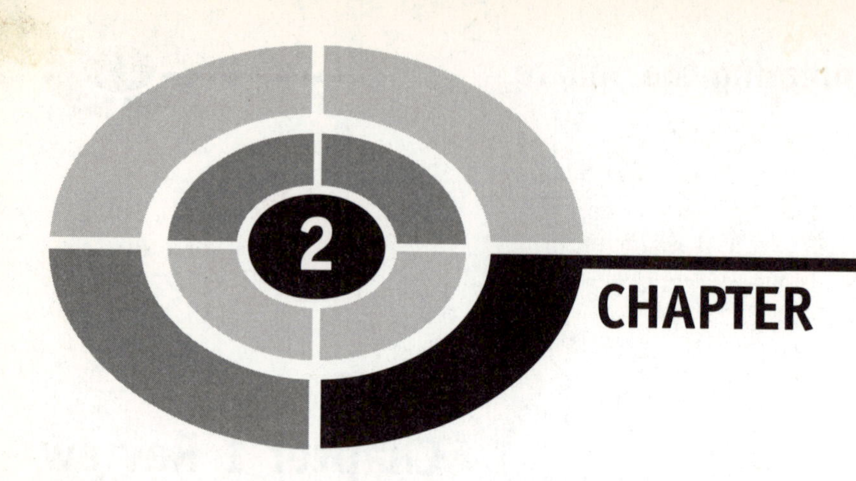

Absolute Value Equations and Inequalities

The *absolute value* of a number is its distance from 0 on the number line. Because distances are not negative, the absolute value of a number is never negative. The symbol for the absolute value is a pair of absolute value bars, "$|\ |$." Hence $|-3| = 3$ because -3 is 3 units away from 0 on the number line. A number written without absolute value bars gives both the distance from 0 as well as the direction. For example, -3 is 3 units to the *left* of 0 and 3 is 3 units to the *right* of 0, but $|-3| = 3$ simply means 3 units *away* from 0. Because 0 is no distance from 0, $|0| = 0$.

EXAMPLES

- $|100| = 100$
- $|-\frac{5}{2}| = \frac{5}{2}$
- $|10 - 1| = 9$
- $|-83| = 83$
- $|5 - 11| = 6$
- $|68 - 90| = 22$

PRACTICE

1. $|-6.75| =$
2. $|8| =$
3. $|-4| =$
4. $|8 - 19| =$
5. $|13 - 25| =$

SOLUTIONS

1. $|-6.75| = 6.75$
2. $|8| = 8$
3. $|-4| = 4$
4. $|8 - 19| = 11$
5. $|13 - 25| = 12$

Technically $\sqrt{x^2} = |x|$ instead of x, unless we have reason to believe x is not negative. For example, suppose $x = -4$. Then $\sqrt{x^2} = \sqrt{(-4)^2} = \sqrt{16} = 4$, not -4. But $|-4| = 4$, so $\sqrt{(-4)^2} = |-4|$ is a true statement. For any even number n and any real number a, $\sqrt[n]{a^n} = |a|$.

Absolute Value Equations

The equation $|x| = 5$ is really the question, "What numbers are 5 units away from 0?" Two numbers are 5 units from 0, 5 and -5, so there are two solutions, $x = 5$ and $x = -5$. Absolute value equations often have two solutions. One can solve an equation of the type |expression| = positive number by solving the two equations: expression = negative number and expression = positive number. Equations such as $|x| = -6$ have no solution because no number has a negative distance from 0. However, $-|x| = -6$, which is equivalent to $|x| = 6$, does have solutions.

EXAMPLES

- $|x| = 16$

  The solutions are $x = 16, -16$.

- $|x+3|=5$

$$x+3=-5 \qquad x+3=5$$
$$x=-8 \qquad x=2$$

The solutions are $x=-8$ and $x=2$.

- $|6-8x|=0$
Because 0 and -0 are the same number, there is only one equation to solve.

$$6-8x=0$$
$$-8x=-6$$
$$x=\frac{-6}{-8}=\frac{3}{4}$$

PRACTICE

Solve for x.

1. $|-5x+1|=6$
2. $|\frac{3}{4}x-8|=1$
3. $|x+\frac{1}{2}|=\frac{2}{3}$
4. $-|3x+4|=-2$
5. $|2x-9|=0$
6. $\left|\frac{3x-2}{5}\right|=7$

SOLUTIONS

1.

$$-5x+1=6 \qquad -5x+1=-6$$
$$-5x=5 \qquad -5x=-7$$
$$x=-1 \qquad x=\frac{7}{5}$$

2.

$$\frac{3}{4}x-8=1 \qquad \frac{3}{4}x-8=-1$$
$$\frac{3}{4}x=9 \qquad \frac{3}{4}x=7$$
$$x=\frac{4}{3}\cdot 9=12 \qquad x=\frac{4}{3}\cdot 7=\frac{28}{3}$$

3.

$$x + \frac{1}{2} = \frac{2}{3} \qquad x + \frac{1}{2} = -\frac{2}{3}$$

$$x = -\frac{1}{2} + \frac{2}{3} \qquad x = -\frac{1}{2} - \frac{2}{3}$$

$$x = -\frac{3}{6} + \frac{4}{6} = \frac{1}{6} \qquad x = -\frac{3}{6} - \frac{4}{6} = -\frac{7}{6}$$

4. $-|3x + 4| = -2$ becomes $|3x + 4| = 2$

$$3x + 4 = 2 \qquad 3x + 4 = -2$$

$$3x = -2 \qquad 3x = -6$$

$$x = -\frac{2}{3} \qquad x = -2$$

5.

$$2x - 9 = 0$$

$$2x = 9$$

$$x = \frac{9}{2}$$

6.

$$\frac{3x - 2}{5} = 7 \qquad \frac{3x - 2}{5} = -7$$

$$5 \cdot \frac{3x - 2}{5} = 5(7) \qquad 5 \cdot \frac{3x - 2}{5} = 5(-7)$$

$$3x - 2 = 35 \qquad 3x - 2 = -35$$

$$3x = 37 \qquad 3x = -33$$

$$x = \frac{37}{3} \qquad x = -\frac{33}{3} = -11$$

Sometimes an absolute value expression is part of a more complex equation. We need to isolate the absolute value expression on one side of the equation, then we can solve it as before.

EXAMPLES

- $2|x-4|+7=13$

$$\begin{aligned} 2|x-4|+7&=13 \\ -7\quad&\;\;-7 \end{aligned}$$

$$2|x-4|=6$$

$$\frac{2|x-4|}{2}=\frac{6}{2}$$

$$|x-4|=3$$

$$x-4=3 \qquad x-4=-3$$

$$x=7 \qquad x=1$$

- $\frac{1}{3}|2x+8|-10=-2$

$$+10 \quad +10$$

$$\frac{1}{3}|2x+8|=8$$

$$3\left(\frac{1}{3}|2x+8|\right)=3(8)$$

$$|2x+8|=24$$

$$2x+8=24 \qquad 2x+8=-24$$

$$2x=16 \qquad 2x=-32$$

$$x=8 \qquad x=-16$$

PRACTICE

Solve for x.

1. $4|5x-2|+3=11$
2. $3-2|x-9|=1$
3. $\frac{2}{5}|4x-3|-9=-1$

SOLUTIONS

1.

$$4|5x-2|+3=11$$

$$-3 \quad -3$$

$$4|5x-2| = 8$$

$$\frac{4|5x-2|}{4} = \frac{8}{4}$$

$$|5x-2| = 2$$

$$5x - 2 = 2 \qquad 5x - 2 = -2$$

$$5x = 4 \qquad 5x = 0$$

$$x = \frac{4}{5} \qquad x = \frac{0}{5} = 0$$

2.

$$3 - 2|x-9| = 1$$

$$-3 \qquad\qquad -3$$

$$-2|x-9| = -2$$

$$\frac{-2|x-9|}{-2} = \frac{-2}{-2}$$

$$|x-9| = 1$$

$$x - 9 = 1 \qquad x - 9 = -1$$

$$x = 10 \qquad x = 8$$

3.

$$\frac{2}{5}|4x-3| - 9 = -1$$

$$+9 \quad +9$$

$$\frac{2}{5}|4x-3| = 8$$

$$\frac{5}{2}\cdot\frac{2}{5}|4x-3| = \frac{5}{2}\cdot 8$$

$$|4x-3| = 20$$

$$4x - 3 = 20 \qquad 4x - 3 = -20$$
$$4x = 23 \qquad 4x = -17$$
$$x = \frac{23}{4} \qquad x = -\frac{17}{4}$$

Absolute Value Inequalities

The inequality $|x| < 4$ is, in mathematical symbols, the question,"What real numbers are closer to 0 than 4 is?" A look at the number line might help with this question.

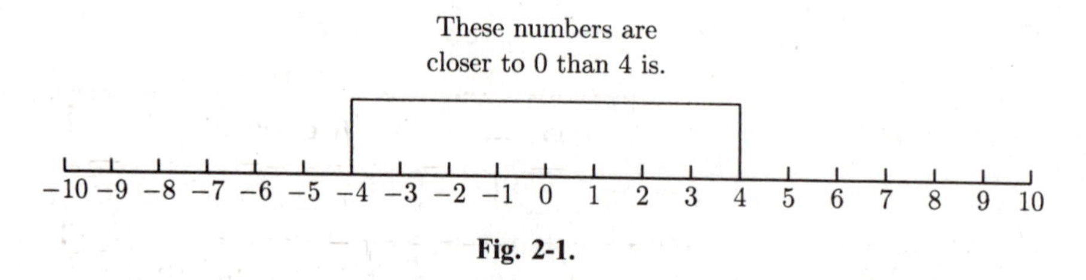

Fig. 2-1.

From the number line we can see that the numbers between -4 and 4 have an absolute value less than 4. The solution to $|x| < 4$ is the interval $(-4, 4)$. In inequality notation, the solution is $-4 < x < 4$. (A double inequality of the form smaller number $< x <$ larger number is shorthand for $x >$ smaller number *and* $x <$ larger number.)

Similarly, the solution to the inequality $|x| > 3$ is all numbers further from 0 than 3 is.

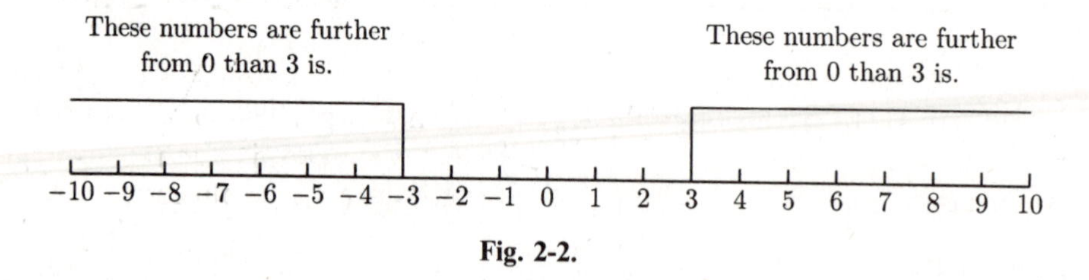

Fig. 2-2.

The solution is all numbers smaller than -3 or larger than 3. In interval notation, the solution is $(-\infty, -3) \cup (3, \infty)$. The "$\cup$" symbol means "or." In inequality notation, the solution is $x < -3$ *or* $x > 3$. The notation "$3 < x < -3$" has no meaning because no number x is *both* larger than 3 and smaller than -3.

Absolute value	Inequalities	Interval(s)
$\lvert x\rvert <$ positive number	neg. number $< x <$ pos. number	(neg. no., pos. no.)
$\lvert x\rvert \leq$ positive number	neg. number $\leq x \leq$ pos. number	[neg. no., pos. no.]
$\lvert x\rvert >$ positive number	$x <$ neg. number or $x >$ pos. number	$(-\infty,$ neg. no.) $\cup$ (pos. no., $\infty)$
$\lvert x\rvert \geq$ positive number	$x \leq$ neg. number or $x \geq$ pos. number	$(-\infty,$ neg. no.] $\cup$ [pos. no., $\infty)$

EXAMPLES

Absolute value	Inequalities	Interval(s)
$\lvert x\rvert < 1$	$-1 < x < 1$	$(-1, 1)$
$\lvert x\rvert > 16$	$x < -16$ or $x > 16$	$(-\infty, -16) \cup (16, \infty)$
$\lvert x\rvert \geq 3$	$x \leq -3$ or $x \geq 3$	$(-\infty, -3] \cup [3, \infty)$
$\lvert x\rvert \leq 5$	$-5 \leq x \leq 5$	$[-5, 5]$
$8 \geq \lvert x\rvert$ is equivalent to $\lvert x\rvert \leq 8$	$-8 \leq x \leq 8$	$[-8, 8]$
$21 < \lvert x\rvert$ is equivalent to $\lvert x\rvert > 21$	$x < -21$ or $x > 21$	$(-\infty, -21) \cup (21, \infty)$

Some absolute value inequalities, like absolute value equations, have no solution: $|x| < -6$. Because absolute values are not negative, no number has an absolute value smaller than -6. If we switch the inequality sign, $|x| > -6$, then we get an inequality for which *every* real number is a solution.

PRACTICE

Solve the inequality and give the solution in inequality and interval notation.

1. $|x| > 12$
2. $|x| \leq 9$
3. $|x| < 10$
4. $|x| \geq 25$

SOLUTIONS

1. $|x| > 12$ $\quad x < -12$ or $x > 12$ $\quad (-\infty, -12) \cup (12, \infty)$
2. $|x| \leq 9$ $\quad -9 \leq x \leq 9$ $\quad [-9, 9]$
3. $|x| < 10$ $\quad -10 < x < 10$ $\quad (-10, 10)$
4. $|x| \geq 25$ $\quad x \leq -25$ or $x \geq 25$ $\quad (-\infty, -25] \cup [25, \infty)$

For some absolute value inequalities, finding the inequalities is only the first step toward finding the solution.

EXAMPLES

- $|4x - 5| \geq 9$

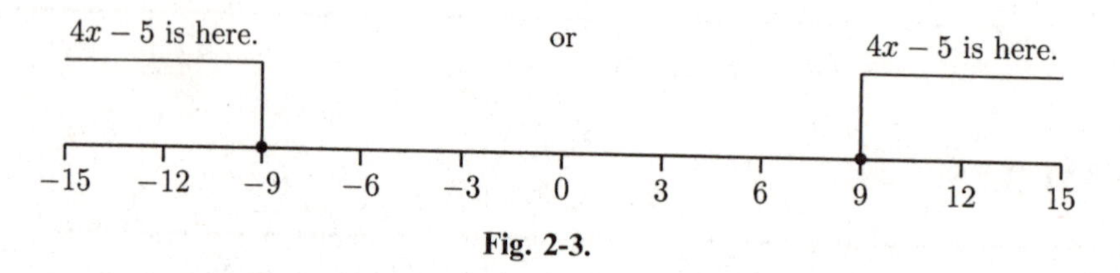

Fig. 2-3.

From the number line, we can see that either $4x - 5 \leq -9$ or $4x - 5 \geq 9$. These are the inequalities we need to solve.

$$\begin{aligned} 4x - 5 &\leq -9 & \qquad 4x - 5 &\geq 9 \\ 4x &\leq -4 & 4x &\geq 14 \\ x &\leq -1 & x &\geq \frac{14}{4} = \frac{7}{2} \end{aligned}$$

The solution in interval notation is $(-\infty, -1] \cup \left[\frac{7}{2}, \infty\right)$.

- $|2x + 5| < 11$

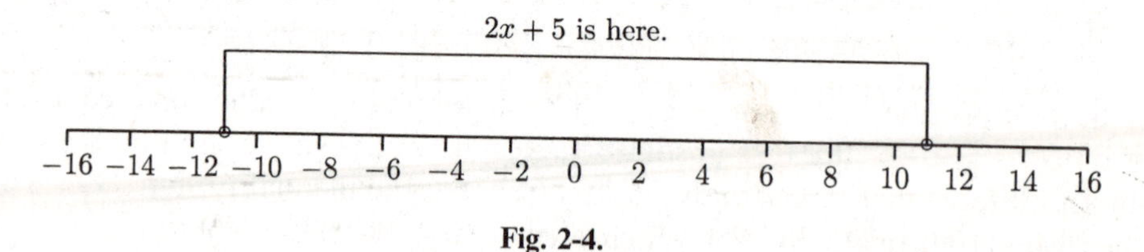

Fig. 2-4.

From the number line, we can see that $2x + 5$ is between -11 and 11. This means that $-11 < 2x + 5 < 11$.

$$\begin{aligned} -11 &< 2x + 5 < 11 \\ -16 &< 2x < 6 \\ -8 &< x < 3 \qquad (-8, 3) \end{aligned}$$

- $|9 - 3x| < 12$

$$-12 < 9 - 3x < 12$$

$$-21 < -3x < 3$$

$$\frac{-21}{-3} > \frac{-3}{-3}x > \frac{3}{-3} \qquad \text{Reverse the signs at this step.}$$

$$7 > x > -1 \quad \text{or} \quad -1 < x < 7 \qquad (-1, 7)$$

- $|(2 - 3x)/4| > 1$

$$\frac{2-3x}{4} < -1 \qquad \frac{2-3x}{4} > 1$$

$$2 - 3x < -4 \qquad 2 - 3x > 4$$

$$-3x < -6 \qquad -3x > 2$$

$$x > 2 \qquad x < -\frac{2}{3} \qquad \left(-\infty, -\frac{2}{3}\right) \cup (2, \infty)$$

Tables 2-1 and 2-2 should help to set up the inequalities for an absolute value inequality.

Table 2-1

Absolute value inequality	**Solve these inequalities**
$\lvert\text{Expression}\rvert >$ pos. number	Expression $<$ neg. number or Expression $>$ pos. number
$\lvert\text{Expression}\rvert \geq$ pos. number	Expression $\leq$ neg. number or Expression $\geq$ pos. number
$\lvert\text{Expression}\rvert <$ pos. number	neg. number $<$ Expression $<$ pos. number
$\lvert\text{Expression}\rvert \leq$ pos. number	neg. number $\leq$ Expression $\leq$ pos. number

Table 2-2

Absolute value inequality	**Interval notation**
$\lvert\text{algebraic expression}\rvert >$ positive number	$(-\infty, a) \cup (b, \infty)$
$\lvert\text{algebraic expression}\rvert \geq$ positive number	$(-\infty, a] \cup [b, \infty)$
$\lvert\text{algebraic expression}\rvert <$ positive number	(a, b)
$\lvert\text{algebraic expression}\rvert \leq$ positive number	$[a, b]$

PRACTICE

Solve the inequality and give the solution in interval notation.

1. $|3x+4| < 5$
2. $|x-2| > 4$
3. $|6-2x| \le 4$
4. $\left|\dfrac{x-4}{3}\right| < 2$
5. $|8-3x| \ge 5$
6. $|\frac{1}{3}x+2| < 4$

SOLUTIONS

1.

$$-5 < 3x+4 < 5$$
$$-9 < 3x < 1$$
$$-3 < x < \frac{1}{3} \quad \left(-3, \frac{1}{3}\right)$$

2.

$$x-2 < -4 \qquad x-2 > 4$$
$$x < -2 \qquad x > 6 \quad (-\infty, -2) \cup (6, \infty)$$

3.

$$-4 \le 6-2x \le 4$$
$$-10 \le -2x \le -2$$
$$\frac{-10}{-2} \ge \frac{-2}{-2}x \ge \frac{-2}{-2}$$
$$5 \ge x \ge 1$$
$$1 \le x \le 5 \quad [1, 5]$$

4.

$$-2 < \frac{x-4}{3} < 2$$
$$3(-2) < 3\left(\frac{x-4}{3}\right) < 3(2)$$
$$-6 < x-4 < 6$$
$$-2 < x < 10 \quad (-2, 10)$$

5.

$$8 - 3x \le -5 \qquad 8 - 3x \ge 5$$

$$-3x \le -13 \qquad -3x \ge -3$$

$$\frac{-3}{-3}x \ge \frac{-13}{-3} \qquad \frac{-3}{-3}x \le \frac{-3}{-3}$$

$$x \ge \frac{13}{3} \qquad x \le 1 \quad (-\infty, 1] \cup \left[\frac{13}{3}, \infty\right)$$

6.

$$-4 < \frac{1}{3}x + 2 < 4$$

$$-6 < \frac{1}{3}x < 2$$

$$3(-6) < 3\left(\frac{1}{3}x\right) < 3(2)$$

$$-18 < x < 6 \quad (-18, 6)$$

Sometimes absolute value expressions are part of more complicated inequalities. As before, we will isolate the absolute value expression on one side of the inequality, then solve the inequality.

EXAMPLES

- $3|x + 4| - 7 \ge 5$

$$3|x + 4| \ge 12$$

$$|x + 4| \ge 4$$

$$x + 4 \le -4 \qquad x + 4 \ge 4$$

$$x \le -8 \qquad x \ge 0 \quad (-\infty, -8] \cup [0, \infty)$$

- $8 - 4\left|\frac{1}{2}x - 1\right| \ge 3$

$$-4\left|\frac{1}{2}x - 1\right| \ge -5$$

$$\frac{-4|(1/2)x - 1|}{-4} \le \frac{-5}{-4}$$

$$\left|\frac{1}{2}x - 1\right| \le \frac{5}{4}$$

$$-\frac{5}{4} \le \frac{1}{2}x - 1 \le \frac{5}{4}$$

$$-\frac{1}{4} \le \frac{1}{2}x \le \frac{9}{4}$$

$$2\left(-\frac{1}{4}\right) \le 2\left(\frac{1}{2}x\right) \le 2\left(\frac{9}{4}\right)$$

$$-\frac{1}{2} \le x \le \frac{9}{2} \quad \left[-\frac{1}{2}, \frac{9}{2}\right]$$

PRACTICE

Solve the inequality and give the solution in interval notation.

1. $\frac{1}{3}|x+4| - 5 < 2$
2. $4 + 2|3x+5| \ge 6$
3. $9 + |6 - 2x| \le 11$
4. $-3|7x+4| - 4 < -13$

SOLUTIONS

1.

$$\frac{1}{3}|x+4| - 5 < 2$$

$$\frac{1}{3}|x+4| < 7$$

$$3\left(\frac{1}{3}|x+4|\right) < 3(7)$$

$$|x+4| < 21$$

$$-21 < x+4 < 21$$

$$-25 < x < 17 \quad (-25, 17)$$

2.

$$4 + 2|3x+5| \ge 6$$

$$2|3x+5| \ge 2$$

$$\frac{2|3x+5|}{2} \ge \frac{2}{2}$$

$$|3x+5| \ge 1$$

$$3x+5 \le -1 \qquad 3x+5 \ge 1$$
$$3x \le -6 \qquad 3x \ge -4$$
$$x \le -2 \qquad x \ge -\frac{4}{3} \quad (-\infty, -2] \cup \left[-\frac{4}{3}, \infty\right)$$

3.

$$9+|6-2x| \le 11$$
$$|6-2x| \le 2$$

$$-2 \le 6-2x \le 2$$
$$-8 \le -2x \le -4$$
$$\frac{-8}{-2} \ge \frac{-2}{-2}x \ge \frac{-4}{-2}$$
$$4 \ge x \ge 2 \quad \text{or} \quad 2 \le x \le 4 \quad [2, 4]$$

4.

$$-3|7x+4|-4 < -13$$
$$-3|7x+4| < -9$$
$$\frac{-3|7x+4|}{-3} > \frac{-9}{-3}$$
$$|7x+4| > 3$$

$$7x+4 < -3 \qquad 7x+4 > 3$$
$$7x < -7 \qquad 7x > -1$$
$$x < -1 \qquad x > -\frac{1}{7} \quad (-\infty, -1) \cup \left(-\frac{1}{7}, \infty\right)$$

Chapter 2 Review

1. What is the solution for $|x-7| = 1$?
   a) $x = 8$ b) $x = 6$ and $x = 8$ c) $x = 8$ and $x = -8$ d) $x = 6$
2. What is the solution for $|x| > 4$?
   a) $(4, -4)$ b) $(-4, 4)$ c) $(4, \infty)$ d) $(-\infty, -4) \cup (4, \infty)$

3. Solve for x: $\frac{1}{2}|x-6|-4=-1$.
   a) $x=12$ b) $x=8$ and $x=4$ c) $x=12$ and $x=0$ d) There is no solution

4. What is the solution for $|2x-3| \leq 1$?
   a) $(-\infty, 2]$ b) $(-\infty, 1] \cup [2, \infty)$ c) $(-\infty, 1] \cup [-\infty, 2]$ d) $[1, 2]$

5. What is the solution for $|-\frac{1}{2}x+5| > 1$?
   a) $(-\infty, 8) \cup (12, \infty)$ b) $(8, 12)$ c) $(-\infty, 12) \cup (8, \infty)$ d) $(12, 8)$

6. Solve for x: $|3x-1|=2$.
   a) $x=1$ b) $x=1$ and $x=-\frac{1}{3}$ c) $x=1$ and $x=-1$ d) $x=1$ and $x=\frac{1}{3}$

SOLUTIONS

1. b) 2. d) 3. c) 4. d) 5. a) 6. b)

CHAPTER 3

The *xy* Coordinate Plane

The *xy* coordinate plane (or *plane*) is made from two number lines. The vertical number line is called the *y*-axis, and the horizontal number line is called the *x*-axis. The number lines cross at 0. This point is called the *origin*. Points on the plane can be located and identified by *coordinates*: (x, y). The first number is called the *x-coordinate*. This number describes how far left or right to go from the origin to locate the point. A negative number tells us that we need to move to the left, and a positive number tells us that we need to move to the right. The second number is called the *y-coordinate*. This number describes how far up or down to go from the origin to locate the point. A negative number tells us that we need to move down, and a positive number tells us that we need to move up.

(+right, +up)   (−left, +up)   (+right, −down)   (−left, −down)

EXAMPLES

- (4, 1) Right 4, up 1
- (−1, −5) Left 1, down 5
- (0, 2) No horizontal movement, up 2
- (−2, 5) Left 2, up 5
- (5, −3) Right 5, down 3
- (−3, 0) Left 3, no vertical movement

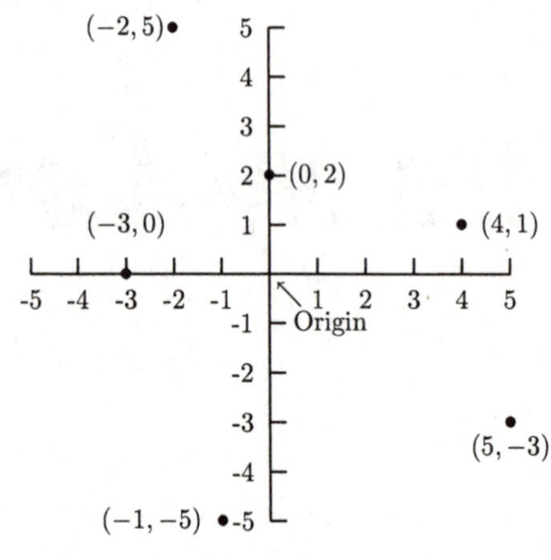

Fig. 3-1.

PRACTICE

Describe the horizontal and vertical movement and locate the point on the plane.

1. (4, 5)
2. (−1, −3)
3. (1, −4)
4. (2, 2)
5. (−6, 4)
6. (0, 3)
7. (−4, 0)
8. (0, 0)

SOLUTIONS

1. (4, 5) Right 4, up 5

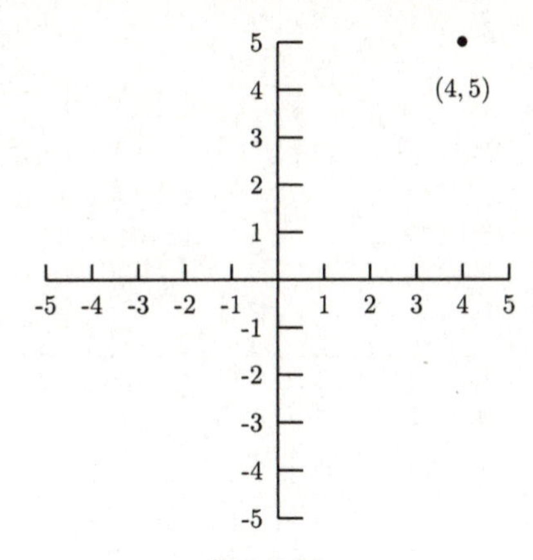

Fig. 3-2.

2. $(-1, -3)$ Left 1, down 3

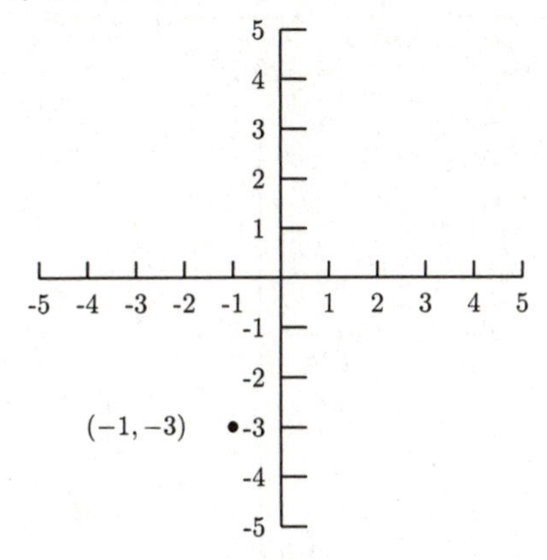

Fig. 3-3.

3. $(1, -4)$ Right 1, down 4

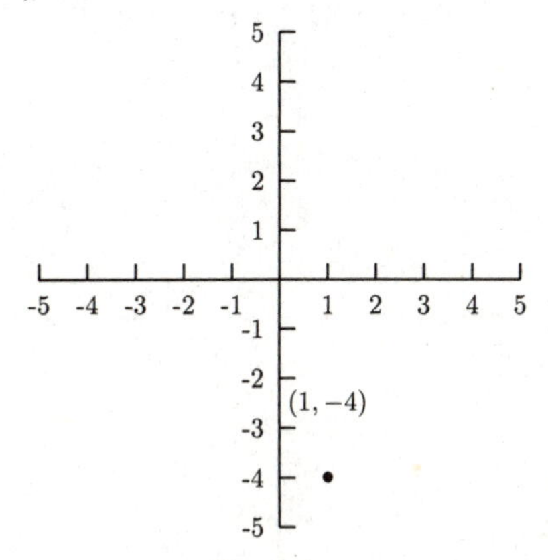

Fig. 3-4.

4. $(2, 2)$ Right 2, up 2

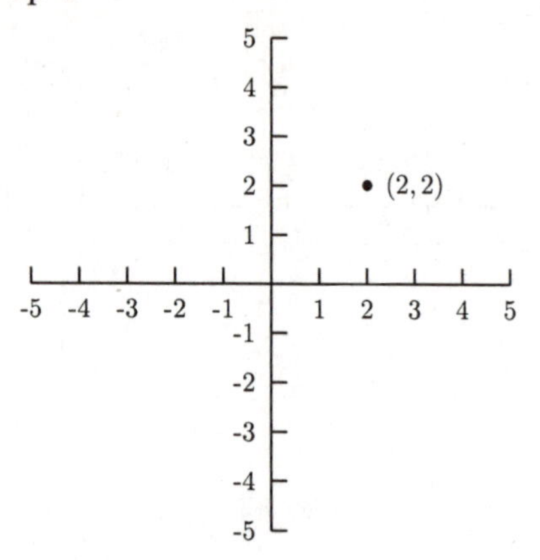

Fig. 3-5.

5. $(-6, 4)$ Left 6, up 4

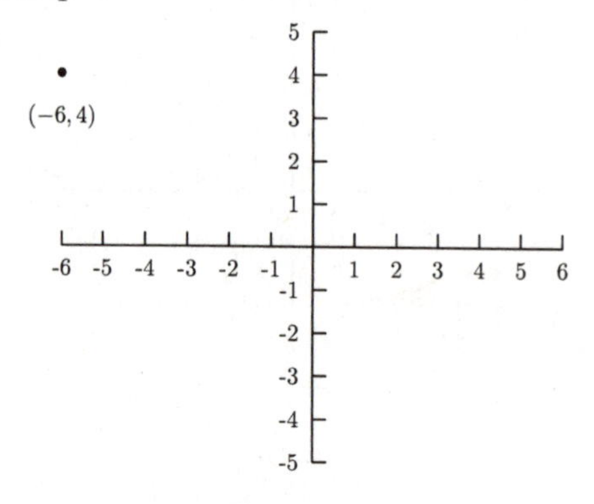

Fig. 3-6.

6. $(0, 3)$ No horizontal movement, up 3

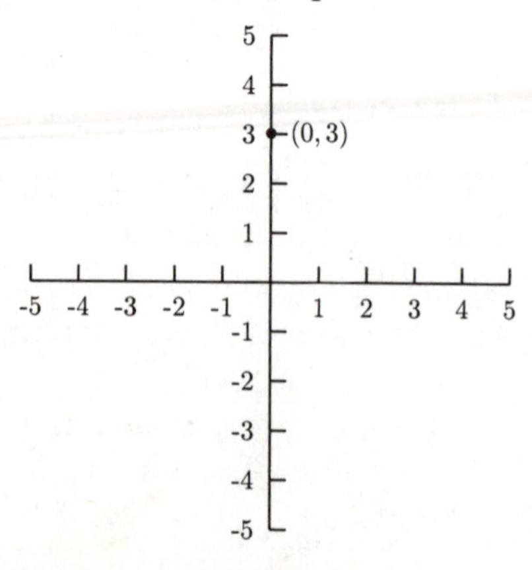

Fig. 3-7.

7. $(-4, 0)$ Left 4, no vertical movement

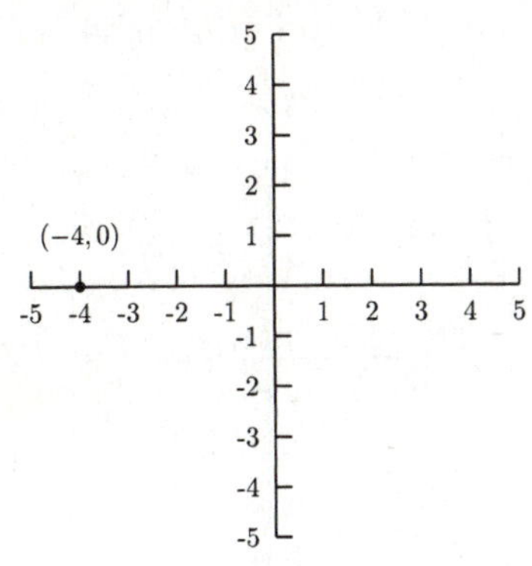

Fig. 3-8.

8. $(0, 0)$ No horizontal movement, no vertical movement. These are the coordinates of the origin.

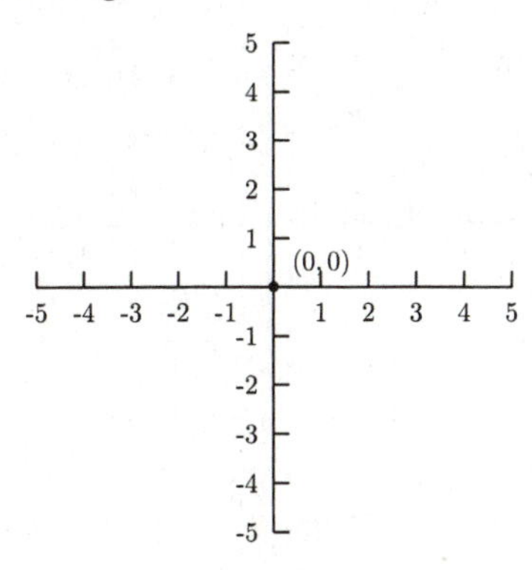

Fig. 3-9.

The Distance Between Two Points

At times we need to find the distance between two points. If the points are on the same vertical line (the x-coordinates are the same), the distance between the points is the *absolute value* of the difference between the y-coordinates. If the points are on the same horizontal line (the y-coordinates are the same), the distance between the points is the absolute value of the difference between the x-coordinates.

EXAMPLES

- The distance between $(1, 4)$ and $(1, 2)$ is $|4 - 2| = |2| = 2$.

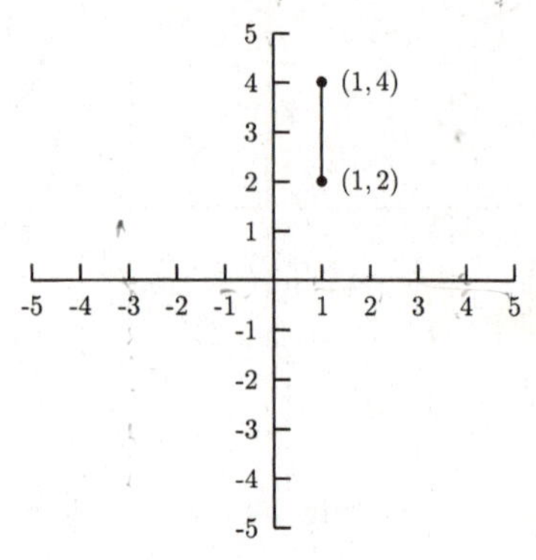

Fig. 3-10.

- The distance between $(2, 3)$ and $(2, -4)$ is $|-4 - 3| = |-7| = 7$.

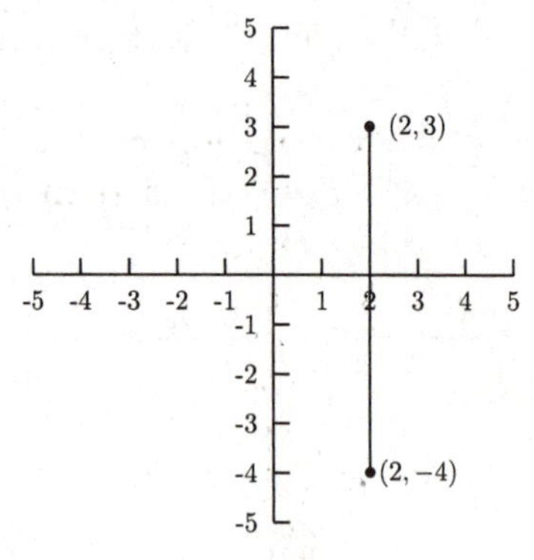

Fig. 3-11.

- The distance between $(-5, 3)$ and $(-1, 3)$ is $|-5 - (-1)| = |-5 + 1| = |-4| = 4$.

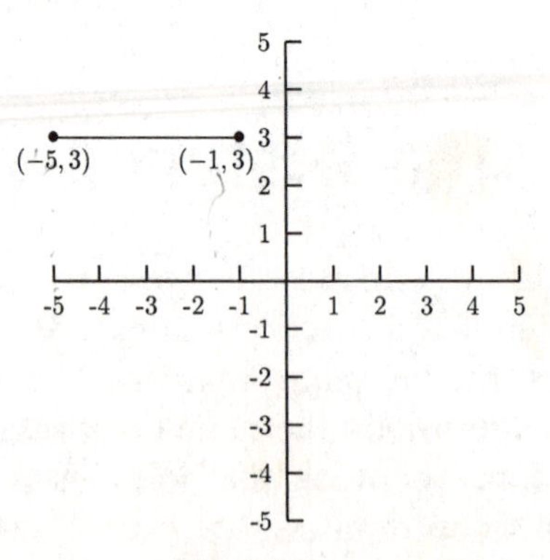

Fig. 3-12.

PRACTICE

Find the distance between the two points.

1. $(5, 6)$ and $(1, 6)$
2. $(4, 3)$ and $(4, 8)$
3. $(-1, 2)$ and $(-1, 9)$
4. $(-3, 1)$ and $(-3, -4)$
5. $(6, 0)$ and $(2, 0)$

SOLUTIONS

1. $|5 - 1| = |4| = 4$
2. $|3 - 8| = |-5| = 5$
3. $|2 - 9| = |-7| = 7$
4. $|1 - (-4)| = |1 + 4| = |5| = 5$
5. $|6 - 2| = |4| = 4$

In the rest of this chapter, we will be working with formulas involving two points. We will call one of them Point 1 with coordinates (x_1, y_1) and the other Point 2 with coordinates (x_2, y_2).

Suppose we have two points that are not on the same vertical line or the same horizontal line. By using Pythagoras' theorem in a clever way, we can find the distance between *any* two points.

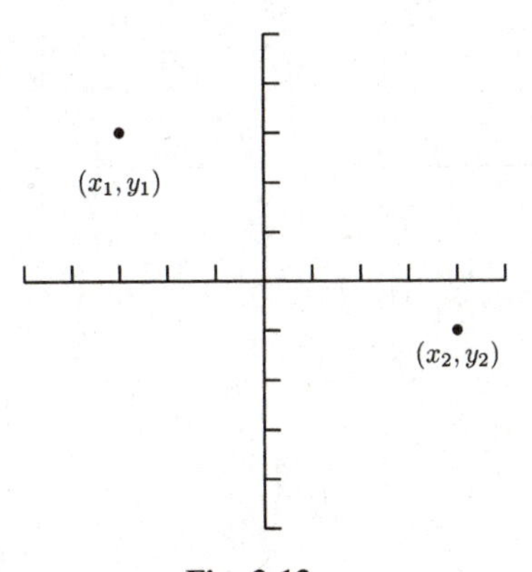

Fig. 3-13.

Draw a vertical line through one of the points and a horizontal line through the other. The point where these lines cross will have the x-coordinate of one of the points and the y-coordinate of the other.

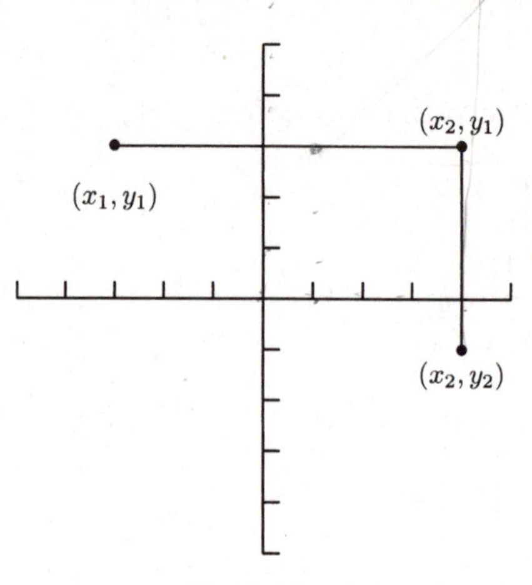

Fig. 3-14.

The three points form a right triangle. The length of the hypotenuse of this triangle is the distance between (x_1, y_1) and (x_2, y_2). We can find this length using Pythagoras' theorem: $a^2 + b^2 = c^2$.

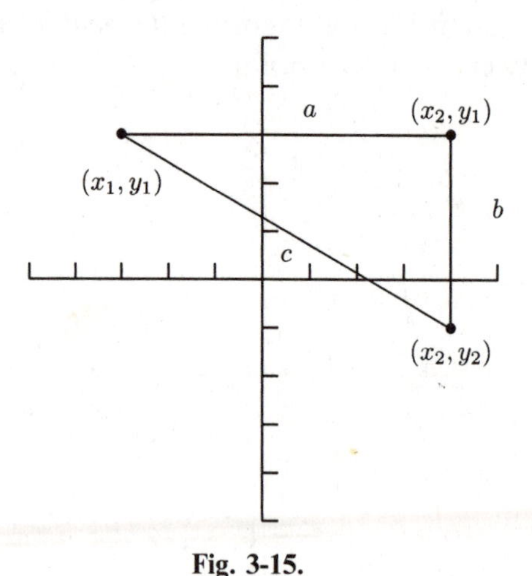

Fig. 3-15.

$a =$ distance between (x_1, y_1) and $(x_2, y_1) = |x_2 - x_1|$; $a^2 = |x_2 - x_1|^2 = (x_2 - x_1)^2$; $b =$ distance between (x_2, y_1) and $(x_2, y_2) = |y_2 - y_1|$; $b^2 = |y_2 - y_1|^2 = (y_2 - y_1)^2$; $c =$ distance between (x_1, y_1) and (x_2, y_2); $c^2 = a^2 + b^2$ or $c^2 = (x_2 - x_1)^2 + (y_2 - y_1)^2$. This means $c = \sqrt{(x_2 - x_1)^2 + (y_2 - y_1)^2}$. The formula, then, for the distance between two points (x_1, y_1) and (x_2, y_2) is $d = \sqrt{(x_2 - x_1)^2 + (y_2 - y_1)^2}$.

EXAMPLES

Use the distance formula to find the distance between the two points.

- $(1,3)$ and $(4,7)$ $x_1 = 1,\ y_1 = 3,\ x_2 = 4,\ y_2 = 7.$

$$d = \sqrt{(4-1)^2 + (7-3)^2} = \sqrt{3^2 + 4^2} = \sqrt{25} = 5$$

- $(-2,5)$ and $(1,4)$ $x_1 = -2,\ y_1 = 5,\ x_2 = 1,\ y_2 = 4$

$$d = \sqrt{(1-(-2))^2 + (4-5)^2} = \sqrt{(1+2)^2 + (4-5)^2}$$

$$= \sqrt{3^2 + (-1)^2} = \sqrt{10}$$

- $(0,-7)$ and $(-2,4)$ $x_1 = 0,\ y_1 = -7,\ x_2 = -2,\ y_2 = 4$

$$d = \sqrt{(-2-0)^2 + (4-(-7))^2} = \sqrt{(-2)^2 + (4+7)^2}$$

$$= \sqrt{125} = 5\sqrt{5}$$

- $(\frac{1}{2},4)$ and $(2,\frac{1}{3})$ $x_1 = \frac{1}{2},\ y_1 = 4,\ x_2 = 2,\ y_2 = \frac{1}{3}$

$$d = \sqrt{\left(2-\frac{1}{2}\right)^2 + \left(\frac{1}{3}-4\right)^2} = \sqrt{\left(\frac{4}{2}-\frac{1}{2}\right)^2 + \left(\frac{1}{3}-\frac{12}{3}\right)^2}$$

$$= \sqrt{\left(\frac{3}{2}\right)^2 + \left(-\frac{11}{3}\right)^2} = \sqrt{\frac{9}{4} + \frac{121}{9}}$$

$$= \sqrt{\frac{81}{36} + \frac{484}{36}} = \sqrt{\frac{565}{36}}$$

$$= \frac{\sqrt{565}}{\sqrt{36}} = \frac{\sqrt{565}}{6}$$

This formula even works for two points on the same horizontal or vertical line. For example, we know the distance between $(3,8)$ and $(3,6)$ is 2. Let us see what happens in the formula.

$$d = \sqrt{(3-3)^2 + (6-8)^2} = \sqrt{0^2 + (-2)^2} = \sqrt{4} = 2$$

PRACTICE

Find the distance between the points.

1. $(-1,4)$, $(3,3)$
2. $(6,-4)$, $(-2,-5)$

3. $(0, 8)$, $(2, 1)$
4. $(7, -3)$, $(5, -3)$
5. $(2, \frac{2}{3})$, $(\frac{1}{5}, \frac{1}{4})$

SOLUTIONS

1.
$$d = \sqrt{(3-(-1))^2 + (3-4)^2} = \sqrt{(3+1)^2 + (-1)^2} = \sqrt{17}$$

2.
$$d = \sqrt{(-2-6)^2 + (-5-(-4))^2} = \sqrt{(-8)^2 + (-5+4)^2} = \sqrt{65}$$

3.
$$d = \sqrt{(2-0)^2 + (1-8)^2} = \sqrt{2^2 + (-7)^2} = \sqrt{53}$$

4.
$$d = \sqrt{(5-7)^2 + (-3-(-3))^2} = \sqrt{(-2)^2 + (-3+3)^2} = 2$$

5.
$$d = \sqrt{\left(\frac{1}{5} - 2\right)^2 + \left(\frac{1}{4} - \frac{2}{3}\right)^2} = \sqrt{\left(\frac{1}{5} - \frac{10}{5}\right)^2 + \left(\frac{3}{12} - \frac{8}{12}\right)^2}$$
$$= \sqrt{\left(-\frac{9}{5}\right)^2 + \left(-\frac{5}{12}\right)^2} = \sqrt{\frac{81}{25} + \frac{25}{144}}$$
$$= \sqrt{\frac{11{,}664}{3600} + \frac{625}{3600}} = \sqrt{\frac{12{,}289}{3600}} = \frac{\sqrt{12{,}289}}{\sqrt{3600}}$$
$$= \frac{\sqrt{12{,}289}}{60}$$

Sometimes we are asked to show that groups of points form shapes such as squares and right triangles. We can use the distance formula to show that the distance between the vertices of a square are equal or that the distances between the vertices of a right triangle follow Pythagoras' theorem.

EXAMPLES

- Show that the points $(-2, \frac{13}{2})$, $(1, 2)$, and $(4, 4)$ are the vertices of a right triangle.

  To use the distance formula on this problem, we need to show that if we square then add the lengths of the two legs (the sides that are not the hypotenuse), this will equal the square of the hypotenuse. While not really necessary, we should plot the points to see which two sides are the legs and which side is the hypotenuse.

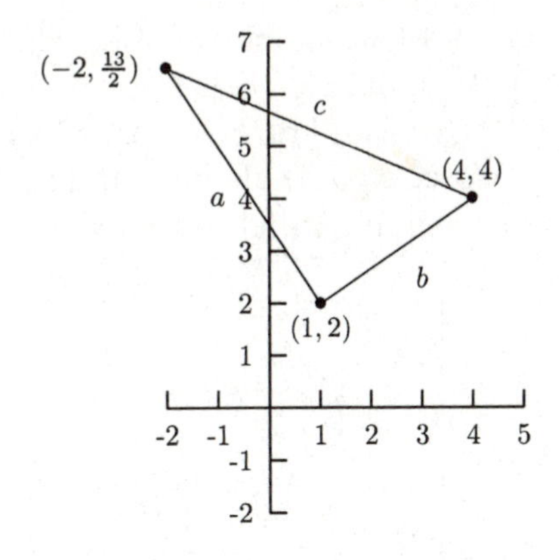

Fig. 3-16.

From the graph we can see that a is the distance from $(-2, \frac{13}{2})$ to $(1, 2)$; b is the distance from $(1, 2)$ to $(4, 4)$; and c is the distance from $(-2, \frac{13}{2})$ to $(4, 4)$.

$$a = \sqrt{(1-(-2))^2 + \left(2 - \frac{13}{2}\right)^2} = \sqrt{9 + \frac{81}{4}} = \sqrt{\frac{117}{4}}$$

$$b = \sqrt{(4-1)^2 + (4-2)^2} = \sqrt{9+4} = \sqrt{13}$$

$$c = \sqrt{(4-(-2))^2 + \left(4 - \frac{13}{2}\right)^2} = \sqrt{36 + \frac{25}{4}} = \sqrt{\frac{169}{4}}$$

$$a^2 = \left(\sqrt{\frac{117}{4}}\right)^2 = \frac{117}{4}$$

$$b^2 = \left(\sqrt{13}\right)^2 = 13$$

$$c^2 = \left(\sqrt{\frac{169}{4}}\right)^2 = \frac{169}{4}$$

Is it true that $a^2 + b^2 = c^2$?

$$a^2 + b^2 = \frac{117}{4} + 13 = \frac{117}{4} + \frac{52}{4} = \frac{169}{4} = c^2$$

Because $a^2 + b^2 = c^2$ is a true statement, $(-2, \frac{13}{2})$, $(1, 2)$, and $(4, 4)$ are the vertices of a right triangle.

- Show that $(2, 5)$, $(6, 3)$, and $(2, 1)$ are the vertices of an isosceles triangle.
  Two sides of an isosceles triangle have the same length. If we plot the points, we should be able to tell which sides have equal length.

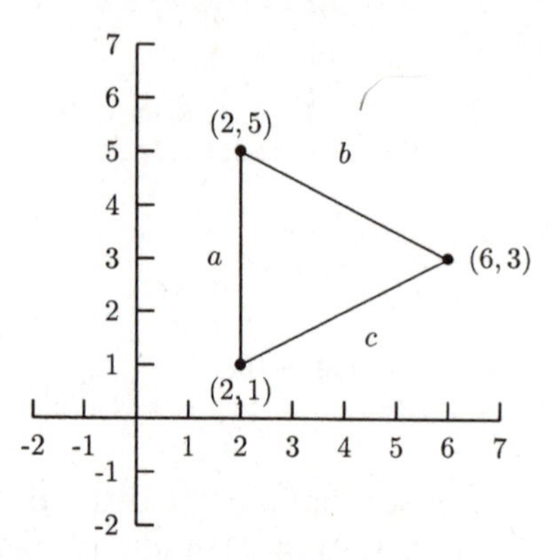

Fig. 3-17.

It appears that sides b and c have equal length, where $b =$ the distance between $(2, 5)$ and $(6, 3)$ and $c =$ the distance between $(2, 1)$ and $(6, 3)$.

$$b = \sqrt{(6-2)^2 + (3-5)^2} = \sqrt{16+4} = \sqrt{20}$$

$$c = \sqrt{(6-2)^2 + (3-1)^2} = \sqrt{16+4} = \sqrt{20}$$

Because $b = c$, the points $(2, 5)$, $(6, 3)$, and $(2, 1)$ are the vertices of an isosceles triangle.

PRACTICE

1. Show that $(3, 1)$, $(10, 0)$, and $(9, 7)$ are the vertices of an isosceles triangle.
2. Show that $(-3, 2)$, $(4, 0)$, and $(-1, 9)$ are the vertices of a right triangle.
3. Show that $(-21, 4)$, $(-4, -3)$, $(-9, 9)$, and $(-16, -8)$ are the vertices of a square.

SOLUTIONS

1. It appears that sides b and c are equal.

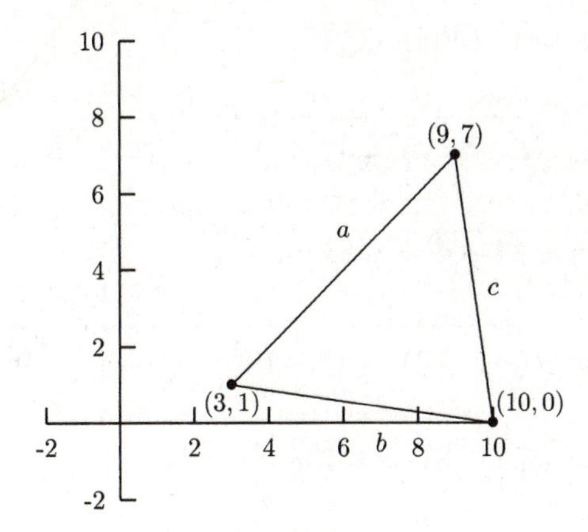

Fig. 3-18.

$$b = \sqrt{(10-3)^2 + (0-1)^2} = \sqrt{49+1} = \sqrt{50}$$

$$c = \sqrt{(9-10)^2 + (7-0)^2} = \sqrt{1+49} = \sqrt{50}$$

Because $b = c$, $(3, 1)$, $(10, 0)$, and $(9, 7)$ are the vertices of an isosceles triangle.

2. We want to show that $a^2 + b^2 = c^2$.

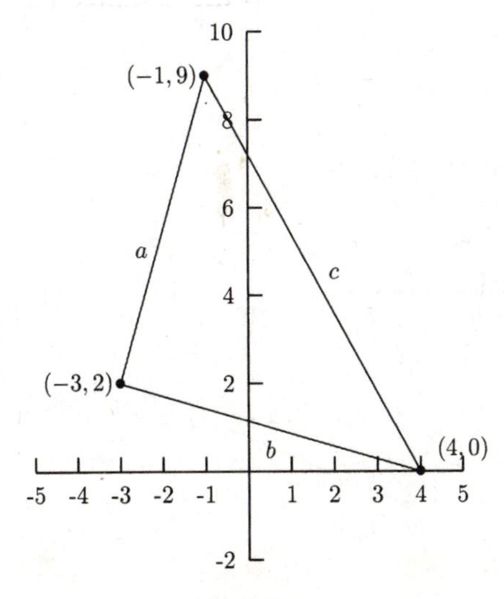

Fig. 3-19.

$$a = \sqrt{(-1-(-3))^2 + (9-2)^2} = \sqrt{(-1+3)^2 + 7^2}$$

$$= \sqrt{4+49} = \sqrt{53}$$

$$b = \sqrt{(-3-4)^2 + (2-0)^2} = \sqrt{(-7)^2 + 2^2}$$

$$= \sqrt{49+4} = \sqrt{53}$$

$$c = \sqrt{(-1-4)^2 + (9-0)^2} = \sqrt{(-5)^2 + 9^2}$$

$$= \sqrt{25+81} = \sqrt{106}$$

$$a^2 + b^2 = \left(\sqrt{53}\right)^2 + \left(\sqrt{53}\right)^2$$

$$= 106 = \left(\sqrt{106}\right)^2 = c^2$$

This means that $(-3, 2)$, $(4, 0)$, and $(-1, 9)$ are the vertices of a right triangle.

3. We want to show that $a = b = c = d$.

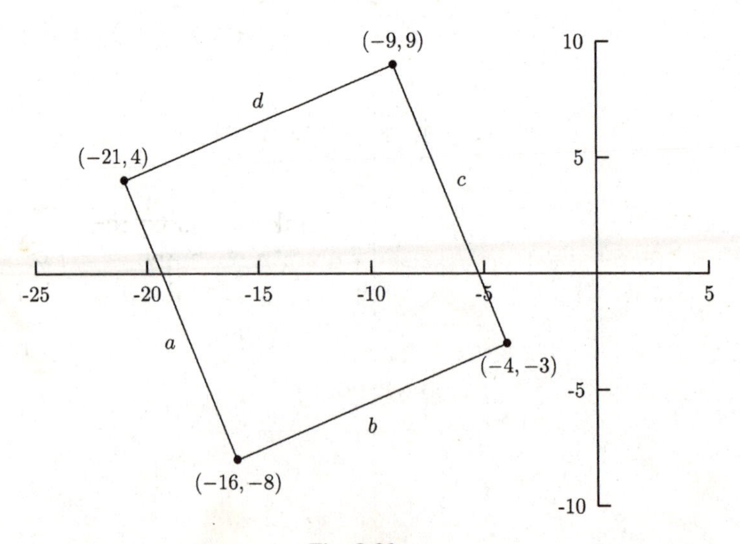

Fig. 3-20.

$$a = \sqrt{(-16 - (-21))^2 + (-8 - 4)^2}$$

$$= \sqrt{(-16 + 21)^2 + (-12)^2} = \sqrt{25 + 144} = 13$$

$$b = \sqrt{(-4 - (-16))^2 + (-3 - (-8))^2}$$

$$= \sqrt{(-4 + 16)^2 + (-3 + 8)^2} = \sqrt{144 + 25} = 13$$

$$c = \sqrt{(-9 - (-4))^2 + (9 - (-3))^2}$$

$$= \sqrt{(-9 + 4)^2 + (9 + 3)^2} = \sqrt{25 + 144} = 13$$

$$d = \sqrt{(-21 - (-9))^2 + (4 - 9)^2}$$

$$= \sqrt{(-21 + 9)^2 + (-5)^2} = \sqrt{144 + 25} = 13$$

Because $a = b = c = d$, $(-21, 4)$, $(-4, -3)$, $(-9, 9)$ and $(-16, -8)$ are the vertices of a square.

The Midpoint Formula

To find the *midpoint* between two points (x_1, y_1) and (x_2, y_2) use the midpoint formula:

$$\left(\frac{x_1 + x_2}{2}, \frac{y_1 + y_2}{2}\right).$$

This formula is easy to remember if we think of finding the *average* of the x-values and the *average* of the y-values. As with the distance formula, it does not matter which point is called (x_1, y_1) and which is called (x_2, y_2).

EXAMPLES

Find the midpoint between the given points.

- $(1, 3)$ and $(4, 7)$

$$\left(\frac{1+4}{2}, \frac{3+7}{2}\right) = \left(\frac{5}{2}, 5\right)$$

- $(-2, 5)$ and $(1, 4)$

$$\left(\frac{-2+1}{2}, \frac{5+4}{2}\right) = \left(-\frac{1}{2}, \frac{9}{2}\right)$$

- $(0, -7)$ and $(-2, 4)$

$$\left(\frac{0+(-2)}{2}, \frac{-7+4}{2}\right) = \left(-1, -\frac{3}{2}\right)$$

- $(\frac{1}{2}, 4)$ and $(2, \frac{1}{3})$

$$\left(\frac{(1/2)+2}{2}, \frac{4+(1/3)}{2}\right) = \left(\frac{5/2}{2}, \frac{13/3}{2}\right)$$
$$= \left(\frac{5}{2} \div 2, \frac{13}{3} \div 2\right) = \left(\frac{5}{2} \cdot \frac{1}{2}, \frac{13}{3} \cdot \frac{1}{2}\right)$$
$$= \left(\frac{5}{4}, \frac{13}{6}\right)$$

PRACTICE

Find the midpoint between the points.

1. $(3, 5)$ and $(1, 2)$
2. $(-1, 4)$ and $(3, 3)$
3. $(6, -4)$ and $(-2, -5)$
4. $(0, 8)$ and $(2, 1)$
5. $(7, -3)$ and $(5, -3)$

SOLUTIONS

1.

$$\left(\frac{3+1}{2}, \frac{5+2}{2}\right) = \left(2, \frac{7}{2}\right)$$

2.

$$\left(\frac{-1+3}{2}, \frac{4+3}{2}\right) = \left(1, \frac{7}{2}\right)$$

3.

$$\left(\frac{6+(-2)}{2}, \frac{-4+(-5)}{2}\right) = \left(2, -\frac{9}{2}\right)$$

4.

$$\left(\frac{0+2}{2}, \frac{8+1}{2}\right) = \left(1, \frac{9}{2}\right)$$

5.

$$\left(\frac{7+5}{2}, \frac{-3+(-3)}{2}\right) = (6, -3)$$

Circles

An equation with two variables can be *graphed* on the *xy*-plane. Think of a graph as a "picture" of all solutions to the equation. Every point on the graph is a solution to the equation and every solution to the equation is a point on the graph.

For example, in the equation $x + y = 5$, *any* pair of numbers whose sum is 5 will be on the graph of the equation. Some of those pairs of numbers are $(0, 5)$, $(1, 4)$, $(3, 2)$, $(4, 1)$, $(5, 0)$, $(6, -1)$, $(7, -2)$, $(-4, 9)$, $(10, -5)$, $(4\frac{1}{2}, \frac{1}{2})$, and $(-\frac{1}{2}, 5\frac{1}{2})$. Now let us plot them.

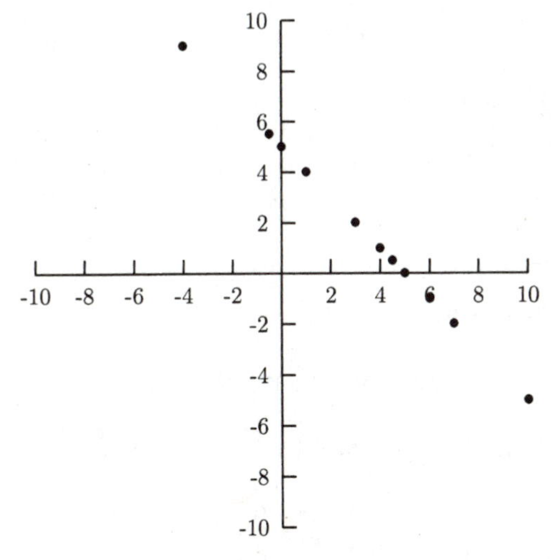

Fig. 3-21.

All of these points lie on a line. For this reason, equations like $x + y = 5$ are called *linear equations*. If we were to draw a line through these points, the sum of the coordinates for every point on the line would be 5.

Graphing equations will be a major part of this book. We will learn how to choose points to plot so that we get a good idea of what an equation's graph looks like with a minimum amount of work.

Every point on a circle is the same distance from the center of the circle. That distance is called the *radius*. This fact along with the distance formula will allow us to discover a formula for the circle in the xy-plane. Call the center of the circle (h, k). That is, the x-coordinate of the circle is h and the y-coordinate is k. Call the radius of the circle r. A point (x, y) is on the circle if its distance from (h, k) is r. When we put this information in the distance formula, it becomes

$$\sqrt{(x-h)^2 + (y-k)^2} = r.$$

If we square both sides of this equation, we get

$$(x-h)^2 + (y-k)^2 = r^2.$$

This is the formula for a circle in the xy-plane with radius r and center (h, k). When we are given the center and radius of a circle, we only need to put these three numbers, h, k, r, in the formula to get an equation for that circle.

EXAMPLES

Find an equation of the circle with the given radius and center.

- Center $(1, 4)$, radius 3.
  Here $h = 1$, $k = 4$, $r = 3$, $r^2 = 9$ and $(x-h)^2 + (y-k)^2 = r^2$ becomes

$$(x-1)^2 + (y-4)^2 = 9$$

- Center $(0, 9)$, radius 4.
  $h = 0$, $k = 9$, $r = 4$, $r^2 = 16$

$$(x-0)^2 + (y-9)^2 = 16$$

  We want to simplify $(x-0)^2$ to x^2.

$$x^2 + (y-9)^2 = 16$$

- Center $(3, 2)$, radius $\frac{1}{2}$.
  $h = 3$, $k = 2$, $r = \frac{1}{2}$, $r^2 = \frac{1}{4}$

$$(x-3)^2 + (y-2)^2 = \frac{1}{4}$$

- Center $(-2, 1)$, radius $r = 6$
  $h = -2,\ k = 1,\ r = 6,\ r^2 = 36$

$$(x - (-2))^2 + (y - 1)^2 = 36$$

We want to simplify $(x - (-2))^2$ to $(x + 2)^2$.

$$(x + 2)^2 + (y - 1)^2 = 36$$

PRACTICE

Find an equation of the circle with the given center and radius.

1. Center $(5, 3)$, radius 2
2. Center $(4, 1)$, radius 7
3. Center $(3, 6)$, radius 1
4. Center $(0, 2)$, radius 3
5. Center $(0, 0)$, radius 5
6. Center $(4, -1)$, radius 2
7. Center $(-5, -2)$, radius 8

SOLUTIONS

1. $h = 5,\ k = 3,\ r = 2,\ r^2 = 4$

$$(x - 5)^2 + (y - 3)^2 = 4$$

2. $h = 4,\ k = 1,\ r = 7,\ r^2 = 49$

$$(x - 4)^2 + (y - 1)^2 = 49$$

3. $h = 3,\ k = 6,\ r = 1,\ r^2 = 1$

$$(x - 3)^2 + (y - 6)^2 = 1$$

4. $h = 0,\ k = 2,\ r = 3,\ r^2 = 9$

$$(x - 0)^2 + (y - 2)^2 = 9$$
$$x^2 + (y - 2)^2 = 9$$

5. $h = 0,\ k = 0,\ r = 5,\ r^2 = 25$

$$(x - 0)^2 + (y - 0)^2 = 25$$
$$x^2 + y^2 = 25$$

6. $h = 4,\ k = -1,\ r = 2,\ r^2 = 4$

$$(x - 4)^2 + (y - (-1))^2 = 4$$
$$(x - 4)^2 + (y + 1)^2 = 4$$

7. $h = -5,\ k = -2,\ r = 8,\ r^2 = 64$

$$(x - (-5))^2 + (y - (-2))^2 = 64$$
$$(x + 5)^2 + (y + 2)^2 = 64$$

When the equation of a circle is in the form $(x - h)^2 + (y - k)^2 = r^2$, we have a good idea of what it looks like. If h, k, and r are integers we can even graph the circle with practically no work. We can mark the center and go up, down, left, and right r units to get four points on the circle. Next draw a circle through these four points. Then we will erase the mark for the center because the center is not really on the circle.

EXAMPLE
Consider the equation $(x - 2)^2 + (y + 1)^2 = 4$. The center of the circle is at $(2, -1)$ and the radius is 2.

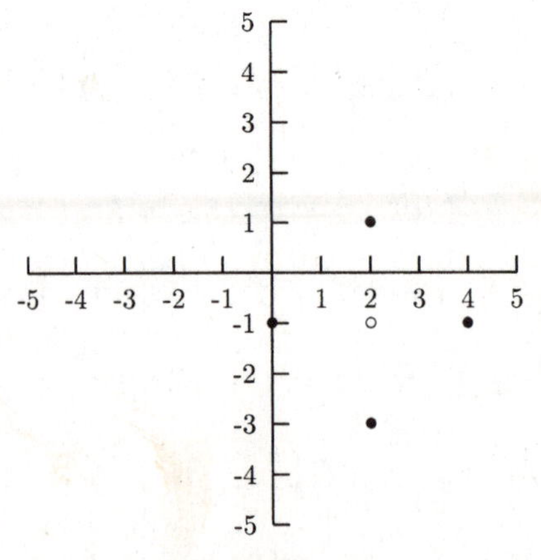

Fig. 3-22.

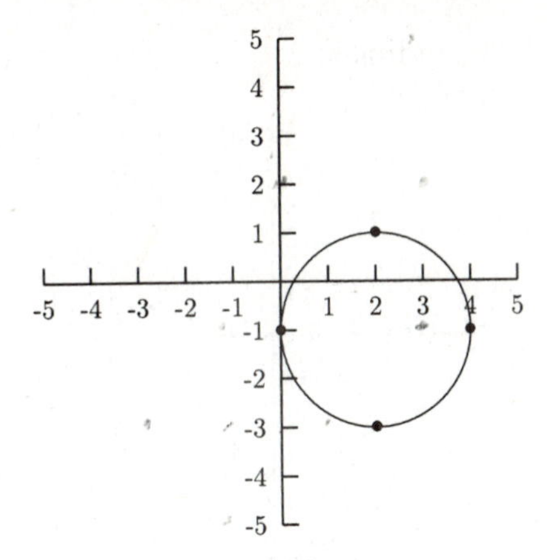

Fig. 3-23.

PRACTICE

Identify the center and radius and graph the circle.

1. $(x-3)^2+(y-2)^2=9$
2. $(x+1)^2+(y+3)^2=4$
3. $(x-4)^2+y^2=1$
4. $x^2+y^2=16$

SOLUTIONS

1. Center = $(3, 2)$, radius = 3

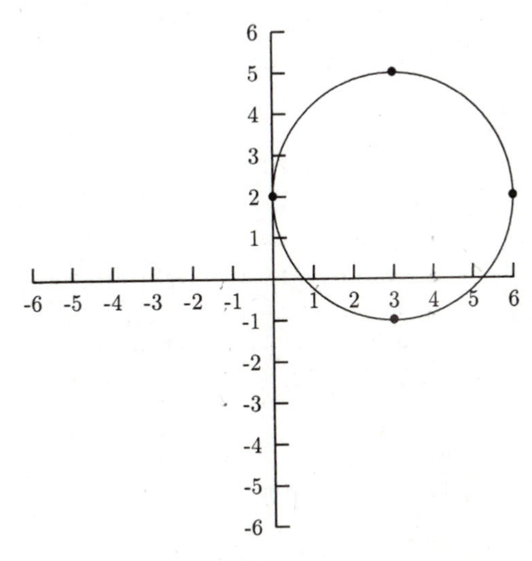

Fig. 3-24.

2. Center = $(-1, -3)$, radius = 2

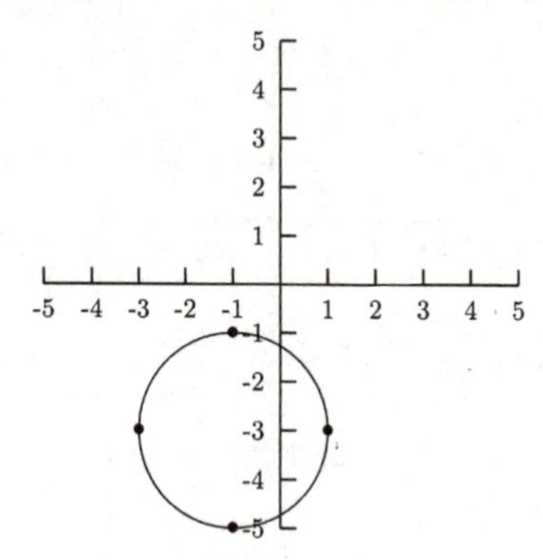

Fig. 3-25.

3. Center = $(4, 0)$, radius = 1

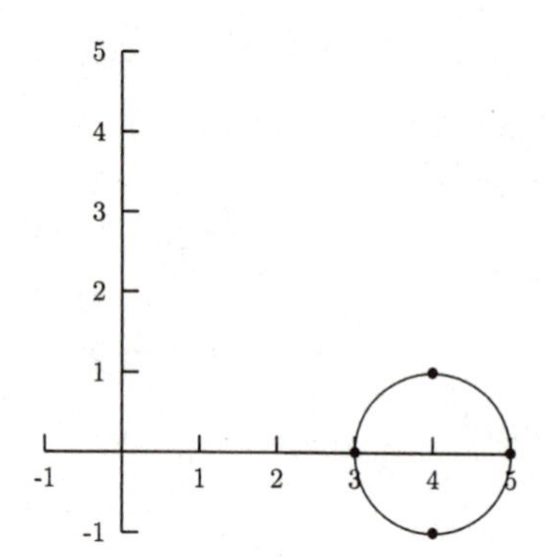

Fig. 3-26.

4. Center = $(0, 0)$, radius = 4

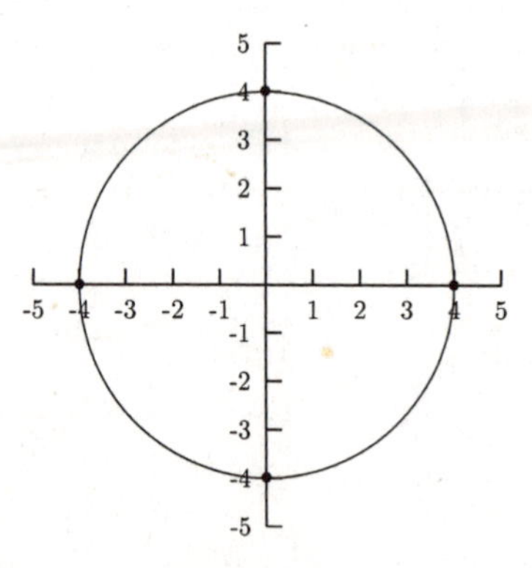

Fig. 3-27.

We can find an equation of a circle without directly knowing its center and radius. When given the endpoints of a diameter (a line segment that stretches the full width of a circle), we can find the center of the circle by finding the midpoint of the diameter. Once we know (h, k), we can use the coordinates of one of the points for x and y in the equation $(x-h)^2 + (y-k)^2 = r^2$ to find r^2.

EXAMPLES

- The endpoints of a diameter of a circle are $(2, 4)$ and $(8, 12)$. Find an equation of the circle.

  The center of the circle can be found by finding the midpoint of $(2, 4)$ and $(8, 12)$.

$$(h, k) = \left(\frac{x_1 + x_2}{2}, \frac{y_1 + y_2}{2}\right) = \left(\frac{2+8}{2}, \frac{4+12}{2}\right) = (5, 8)$$

  So far, we know that the equation is $(x-5)^2 + (y-8)^2 = r^2$. To find r^2, we will use the endpoint $(2, 4)$. (The endpoint $(8, 12)$ would also work.) We will substitute $x = 2$ and $y = 4$ in the equation and solve for r^2.

$$(x-5)^2 + (y-8)^2 = r^2$$
$$(2-5)^2 + (4-8)^2 = r^2$$
$$(-3)^2 + (-4)^2 = r^2$$
$$9 + 16 = r^2$$
$$25 = r^2$$

  An equation of the circle is $(x-5)^2 + (y-8)^2 = 25$.

- A circle has center $(-4, 3)$ and the point $(2, 11)$ is on the circle. Find an equation of the circle.

  Because $(-4, 3)$ is the center, we already know the equation is $(x+4)^2 + (y-3)^2 = r^2$. Let $x = 2$ and $y = 11$ in this equation to find r^2.

$$(2+4)^2 + (11-3)^2 = r^2$$
$$6^2 + 8^2 = r^2$$
$$100 = r^2$$

  An equation for the circle is $(x+4)^2 + (y-3)^2 = 100$.

PRACTICE

Use the information given about the circle to find its equation.

1. A diameter to the circle has endpoints $(2, 1)$ and $(-4, 9)$.
2. A diameter to the circle has endpoints $(0, 4)$ and $(-12, 9)$.
3. The center of the circle has coordinates $(1, 8)$ and the point $(13, 13)$ is on the circle.
4. The center of the circle is $(5, 0)$ and $(-5, 6)$ is on the circle.

SOLUTIONS

1. The midpoint will be the center of the circle.

$$(h, k) = \left(\frac{2 + (-4)}{2}, \frac{1 + 9}{2}\right) = (-1, 5)$$

So far, we know the equation is $(x + 1)^2 + (y - 5)^2 = r^2$. We could use either $(2, 1)$ or $(-4, 9)$ in the equation to find r^2. Here, $(2, 1)$ will be used.

$$(2 + 1)^2 + (1 - 5)^2 = r^2$$
$$9 + 16 = r^2$$
$$25 = r^2$$

The equation is $(x + 1)^2 + (y - 5)^2 = 25$.

2. The midpoint will be the center of the circle.

$$(h, k) = \left(\frac{0 + (-12)}{2}, \frac{4 + 9}{2}\right) = \left(-6, \frac{13}{2}\right)$$

So far, we know the equation is $(x + 6)^2 + (y - \frac{13}{2})^2 = r^2$. We will use $(0, 4)$ to find r^2.

$$(0 + 6)^2 + \left(4 - \frac{13}{2}\right) = r^2$$
$$6^2 + \left(-\frac{5}{2}\right)^2 = r^2$$
$$36 + \frac{25}{4} = r^2$$
$$\frac{169}{4} = r^2$$

The equation is $(x + 6)^2 + (y - \frac{13}{2})^2 = \frac{169}{4}$.

3. The center of the circle is $(1, 8)$. This means that the circle equation begins as $(x - 1)^2 + (y - 8)^2 = r^2$. We will use $(13, 13)$ to find r^2.

$$(13 - 1)^2 + (13 - 8)^2 = r^2$$
$$144 + 25 = r^2$$
$$169 = r^2$$

The equation is $(x - 1)^2 + (y - 8)^2 = 169$.

4. Because the center is $(5, 0)$, we know the equation begins as $(x - 5)^2 + y^2 = r^2$. We will use $(-5, 6)$ to find r^2.

$$(-5 - 5)^2 + 6^2 = r^2$$
$$100 + 36 = r^2$$
$$136 = r^2$$

The equation is $(x - 5) + y^2 = 136$.

Equations of circles are not always written in the form $(x - h)^2 + (y - k)^2 = r^2$. For example, the equation $(x - 2)^2 + (y + 3)^2 = 16$ might be written in its expanded form.

$$(x - 2)^2 + (y + 3)^2 = 16$$
$$(x - 2)(x - 2) + (y + 3)(y + 3) = 16$$
$$x^2 - 4x + 4 + y^2 + 6y + 9 = 16 \quad \text{After using the FOIL method}$$
$$x^2 + y^2 - 4x + 6y - 3 = 0$$

In the following, we will be given equations like the one above and use completing the square to rewrite them in the form $(x - h)^2 + (y - k)^2 = r^2$.

EXAMPLES

- $x^2 + y^2 - 12x + 4y + 36 = 0$

  For the first step, move the constant term (the number without a variable) to the right side of the equation, writing the left side with the x-terms together and the y-terms together.

$$x^2 - 12x + y^2 + 4y = -36$$

Next, we will complete the square for the x-terms and the y-terms and will add both numbers to each side of the equation.

$$x^2 - 12x + 36 + y^2 + 4y + 4 = -36 + 36 + 4$$

In the last step, we will write the left side of the equation as the sum of two perfect squares. For $x^2 - 12x + 36 = (x - \_)^2$, we will use $\sqrt{36} = 6$. For $y^2 + 4y + 4 = (y + \_)^2$, we will use $\sqrt{4} = 2$.

$$(x - 6)^2 + (y + 2)^2 = 4$$

Now we can see that this equation is an equation of a circle which has center $(6, -2)$ and radius 2.

- $x^2 + y^2 - 8x - 4y = -11$

$$x^2 - 8x + \_ + y^2 - 4y + \_ = -11 + \_ + \_$$

$$\left(\frac{8}{2}\right)^2 = 16 \quad \text{and} \quad \left(\frac{4}{2}\right)^2 = 4$$

$$x^2 - 8x + 16 + y^2 - 4y + 4 = -11 + 16 + 4$$

For $x^2 - 8x + 16 = (x - \_)^2$, we will use $\sqrt{16} = 4$. For $y^2 - 4y + 4 = (y - \_)^2$, we use $\sqrt{4} = 2$.

$$(x - 4)^2 + (y - 2)^2 = 9$$

- $x^2 + y^2 - 2y - 14 = 0$
Because x^2 already is a perfect square, we only need to complete the square on the y-terms.

$$x^2 + y^2 - 2y + \_ = 14 + \_$$

$$\left(\frac{2}{2}\right)^2 = 1$$

$$x^2 + y^2 - 2y + 1 = 14 + 1$$

For $y^2 - 2y + 1 = (y - \_)^2$ we will use $\sqrt{1} = 1$

$$x^2 + (y - 1)^2 = 15$$

- $x^2 + y^2 - x + \frac{4}{5}y - \frac{59}{100} = 0$
It might be tempting to clear the fraction on this problem (that is, to multiply both sides of the equation by the least common denominator).

This will not work because the coefficients of x^2 and y^2 must be 1 when completing the square.

$$x^2 - x + \frac{1}{4} + y^2 + \frac{4}{5}y + \frac{4}{25} = \frac{59}{100} + \frac{1}{4} + \frac{4}{25}$$

$$\frac{59}{100} + \frac{1}{4} + \frac{4}{25} = \frac{59}{100} + \frac{25}{100} + \frac{16}{100} = \frac{100}{100} = 1$$

$$\left(x - \frac{1}{2}\right)^2 + \left(y + \frac{2}{5}\right)^2 = 1$$

PRACTICE

Complete the square to find the center and radius of the circle.

1. $x^2 + y^2 - 14x - 10y + 68 = 0$
2. $x^2 + y^2 + 4x - 8y + 11 = 0$
3. $x^2 + y^2 - 12x = -21$
4. $x^2 + y^2 - \frac{3}{2}x + 6y - \frac{247}{16} = 0$

SOLUTIONS

1. $x^2 - 14x + \_ + y^2 - 10y + \_ = -68 + \_ + \_$

$$x^2 - 14x + 49 + y^2 - 10y + 25 = -68 + 49 + 25$$

$$(x - 7)^2 + (y - 5)^2 = 6$$

The center is $(7, 5)$ and the radius is $\sqrt{6}$.

2. $x^2 + 4x + \_ + y^2 - 8y + \_ = -11 + \_ + \_$

$$x^2 + 4x + 4 + y^2 - 8y + 16 = -11 + 4 + 16$$

$$(x + 2)^2 + (y - 4)^2 = 9$$

The center is $(-2, 4)$ and the radius is 3.

3. $x^2 - 12x + \_ + y^2 = -21 + \_$

$$x^2 - 12x + 36 + y^2 = -21 + 36$$

$$(x - 6)^2 + y^2 = 15$$

The center is $(6, 0)$ and the radius is $\sqrt{15}$.

4. $x^2 - \frac{3}{2}x + \_ + y^2 + 6y + \_ = \frac{247}{16} + \_ + \_$

$$x^2 - \frac{3}{2}x + \frac{9}{16} + y^2 + 6y + 9 = \frac{247}{16} + \frac{9}{16} + 9$$

$$\frac{247}{16} + \frac{9}{16} + 9 = \frac{247}{16} + \frac{9}{16} + \frac{144}{16} = \frac{400}{16} = 25$$

$$\left(x - \frac{3}{4}\right)^2 + (y + 3)^2 = 25$$

The center is $(\frac{3}{4}, -3)$ and the radius is 5.

Sometimes the coefficient of x^2 and y^2 is not 1. In this case, we must divide both sides of the equation by this number *before* completing the square. It is worth mentioning that in equations of circles, x^2 and y^2 will always have the same coefficient. If the coefficients are different, the graph of the equation will not be a circle.

EXAMPLE

$$3x^2 + 3y^2 - 30x - 12y + 84 = 0$$

$$\frac{1}{3}(3x^2 + 3y^2 - 30x - 12y) = \frac{1}{3}(-84)$$

$$x^2 + y^2 - 10x - 4y = -28$$

$$x^2 - 10x + \_ + y^2 - 4y + \_ = -28 + \_ + \_$$

$$x^2 - 10x + 25 + y^2 - 4y + 4 = -28 + 25 + 4$$

$$(x - 5)^2 + (y - 2)^2 = 1$$

Chapter 3 Review

1. What is the distance between $(-3, 4)$ and $(-2, 9)$?
   a) $\sqrt{24}$ b) 4 c) 24 d) $\sqrt{26}$
2. What are the center and radius of the circle whose equation is $x^2 + (y + 1)^2 = 5$?
   a) The center is $(0, 1)$, and the radius is 5.
   b) The center is $(0, 1)$, and the radius is $\sqrt{5}$.
   c) The center is $(0, -1)$, and the radius is 5.
   d) The center is $(0, -1)$, and the radius is $\sqrt{5}$.

3. What are the center and radius of the circle whose equation is $x^2 + y^2 + 6x - 8y = 75$?
   a) The center is $(-3, 4)$, and the radius is $\sqrt{75}$.
   b) The center is $(-3, 4)$, and the radius is 10.
   c) The center is $(3, -4)$, and the radius is $\sqrt{75}$.
   d) The center is $(3, -4)$, and the radius is 10.

4. TRUE OR FALSE: $(3, 14)$, $(10, -3)$, $(15, 9)$, $(-2, 2)$ are the vertices of a square.
   a) True b) False c) Cannot be determined

5. What is the midpoint between the points $(-3, 2)$ and $(1, 6)$?
   a) $(-\frac{1}{2}, \frac{7}{2})$ b) $(-2, 2)$ c) $(-1, 4)$ d) $(-\frac{5}{2}, -\frac{5}{2})$

6. What is an equation of the circle which has a diameter with endpoints $(6, -1)$ and $(2, -5)$?
   a) $(x - 4)^2 + (y + 3)^2 = 8$ b) $(x - 6)^2 + (y + 1)^2 = 16$
   c) $(x - 2)^2 + (y + 5)^2 = 16$ d) $(x - 2)^2 + (y - 2)^2 = 17$

SOLUTIONS

1. d) 2. d) 3. b) 4. a) 5. c) 6. a)

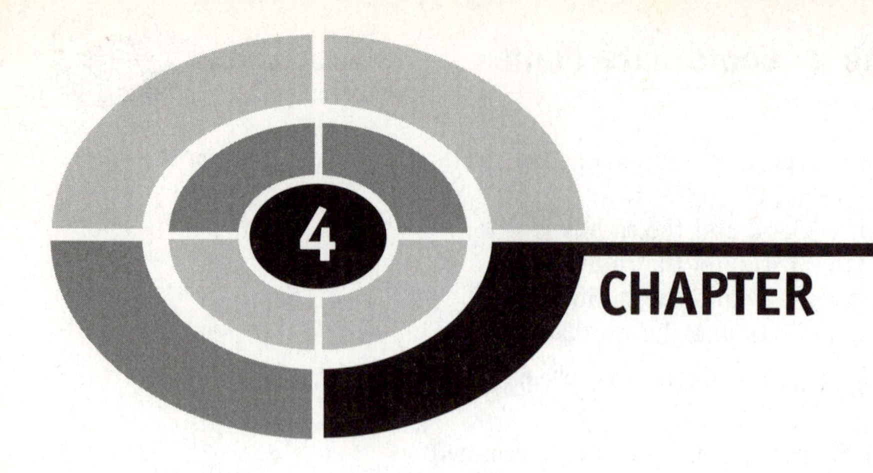

Lines and Parabolas

The graph to an equation of the form $Ax + By = C$ will be a line. An equation that can be put in this form is called a *linear equation*. We only need two points to graph a line. It does not matter which two points, but we will choose points that would be easy to graph.

If A and B are each nonzero, we can pick two x-values at random. We will put them into the equation to compute the y-values.

EXAMPLES

- $2x + 3y = 6$

  We can choose *any* two numbers for x. Here we will use $x = 0$ and $x = 6$.

$$\begin{aligned} 2(0) + 3y &= 6 \\ 3y &= 6 \\ y &= 2 \end{aligned} \qquad \begin{aligned} 2(6) + 3y &= 6 \\ 12 + 3y &= 6 \\ 3y &= -6 \\ y &= -2 \end{aligned}$$

Plot $(0, 2)$ and $(6, -2)$.

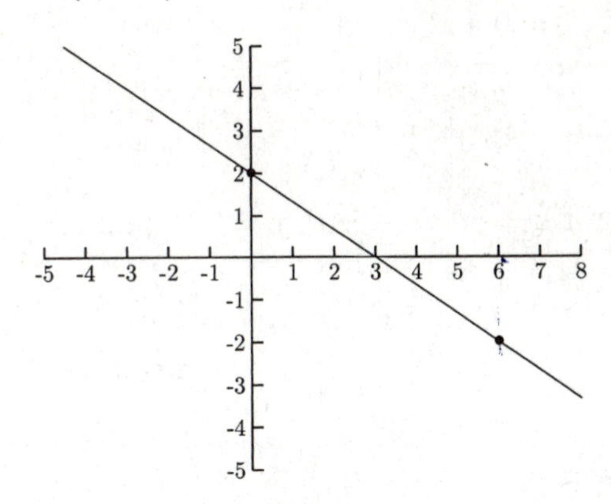

Fig. 4-1.

- $4x - y = 7$
  We will use $x = 0$ and $x = 2$.

$$\begin{aligned} 4(0) - y &= 7 \\ -y &= 7 \\ y &= -7 \end{aligned} \qquad \begin{aligned} 4(2) - y &= 7 \\ 8 - y &= 7 \\ -y &= -1 \\ y &= 1 \end{aligned}$$

Plot $(0, -7)$ and $(2, 1)$.

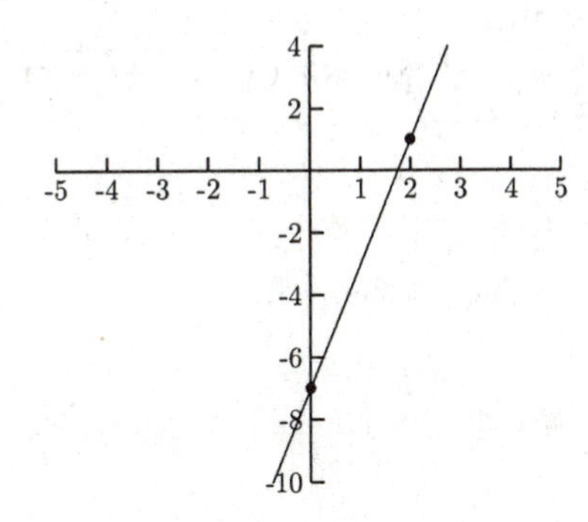

Fig. 4-2.

PRACTICE

Find the y-values for the given x-values and use the two points to plot the line.

1. $-3x + y = 5$, $x = 0$ and $x = -1$

2. $2x + 4y = 8$, $x = 0$ and $x = -2$
3. $x - 4y = 12$, $x = 0$ and $x = 4$
4. $-3x + 4y = -6$, $x = 0$ and $x = 4$

SOLUTIONS

1. Put $x = 0$ and $x = -1$ in $-3x + y = 5$ to find y.

$$-3(0) + y = 5 \qquad -3(-1) + y = 5$$

$$y = 5 \qquad 3 + y = 5$$

$$y = 2$$

Plot $(0, 5)$ and $(-1, 2)$.

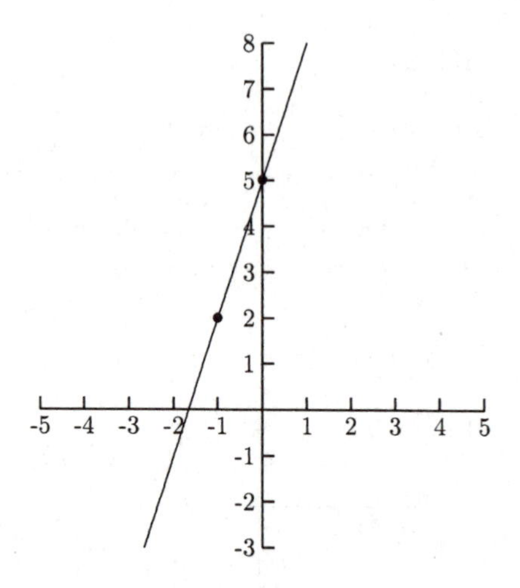

Fig. 4-3.

2. Put $x = 0$ and $x = -2$ in $2x + 4y = 8$ to find y.

$$2(0) + 4y = 8 \qquad 2(-2) + 4y = 8$$

$$4y = 8 \qquad -4 + 4y = 8$$

$$y = 2 \qquad 4y = 12$$

$$y = 3$$

Plot (0, 2) and (−2, 3).

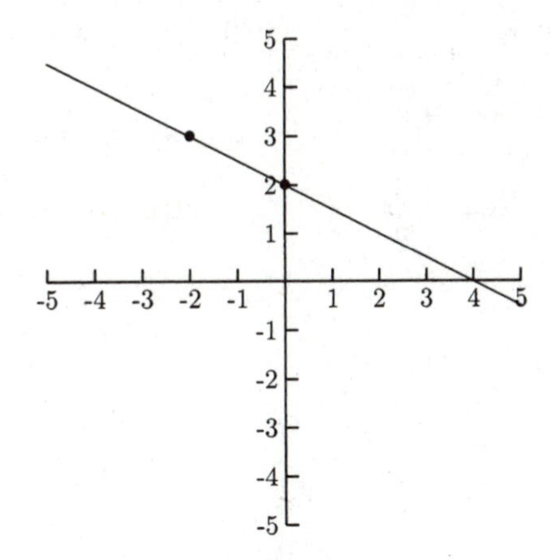

Fig. 4-4.

3. Put $x = 0$ and $x = 4$ in $x - 4y = 12$ to find y.

$$0 - 4y = 12 \qquad 4 - 4y = 12$$
$$y = -3 \qquad -4y = 8$$
$$y = -2$$

Plot (0, −3) and (4, −2)

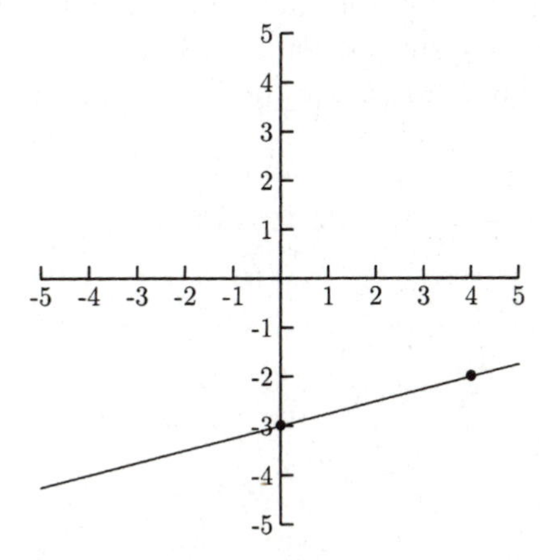

Fig. 4-5.

4. Put $x = 0$ and $x = 4$ in $-3x + 4y = -6$ to find y.

$$-3(0) + 4y = -6 \qquad -3(4) + 4y = -6$$

$$y = \frac{-6}{4} \qquad -12 + 4y = -6$$

$$y = -\frac{3}{2} \qquad 4y = -6 + 12 = 6$$

$$y = \frac{6}{4} = \frac{3}{2}$$

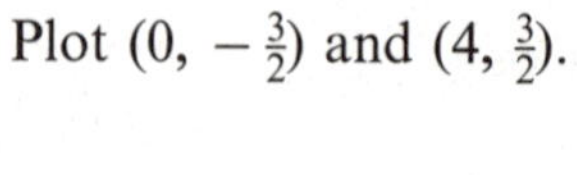

Plot $(0, -\frac{3}{2})$ and $(4, \frac{3}{2})$.

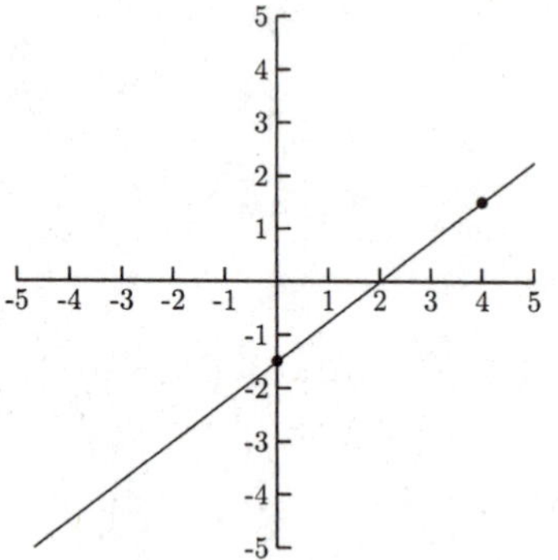

Fig. 4-6.

You might have noticed that $x = 0$ was selected for one of the points in all of the previous examples and practice problems. This point was chosen for two reasons: one, computing y is easier if $x = 0$; and two, it is an important point in its own right. A point on a graph whose x-coordinate is 0 is called a *y-intercept*. This is where the graph touches the y-axis. Many of the graphs in this book will have exactly one y-intercept. Some graphs have more than one y-intercept and some have none. See the figures below.

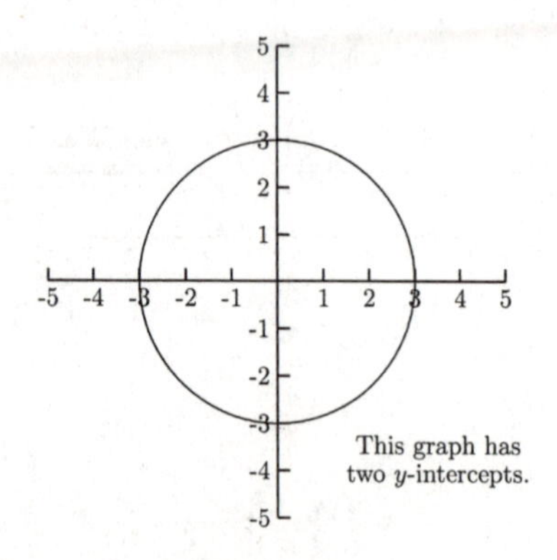

Fig. 4-7.

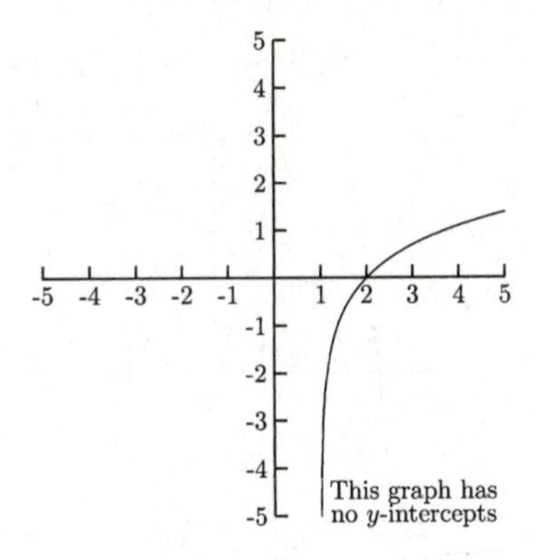

Fig. 4-8.

The y-coordinate is 0 for points on the graph that touch the x-axis. This point is called the *x-intercept*. Some of the graphs in this book will have exactly one x-intercept, some will have more than one, and still others will not have any.

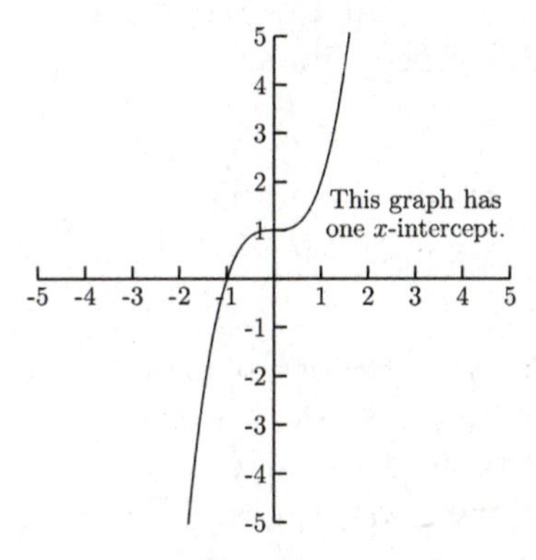

Fig. 4-9.

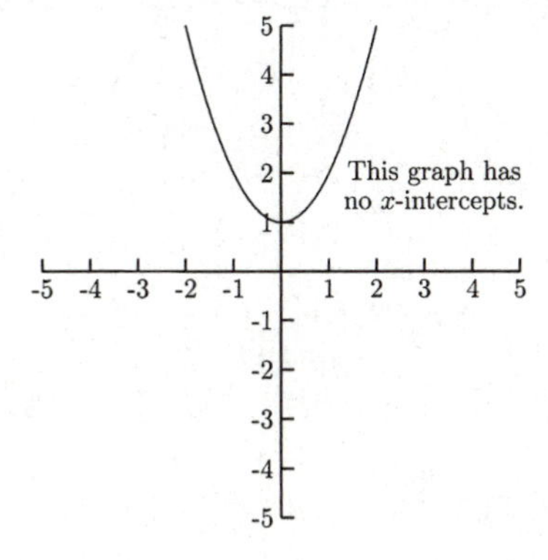

Fig. 4-10.

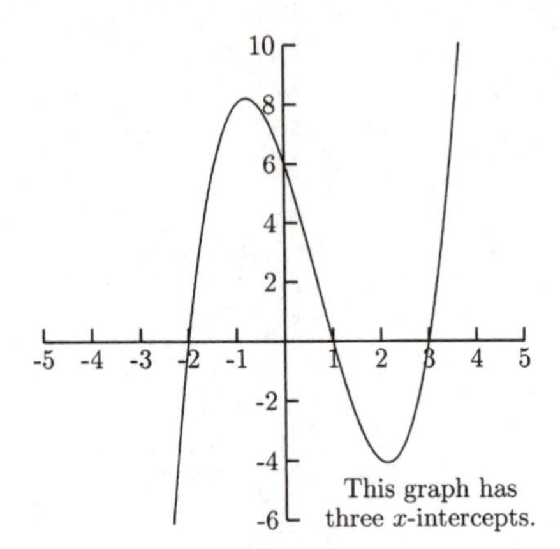

Fig. 4-11.

One way to find the intercepts is by looking at the graph. Rather than say $(a, 0)$ is an x-intercept, we say for short that a is an x-intercept. The x-intercept in Fig. 4-9 is -1 and the y-intercept is 1. The x-intercepts in Fig. 4-11 are -2, 1, and 3. The y-intercept is 6.

PRACTICE

Find the x- and y-intercepts on the graphs.

1.

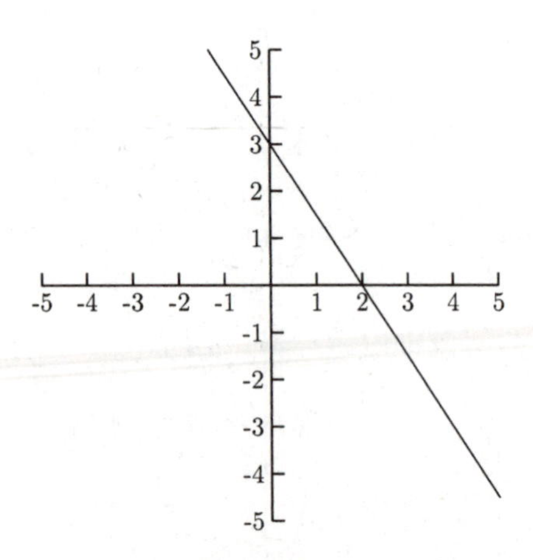

Fig. 4-12.

The x-intercept(s) is/are ___
The y-intercept(s) is/are ___

2.

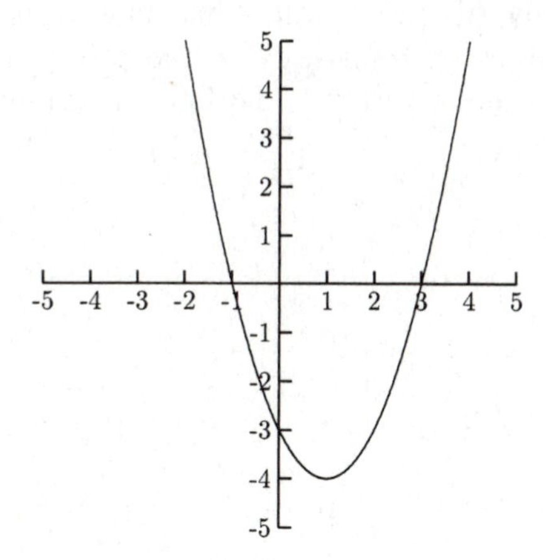

Fig. 4-13.

The x-intercept(s) is/are ___
The y-intercept(s) is/are ___

3.

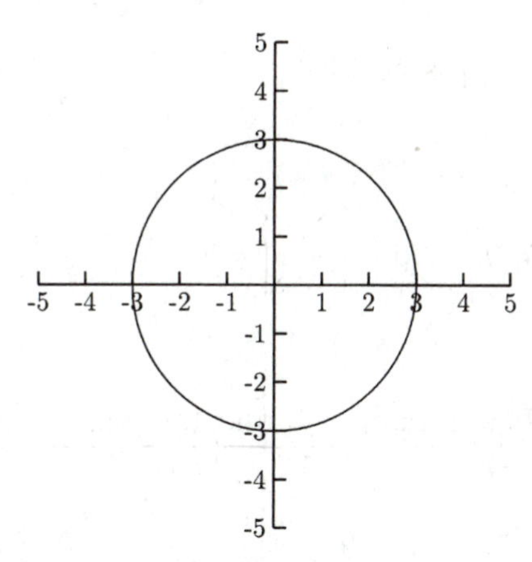

Fig. 4-14.

The x-intercept(s) is/are ___
The y-intercept(s) is/are ___

SOLUTIONS

1. The x-intercept is 2, and the y-intercept is 3.
2. The x-intercepts are -1 and 3, and the y-intercept is -3.
3. The x-intercepts are -3 and 3, and the y-intercepts are -3 and 3.

Intercepts can be found without looking at the graph. We can find the intercepts algebraically (if an equation has intercepts), by substituting 0 for one of the variables and solving for the other variable. To find the x-intercept, let $y = 0$ and solve for x. To find the y-intercept, let $x = 0$ and solve for y.

EXAMPLES

- $2x + 3y = 6$

$$\begin{array}{rr} y = 0 & x = 0 \\ 2x + 3(0) = 6 & 2(0) + 3y = 6 \\ 2x = 6 & 3y = 6 \\ x = 3 & y = 2 \end{array}$$

The x-intercept is 3, and the y-intercept is 2.

- $y = x^2 - x - 2$

$$\begin{array}{rl} y = 0 & x = 0 \\ x^2 - x - 2 = 0 & y = 0^2 - 0 - 2 \\ (x-2)(x+1) = 0 & y = -2 \\ x - 2 = 0 \quad \text{and} \quad x + 1 = 0 & \\ x = 2 \quad \text{and} \quad x = -1 & \end{array}$$

The x-intercepts are 2 and -1, and the y-intercept is -2.

- $x^2 + y^2 = 16$

$$\begin{array}{rr} y = 0 & x = 0 \\ x^2 + 0^2 = 16 & 0^2 + y^2 = 16 \\ x^2 = 16 & y^2 = 16 \\ x = \pm 4 & y = \pm 4 \end{array}$$

The x-intercepts are -4 and 4, and the y-intercepts are -4 and 4.

- $y = (x + 8)/(x - 2)$
  The only way a fraction can be zero is if the numerator is zero. Here the numerator is $x + 8$, so we will solve $x + 8 = 0$ to find the x-intercept.

$$\begin{aligned} y &= 0 & x &= 0 \\ x + 8 &= 0 & y &= \frac{0+8}{0-2} \\ x &= -8 & y &= -4 \end{aligned}$$

  The x-intercept is -8, and the y-intercept is -4.

PRACTICE

Find the x- and y-intercepts algebraically.

1. $x - 2y = 4$
2. $y = 3x - 12$
3. $y = x^2 + 3x - 4$
4. $y = \dfrac{x+6}{x+12}$

SOLUTIONS

1. $x - 2y = 4$

$$\begin{aligned} x - 2(0) &= 4 & 0 - 2y &= 4 \\ x &= 4 & -2y &= 4 \\ & & y &= -2 \end{aligned}$$

   The x-intercept is 4, and the y-intercept is -2.

2. $y = 3x - 12$

$$\begin{aligned} 3x - 12 &= 0 & y &= 3(0) - 12 \\ 3x &= 12 & y &= -12 \\ x &= 4 & & \end{aligned}$$

   The x-intercept is 4, and the y-intercept is -12.

3. $y = x^2 + 3x - 4$

$$x^2 + 3x - 4 = 0 \qquad y = 0^2 + 3(0) - 4$$

$$(x + 4)(x - 1) = 0 \qquad y = -4$$

$$x + 4 = 0 \quad \text{and} \quad x - 1 = 0$$

$$x = -4 \quad \text{and} \quad x = 1$$

The x-intercepts are -4 and 1, and the y-intercept is -4.

4. $y = (x + 6)/(x + 12)$

When finding the x-intercept, we only need to solve $x + 6 = 0$ because the only way a fraction can be zero is if its numerator is zero.

$$x + 6 = 0 \qquad y = \frac{0 + 6}{0 + 12}$$

$$x = -6 \qquad y = \frac{6}{12} = \frac{1}{2}$$

The x-intercept is -6, and the y-intercept is $\frac{1}{2}$.

We can tell whether or not a graph has intercepts by looking at it. What happens if we *do not* have the graph? A graph will not have an x-intercept if when we let $y = 0$ in its equation we do not get a real number solution. A graph will not have a y-intercept if when we let $x = 0$ in its equation we do not get a real number solution.

EXAMPLES

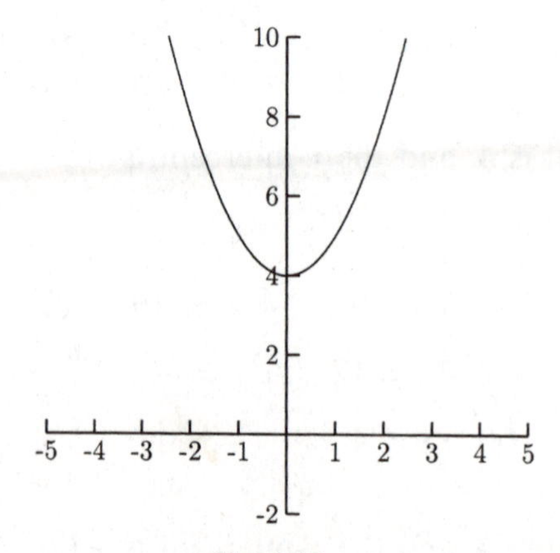

Fig. 4-15.

The graph of $y = x^2 + 4$ in Fig. 4-15 does not have any x-intercepts. Let us see what happens if we try to find the x-intercepts algebraically.

$$x^2 + 4 = 0$$
$$x^2 = -4$$
$$x = \pm\sqrt{-4}$$

$\sqrt{-4}$ is not a real number, so the equation $x^2 + 4 = 0$ does not have any real solution.

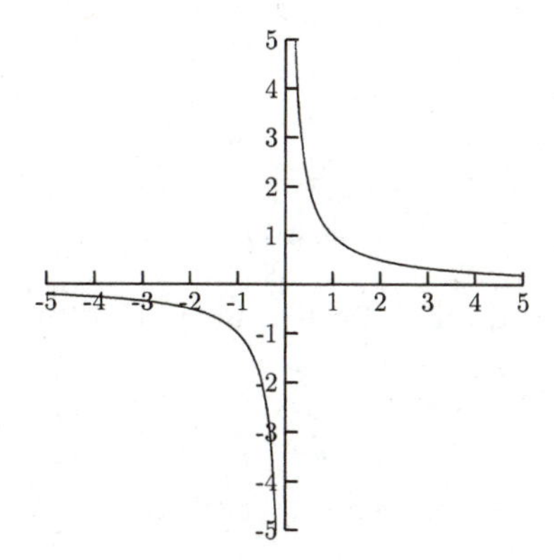

Fig. 4-16.

The graph of $y = 1/x$ in Fig. 4-16 does not have any intercepts at all.

$$y = 0 \qquad x = 0$$
$$\frac{1}{x} = 0 \qquad \frac{1}{0} = y$$

As was mentioned earlier, a fraction can equal zero only if the numerator is zero. The equation $1/x = 0$ has no solution because the fraction is zero but the numerator, 1, is never zero. This shows that the graph of $y = 1/x$ has no x-intercept. The equation $1/0 = y$ has no solution because $1/0$ is not a number. This shows that the graph of $y = 1/x$ has no y-intercept, either.

The Slope of a Line

Another important part of a line is its *slope*. The slope is a measure of a line's tilt. Some lines have steep slopes and others have more gradual slopes. A line that tilts upward will have a different slope than one that tilts downward.

A line has a steep slope if a small horizontal change results in a large vertical change.

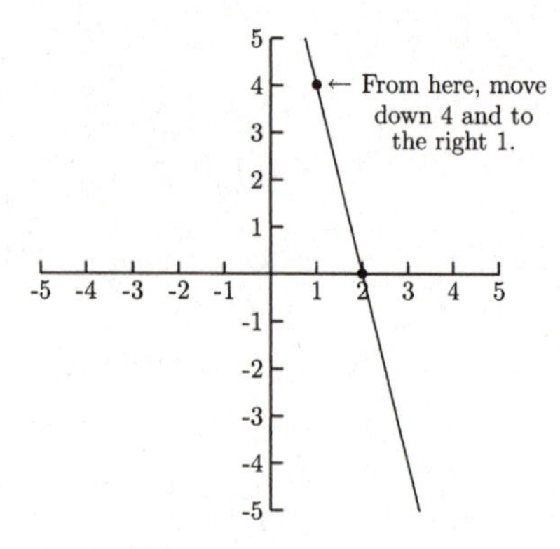

Fig. 4-17.

A line has a more gradual slope if a large horizontal change results in a small vertical change.

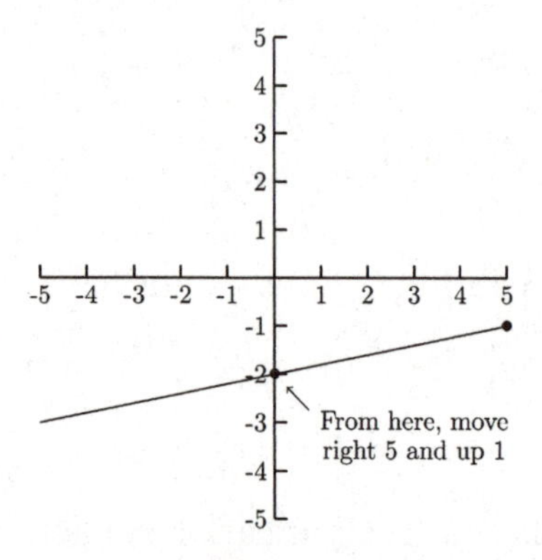

Fig. 4-18.

The slope of a line is measured by a number. This number is a quotient (a fraction) where the vertical change is divided by the horizontal change. In Fig. 4-17, to move from one point to the other, we had a vertical change of down 4 and a horizontal change of 1. This means that the slope of the line $4x + y = 8$ is $\frac{-4}{1}$. In Fig. 4-18, to move from one point to the other, we moved up 1 and to the right 5. The slope to the line $x - 5y = 10$ is $\frac{1}{5}$.

One of the convenient things about the slope of a line is that it does not matter which two points we use—the quotient of the vertical change to the horizontal change will be the same. Suppose we use two other points on the line $4x + y = 8$. If we moved from $(-4, 24)$ to $(2, 0)$, then we would go down

24 (a change of −24) and to the right 6. This quotient is $\frac{-24}{6} = \frac{-4}{1}$, the same as with the two points in Fig. 4-17. This idea leads us to the slope formula. This formula is important and is worth memorizing. If (x_1, y_1) and (x_2, y_2) are two points on a line, then the slope, m, of the line is the number

$$m = \frac{y_2 - y_1}{x_2 - x_1} = \frac{\text{vertical change}}{\text{horizontal change}}.$$

EXAMPLES

Find the slope of the line using the given points.

- $2x + 3y = -6$ $(0, -2)$ and $(-3, 0)$

  Here $(x_1, y_1) = (0, -2)$ and $(x_2, y_2) = (-3, 0)$.

$$m = \frac{0 - (-2)}{-3 - 0} = \frac{2}{-3} = -\frac{2}{3}$$

  It does not matter which point we call (x_1, y_1) and which we call (x_2, y_2). We will compute m with $(x_1, y_1) = (-3, 0)$ and $(x_2, y_2) = (0, -2)$.

$$m = \frac{-2 - 0}{0 - (-3)} = \frac{-2}{3} = -\frac{2}{3}$$

- $3x - y = 4$ $(1, -1)$ and $(-2, -10)$

$$m = \frac{-10 - (-1)}{-2 - 1} = \frac{-9}{-3} = \frac{3}{1} = 3$$

  Normally, when the denominator is 1, we write the slope as an integer.

- $x - 2y = -2$ $(4, 3)$ and $(-2, 0)$

$$m = \frac{0 - 3}{-2 - 4} = \frac{-3}{-6} = \frac{1}{2}$$

PRACTICE

Find the slope of the line using the given points.

1. $2x + 3y = -12$ $(0, -4)$ and $(3, -6)$
2. $2x - y = 1$ $(0, -1)$ and $(1, 1)$
3. $x - y = 4$ $(3, -1)$ and $(2, -2)$
4. $x + 2y = 6$ $(2, 2)$ and $(-4, 5)$
5. $3x - 5y = 10$ $(10, 4)$ and $(5, 1)$

SOLUTIONS

1.

$$m = \frac{-6-(-4)}{3-0} = \frac{-2}{3} = -\frac{2}{3}$$

2.

$$m = \frac{1-(-1)}{1-0} = \frac{2}{1} = 2$$

3.

$$m = \frac{-2-(-1)}{2-3} = \frac{-1}{-1} = 1$$

4.

$$m = \frac{5-2}{-4-2} = \frac{3}{-6} = -\frac{1}{2}$$

5.

$$m = \frac{1-4}{5-10} = \frac{-3}{-5} = \frac{3}{5}$$

Horizontal and Vertical Lines

The y-values of a horizontal line are the same number. The equation of a horizontal line is in the form $y =$ number.

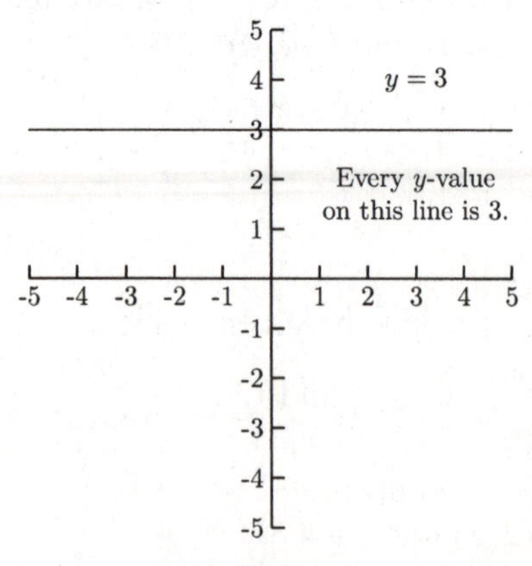

Fig. 4-19.

What would the slope of a horizontal line be? No matter which two points we choose, their y-values will be the same. This means that y_1 and y_2 will be equal, so $y_2 - y_1 = 0$.

$$m = \frac{y_2 - y_1}{x_2 - x_1} = \frac{0}{x_2 - x_1} = 0$$

The slope of *any* horizontal line is 0.

The x-values of a vertical line are the same number. The equation of a vertical line is in the form $x =$ number.

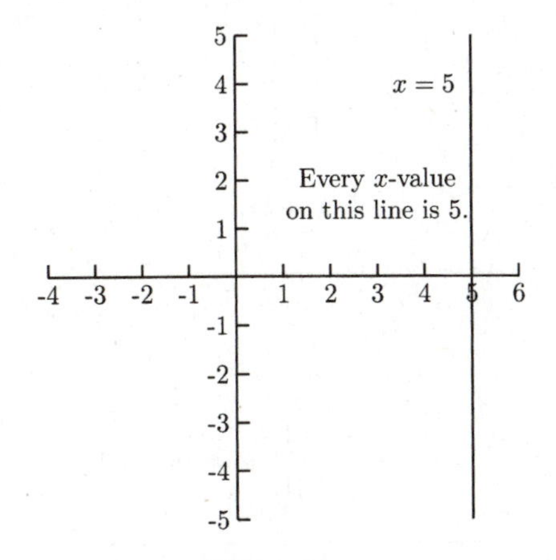

Fig. 4-20.

Because all of the x-values on a vertical line are the same, x_2 and x_1 are the same. This means that the denominator of the slope of a vertical line is 0, so the slope is undefined.

$$m = \frac{y_2 - y_1}{x_2 - x_1} = \frac{y_2 - y_1}{0}$$

In addition to saying that the slope of a vertical line is undefined, we also say it does not exist. To say that the slope of a line does not exist is *not* the same as saying that the slope is 0. The slope of horizontal line is 0; the slope of a vertical line does not exist.

PRACTICE

Graph each line. State whether the slope is zero or does not exist.

1. $x = 1$
2. $y = 4$
3. $x = -\frac{3}{2}$

SOLUTIONS

1.

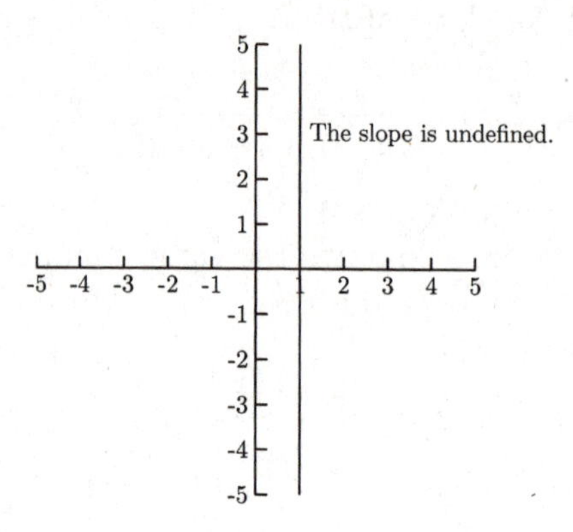

Fig. 4-21.

2.

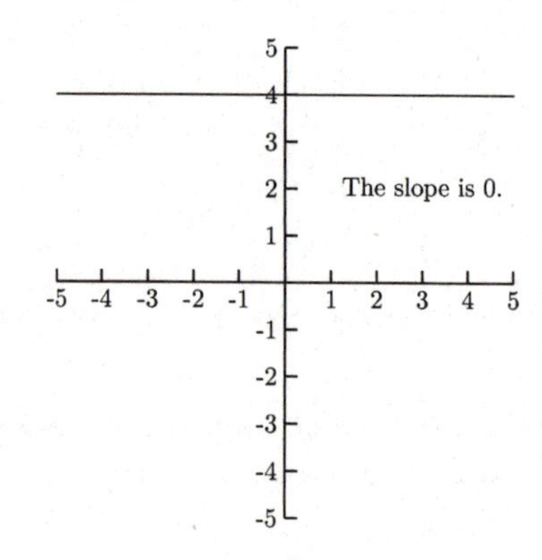

Fig. 4-22.

3.

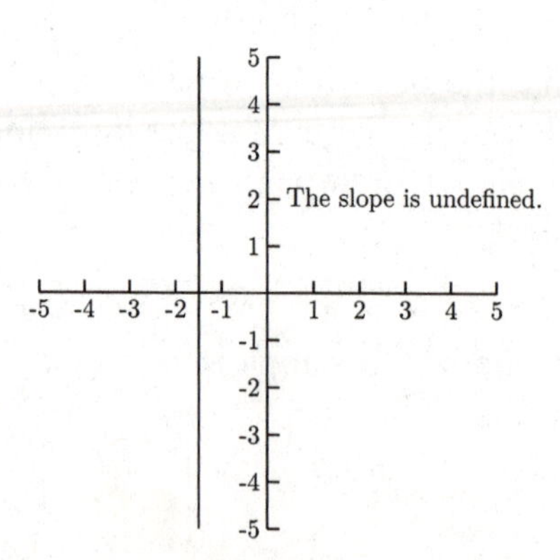

Fig. 4-23.

Two points on a line not only allow us to graph the line, they also give us enough information to find an equation for the line. First, we need to use the slope formula to find the slope of the line. Second, we need to use the slope and one of the points in the *point–slope formula.*

$$y - y_1 = m(x - x_1)$$

This formula comes directly from the slope formula. All that was done to the slope formula was to replace (x_2, y_2) with (x, y) and to clear the fraction.

$$m = \frac{y - y_1}{x - x_1}$$

$$(x - x_1)m = (x - x_1) \cdot \frac{y - y_1}{x - x_1}$$

$$(x - x_1)m = y - y_1$$

$$y - y_1 = m(x - x_1)$$

In the following examples and practice problems, we will be putting linear equations in the general form $Ax + By = C$, where A, B, and C are integers and A is not negative.

EXAMPLES

Find the equation of the line containing the given points.

- $(3, 5)$ and $(-6, -13)$

  First find the slope of the line containing these points.

  $$m = \frac{-13 - 5}{-6 - 3} = \frac{-18}{-9} = 2$$

  We can use either $(3, 5)$ or $(-6, -13)$ as (x_1, y_1) in the point–slope formula. If we use $(3, 5)$, $y - y_1 = m(x - x_1)$ becomes

  $$y - 5 = 2(x - 3)$$

  Now we will put this equation in the general form.

  $$y - 5 = 2(x - 3)$$

  $$y - 5 = 2x - 6 \qquad \text{We need } x \text{ and } y \text{ on the same side.}$$

  $$-2x + y = -1$$

  $$-(-2x + y) = -(-1) \qquad A \text{ needs to be positive.}$$

  $$2x - y = 1$$

To see that it does not matter which point we choose for (x_1, y_1), we will find this equation using $(-6, -13)$.

$$y - (-13) = 2(x - (-6))$$
$$y + 13 = 2x + 12$$
$$-2x + y = -1$$
$$2x - y = 1$$

- $(-4, 0)$ and $(0, 4)$ (These are the intercepts.)

$$m = \frac{4 - 0}{0 - (-4)} = \frac{4}{4} = 1$$

Use $(-4, 0)$ as (x_1, y_1).

$$y - 0 = 1(x - (-4))$$
$$y = x + 4$$
$$-x + y = 4$$
$$x - y = -4$$

- $(-3, -2)$ and $(6, 1)$

$$m = \frac{1 - (-2)}{6 - (-3)} = \frac{3}{9} = \frac{1}{3}$$

Use $(-3, -2)$ as (x_1, y_1).

$$y - (-2) = \frac{1}{3}(x - (-3))$$
$$y + 2 = \frac{1}{3}(x + 3)$$
$$3(y + 2) = \left(\frac{1}{3} \cdot 3\right)(x + 3) \quad \text{Multiply by the lowest common denominator}$$
$$3y + 6 = x + 3$$
$$-x + 3y = -3$$
$$x - 3y = 3$$

- $(4,6)$ and $(-3,6)$
  The y-values are the same, making this a horizontal line. The equation for this line is $y=6$—no work is necessary. The method used above will still work on horizontal lines, though.

$$m=\frac{6-6}{-3-4}=\frac{0}{-7}=0$$

$$y-6=0(x-4)$$

$$y-6=0 \quad \text{or} \quad y=6$$

- $(-2,0)$ and $(-2,5)$
  This line is a vertical line because the x-values are the same. The equation for this line is $x=-2$. No work is necessary (or even possible).

PRACTICE

Find an equation of the line containing the given points. Put the equation in the general form $Ax+By=C$, where A, B, and C are integers and A is not negative, or in the form $x=$ number or $y=$ number.

1. $(1,2)$ and $(5,-2)$
2. $(2,-7)$ and $(-1,5)$
3. $(4,-6)$ and $(4,2)$
4. $(5,-1)$ and $(-10,-10)$
5. $(1,5)$ and $(4,2)$
6. $(4,8)$ and $(-1,8)$
7. $(-1,\frac{3}{2})$ and $(2,-\frac{3}{2})$
8. $(2,6)$ and $(\frac{1}{3},1)$

SOLUTIONS

1.

$$m=\frac{-2-2}{5-1}=\frac{-4}{4}=-1$$

$$y-2=-1(x-1)$$

$$y-2=-x+1$$

$$x+y=3$$

2.

$$m = \frac{5-(-7)}{-1-2} = \frac{12}{-3} = -4$$

$$y-(-7) = -4(x-2)$$

$$y+7 = -4x+8$$

$$4x+y = 1$$

3. The x-values are the same, making this a vertical line. The equation is $x = 4$.

4.

$$m = \frac{-10-(-1)}{-10-5} = \frac{-9}{-15} = \frac{3}{5}$$

$$y-(-1) = \frac{3}{5}(x-5)$$

$$5(y+1) = \left(5 \cdot \frac{3}{5}\right)(x-5)$$

$$5y+5 = 3(x-5)$$

$$5y+5 = 3x-15$$

$$-3x+5y = -20$$

$$3x-5y = 20$$

5.

$$m = \frac{2-5}{4-1} = \frac{-3}{3} = -1$$

$$y-5 = -1(x-1)$$

$$y-5 = -x+1$$

$$x+y = 6$$

6. Because the y-values are the same, this line is horizontal. The equation is $y = 8$.

7.

$$m = \frac{-(3/2) - (3/2)}{2 - (-1)} = \frac{-(6/2)}{3} = \frac{-3}{3} = -1$$

$$y - \frac{3}{2} = -1(x - (-1))$$

$$y - \frac{3}{2} = -(x + 1)$$

$$y - \frac{3}{2} = -x - 1$$

$$x + y = \frac{3}{2} - 1$$

$$x + y = \frac{1}{2}$$

$$2(x + y) = 2 \cdot \frac{1}{2}$$ We want C to be an integer.

$$2x + 2y = 1$$

8.

$$m = \frac{1 - 6}{(1/3) - 2} = \frac{-5}{(1/3) - (6/3)}$$

$$= \frac{-5}{-(5/3)} = -5 \div \frac{-5}{3}$$

$$= -5 \cdot \frac{3}{-5}$$

$$= 3$$

$$y - 6 = 3(x - 2)$$

$$y - 6 = 3x - 6$$

$$-3x + y = 0$$

$$3x - y = 0$$

The Slope–Intercept Form of the Line

Now we are ready to learn a new form of the line. Remember when a circle is in the form $(x-h)^2+(y-k)^2=r^2$ we know the circle's center and radius. There is a form of the line which gives the same kind of information. This form is called the *slope–intercept* form of the line. When an equation is in this form, we know the line's slope and y-intercept.

To discover this form, we will examine a practice problem from an earlier practice set. Two points on the line are $(0, -4)$ (this is the y-intercept) and $(3, -6)$. The slope of the line is $-\frac{2}{3}$ and the general form of the equation is $2x+3y=-12$. Solve this equation for y (this means to isolate y on one side of the equation).

$$2x+3y=-12$$

$$3y=-2x-12$$

$$\frac{3y}{3}=\frac{-2x}{3}-\frac{12}{3}$$

$$y=-\frac{2}{3}x-4$$

The coefficient of x is $-\frac{2}{3}$, which is the slope; and the constant term is -4, the y-intercept. This will happen *every* time a linear equation is solved for y. This is why $y=mx+b$ is called the slope–intercept form of the line. Because a vertical line has no y term and no slope, there is no slope–intercept form for a vertical line.

EXAMPLES

- $y=3x+4$. The slope is 3. The y-intercept is 4.
- $y=x-2$. The slope is 1. The y-intercept is -2.
- $y=-\frac{1}{2}$. This equation could be rewritten as $y=0x-\frac{1}{2}$. The slope is 0. The y-intercept is $-\frac{1}{2}$.
- $y=x$. This equation could be rewritten as $y=1x+0$. The slope is 1. The y-intercept is 0.

PRACTICE

Identify the slope and y-intercept.

1. $y=-2x+6$
2. $y=\frac{3}{4}x-5$
3. $y=x+\frac{2}{3}$

4. $y = 4x$
5. $y = 10$

SOLUTIONS

1. The slope is -2, and the y-intercept is 6.
2. The slope is $\frac{3}{4}$, and the y-intercept is -5.
3. The slope is 1, and the y-intercept is $\frac{2}{3}$.
4. The equation can be rewritten as $y = 4x + 0$. The slope is 4, and the y-intercept is 0.
5. The equation can be rewritten as $y = 0x + 10$. The slope is 0, and the y-intercept is 10.

Graphing a Line Using the Slope and y-Intercept

We can graph a line using the slope and any point on the line. In particular, we can graph a line using the slope and y-intercept. Remember what information the slope is giving: the vertical change over the horizontal change. We will begin by plotting the y-intercept. Then we will use the slope to get another point on the line. Finally, we will draw a line through these two points.

EXAMPLES

- $y = \frac{2}{3}x + 1$
  Plot $(0, 1)$.

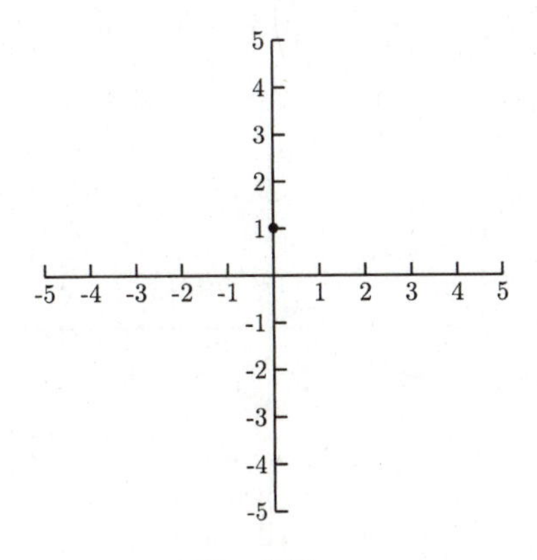

Fig. 4-24.

The slope is $\frac{2}{3}$. From the point already plotted, go up 2 units and to the right 3 units.

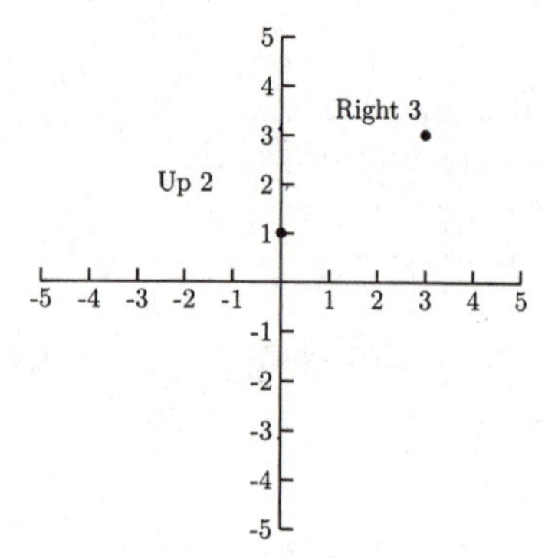

Fig. 4-25.

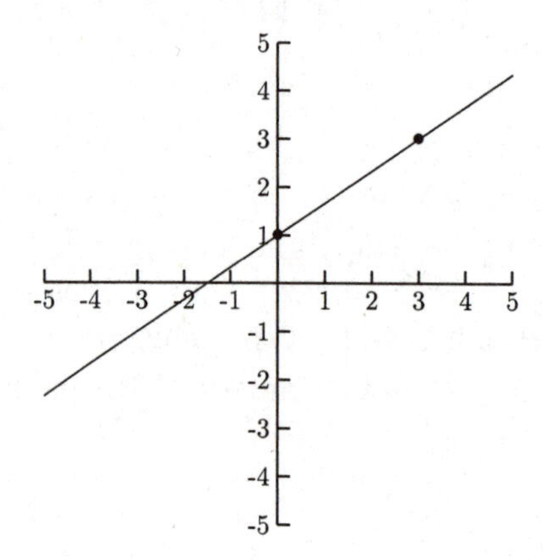

Fig. 4-26.

- $y = \frac{3}{5}x - 2$
  First plot $(0, -2)$. Next, go up 3 units and to the right 5 units.

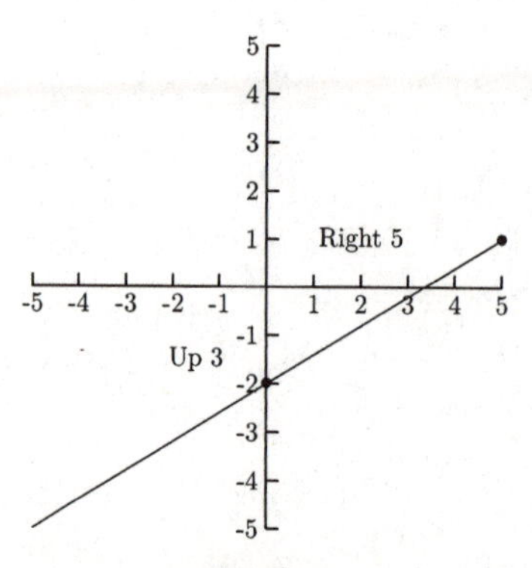

Fig. 4-27.

- $y = 3x + 1 = \frac{3}{1}x + 1$
  Plot (0, 1) then go up 3 units and to the right 1 unit.

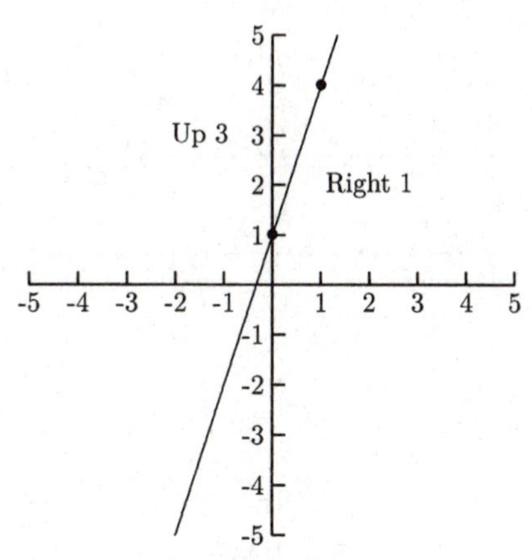

Fig. 4-28.

- $y = -\frac{5}{2}x = -\frac{5}{2}x + 0 = \frac{-5}{2}x + 0 = \frac{5}{-2}x + 0$
  Plot (0, 0). Either go down 5 units then to the right 2 units or go up 5 units then to the left 2 units.

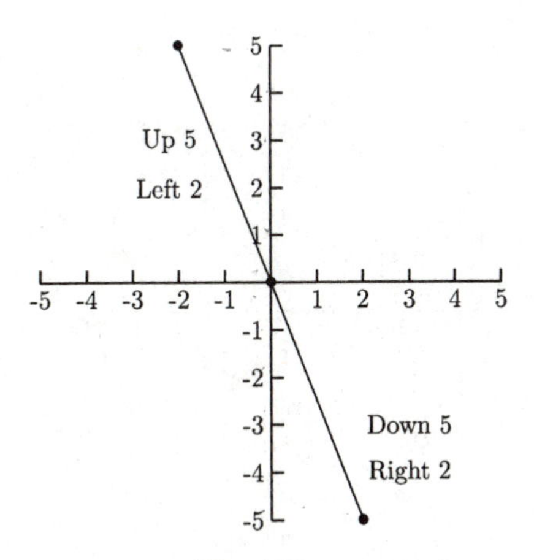

Fig. 4-29.

- $y = 2 = 0x + 2$
  Plot $(0, 2)$. Go to the left or right any distance; do not move up or down.

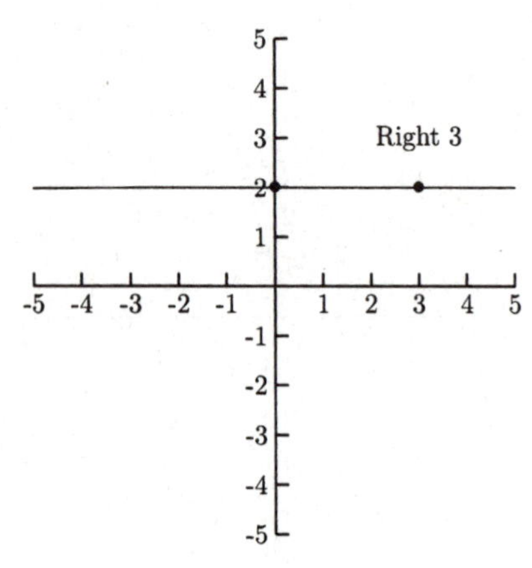

Fig. 4-30.

Parallel and Perpendicular Lines

Two lines are parallel if they have the same slope or if each slope is undefined.

EXAMPLES

-

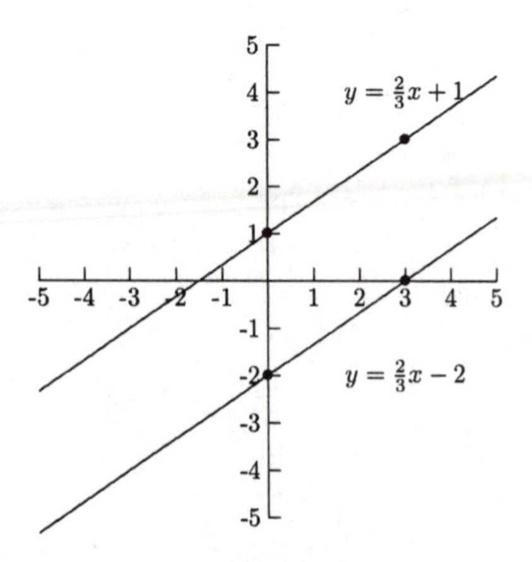

Fig. 4-31.

•

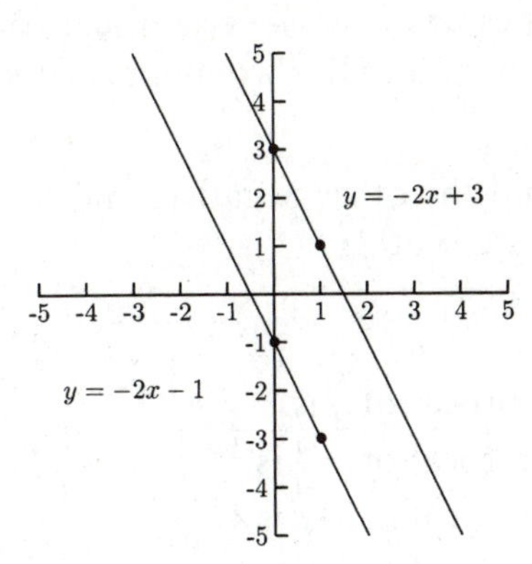

Fig. 4-32.

•

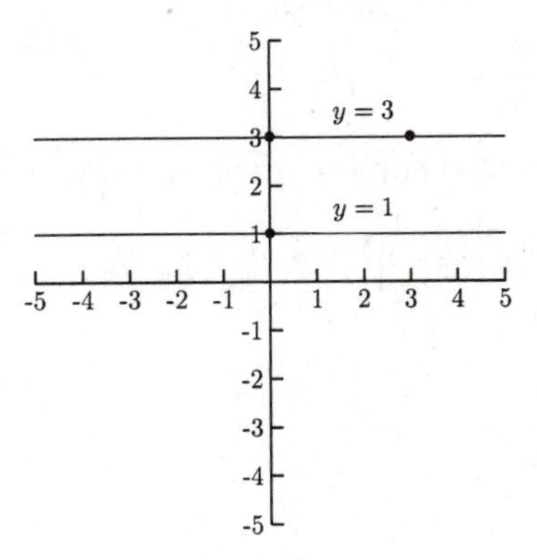

Fig. 4-33.

•

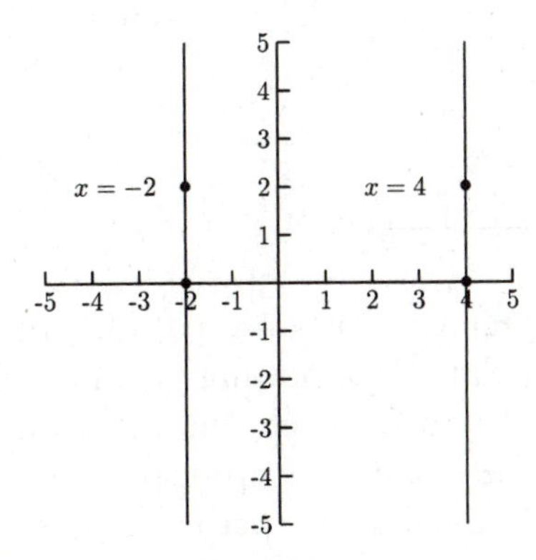

Fig. 4-34.

Two lines are perpendicular (that is, they form a 90° angle where they cross each other) if their slopes are negative reciprocals of each other (or one is vertical and the other is horizontal). Two numbers are negative reciprocals of each other if

- one is positive and the other negative; and
- inverting one gets the other.

EXAMPLES

- The negative reciprocal of $\frac{2}{3}$ is $-\frac{3}{2}$.
- The negative reciprocal of $-\frac{4}{5}$ is $\frac{5}{4}$.
- The negative reciprocal of -2 is $\frac{1}{2}$.
- The negative reciprocal of $\frac{5}{8}$ is $-\frac{8}{5}$.
- The negative reciprocal of $\frac{1}{4}$ is -4.
- The negative reciprocal of 1 is -1.

PRACTICE

Find the negative reciprocal for the given number.

1. $\frac{2}{7}$
2. $-\frac{4}{3}$
3. $\frac{1}{5}$
4. -3
5. -1

SOLUTIONS

1. $-\frac{7}{2}$
2. $\frac{3}{4}$
3. -5
4. $\frac{1}{3}$
5. 1

EXAMPLES

We can determine whether two lines are parallel, perpendicular, or neither parallel nor perpendicular, by putting their equations in the form $y = mx + b$. If m is the same for each line (or both are vertical), the lines are parallel. If one m is the negative reciprocal of the other (or one is vertical and the other horizontal), the lines are perpendicular. Otherwise, the lines are neither parallel nor perpendicular.

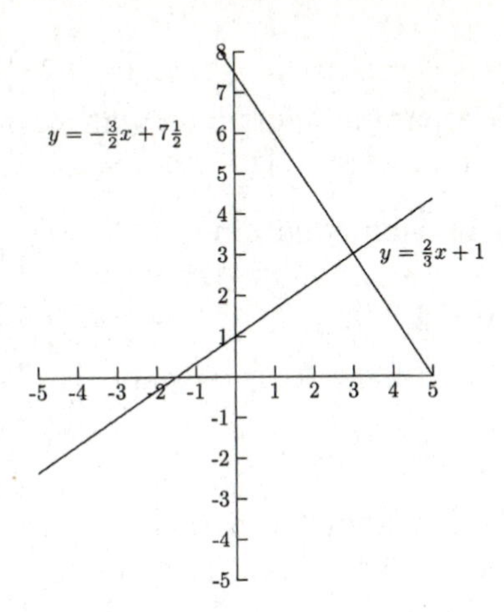

Fig. 4-35.

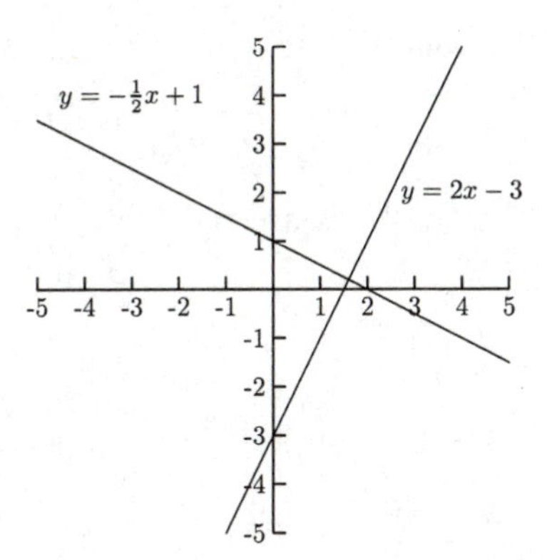

Fig. 4-36.

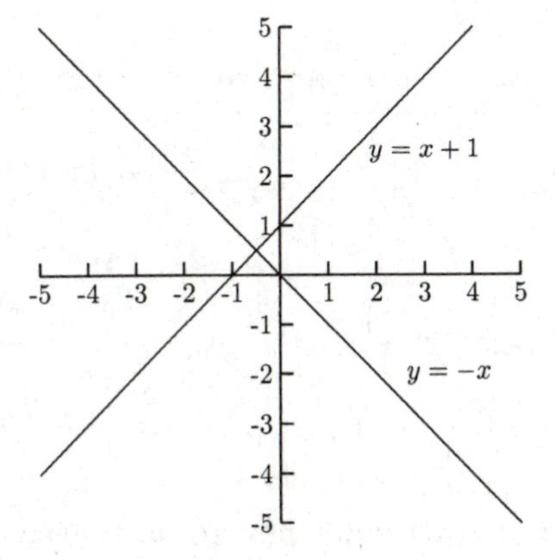

Fig. 4-37.

EXAMPLES

Determine if the lines are parallel, perpendicular, or neither.

- $2x - y = 4$ $\quad\quad$ $2x - y = -5$

  First, solve for y in each equation.

$$\begin{aligned} 2x - y &= 4 \\ -y &= -2x + 4 \\ y &= 2x - 4 \end{aligned} \qquad \begin{aligned} 2x - y &= -5 \\ -y &= -2x - 5 \\ y &= 2x + 5 \end{aligned}$$

  Each slope is 2 so the lines are parallel.

- $4x + y = 6$ $\quad\quad$ $8x + 2y = 4$

$$\begin{aligned} 4x + y &= 6 \\ y &= -4x + 6 \end{aligned} \qquad \begin{aligned} 8x + 2y &= 4 \\ 2y &= -8x + 4 \\ y &= -4x + 2 \end{aligned}$$

  Each slope is -4 so the lines are parallel.

- $x - 3y = -6$ $\quad\quad$ $3x + y = 2$

$$\begin{aligned} x - 3y &= -6 \\ -3y &= -x - 6 \\ y &= \frac{-x}{-3} - \frac{6}{-3} \\ y &= \frac{1}{3}x + 2 \end{aligned} \qquad \begin{aligned} 3x + y &= 2 \\ y &= -3x + 2 \end{aligned}$$

  The slopes are negative reciprocals of each other, so the lines are perpendicular.

- $3x - 4y = 4$ $\quad\quad$ $-4x + 3y = 9$

$$\begin{aligned} 3x - 4y &= 4 \\ -4y &= -3x + 4 \\ y &= \frac{3}{4}x - 1 \end{aligned} \qquad \begin{aligned} -4x + 3y &= 9 \\ 3y &= 4x + 9 \\ y &= \frac{4}{3} + 3 \end{aligned}$$

  The slopes are not equal and they are not negative reciprocals, so the lines are neither parallel nor perpendicular.

- $x = 4$ $y = -2$
  The first line is vertical and the second line is horizontal. These lines are perpendicular.

PRACTICE

Determine if the lines are parallel, perpendicular, or neither.

1. $3x - 8y = 8$ and $3x - 8y = -16$
2. $3x - 4y = -8$ and $4x - 3y = 3$
3. $3x - 5y = -10$ and $5x + 3y = -3$
4. $y = 6$ and $y = -4$
5. $3x - 3y = 2$ and $4x + 4y = 1$
6. $2x - y = 4$ and $6x - 3y = 9$
7. $x = 1$ and $y = 1$

SOLUTIONS

1. Parallel.

$$3x - 8y = 8 \qquad 3x - 8y = -16$$
$$-8y = -3x + 8 \qquad -8y = -3x - 16$$
$$y = \frac{3}{8}x - 1 \qquad y = \frac{3}{8}x + 2$$

2. Neither.

$$3x - 4y = -8 \qquad 4x - 3y = 3$$
$$-4y = -3x - 8 \qquad -3y = -4x + 3$$
$$y = \frac{3}{4}x + 2 \qquad y = \frac{4}{3}x - 1$$

3. Perpendicular.

$$3x - 5y = -10 \qquad 5x + 3y = -3$$
$$-5y = -3x - 10 \qquad 3y = -5x - 3$$
$$y = \frac{3}{5}x + 2 \qquad y = -\frac{5}{3}x - 1$$

4. Both lines are horizontal, so they are parallel.

5. Perpendicular (−1 and 1 are negative reciprocals).

$$3x - 3y = 2 \qquad\qquad 4x + 4y = 1$$
$$-3y = -3x + 2 \qquad\qquad 4y = -4x + 1$$
$$y = 1x - \frac{2}{3} \qquad\qquad y = -1x + \frac{1}{4}$$
$$y = x - \frac{2}{3} \qquad\qquad y = -x + \frac{1}{4}$$

6. Parallel.

$$2x - y = 4 \qquad\qquad 6x - 3y = 9$$
$$-y = -2x + 4 \qquad\qquad -3y = -6x + 9$$
$$y = 2x - 4 \qquad\qquad y = 2x - 3$$

7. One line is vertical and the other is horizontal, so these lines are perpendicular.

There is another way to find an equation of a line when we know the slope and a point that is a little faster than using the point–slope form ($y - y_1 = m(x - x_1)$). We can put the slope and the point we know into the slope–intercept form ($y = mx + b$). The only unknown would then be b.

EXAMPLES

- The slope is 4 and the point $(1, -2)$ is on the line.

  Because $m = 4$, $y = mx + b$ becomes $y = 4x + b$. We can find b by letting $x = 1$ and $y = -2$ in $y = 4x + b$.

$$-2 = 4(1) + b$$
$$-2 = 4 + b$$
$$-6 = b$$

  The equation containing $(1, -2)$ with slope 4 is $y = 4x - 6$.

- The slope is 2 and the x-intercept is 3.

  To say that the x-intercept is 3 is the same as saying $(3, 0)$ is a point on the line.

$$y = 2x + b$$
$$0 = 2(3) + b$$
$$-6 = b$$
$$y = 2x - 6$$

PRACTICE

Find the equation of the line.

1. The slope is 3 and $(-1, -8)$ is on the line.
2. The slope is $\frac{1}{2}$ and $(2, 5)$ is on the line.
3. The slope is $-\frac{4}{9}$ and $(18, -8)$ is on the line.
4. The slope is -5 and the x-intercept is 1.
5. The slope is -2 and the y-intercept is 6.

SOLUTIONS

1. $m = 3,\ x = -1,\ y = -8$

$$\begin{aligned} y &= mx + b \\ -8 &= 3(-1) + b \\ -5 &= b \\ y &= 3x - 5 \end{aligned}$$

2. $m = \frac{1}{2},\ x = 2,\ y = 5$

$$\begin{aligned} y &= mx + b \\ 5 &= \frac{1}{2}(2) + b \\ 5 &= 1 + b \\ 4 &= b \\ y &= \frac{1}{2}x + 4 \end{aligned}$$

3. $m = -\frac{4}{9},\ x = 18,\ y = -8$

$$\begin{aligned} y &= mx + b \\ -8 &= -\frac{4}{9}(18) + b \\ -8 &= -8 + b \\ 0 &= b \\ y &= -\frac{4}{9}x \end{aligned}$$

4. $m = -5$. To say that the x-intercept is 1 is another way of saying $(1, 0)$ is on the line, so $x = 1,\ y = 0$.

$$y = mx + b$$

$$0 = -5(1) + b$$

$$5 = b$$

$$y = -5x + 5$$

5. The y-intercept is 6 and $m = -2$. There is nothing we need to do but write down the equation: $y = -2x + 6$.

The relationship between many pairs of variables can be described by linear equations. These variables are called *linearly related*. For example, if one is paid \$12 per hour, the daily pay (before deductions) would be described by the equation $p = 12h$, where p represents the daily pay, and h represents the number of hours worked for the day. The slope of this line is 12 and the p-intercept (this is like the y-intercept) is 0. We will first use linear equations to answer such questions as, "If you were paid \$60, how many hours did you work?" Later we will use two pairs of numbers to find a linear equation.

EXAMPLES

- An electric company bills y dollars for x kilowatt hours used each month. The equation for each family's electric bill is $y = 0.06x + 20$. If a family's electric bill one month was \$68, how many kilowatt hours were used?

  The information given in the problem is $y = 68$ for $y = 0.06x + 20$. Substitute $y = 68$ in the equation and solve for x, the number of kilowatt hours used.

$$y = 0.06x + 20$$

$$68 = 0.06x + 20$$

$$48 = 0.06x$$

$$\frac{48}{0.06} = x$$

$$800 = x$$

The family used 800 kilowatt hours of electricity.

- The relationship between degrees Celsius and Fahrenheit is $F = \frac{9}{5}C + 32$.
  (a) If the temperature in Fahrenheit is 86 degrees, what is the temperature in Celsius?
  (b) If the temperature is -20 degrees Celsius, what is the temperature on the Fahrenheit scale?

  (a) Substitute $F = 86$ in $F = \frac{9}{5}C + 32$ and solve for C.

$$F = \frac{9}{5}C + 32$$

$$86 = \frac{9}{5}C + 32$$

$$54 = \frac{9}{5}C$$

$$\frac{5}{9} \cdot 54 = C$$

$$30 = C$$

  The temperature is 30 degrees Celsius.

  (b) Substitute $C = -20$ and compute F.

$$F = \frac{9}{5}(-20) + 32$$

$$F = -36 + 32 = -4$$

  The temperature is -4 degrees Fahrenheit.

- For the years 1990–1999, enrollment at a small college is approximated by the equation $y = 75x + 1100$, where y represents the number of students enrolled and x represents the number of years after 1990. Find the approximate enrollment for the years 1990, 1996, and 1999. In what year was enrollment about 1475?

  Because x represents the number of years after 1990, $x = 0$ is the year 1990; $x = 6$ is the year 1996; and $x = 9$ is the year 1999. (Because the equation is only good for the years 1990–1999, the only values of x we can use are $x = 0, 1, 2, \ldots, 9$.) We want to find y for $x = 0$, $x = 6$, $x = 9$.

  When $x = 0$, $y = 75(0) + 1100$. Enrollment for 1990 was about 1100.

When $x = 6$, $y = 75(6) + 1100 = 1550$. Enrollment for 1996 was about 1550.

When $x = 9$, $y = 75(9) + 1100 = 1775$. Enrollment for 1999 was about 1775.

For the question "In what year was enrollment about 1475?" let $y = 1475$ and solve for x.

$$1475 = 75x + 1100$$

$$375 = 75x$$

$$5 = x$$

Enrollment was about 1475 in the year $1990 + 5 = 1995$.

PRACTICE

1. A saleswoman's salary is given by the equation $y = 0.08x + 15{,}000$, where y is her annual salary and x is her annual sales level.

   (a) If her annual sales level was \$190,000, what was her annual salary?
   (b) If her annual salary was \$25,080, what was her annual sales level?

2. The relationship between degrees Celsius and degrees Fahrenheit is given by the equation $C = \frac{5}{9}(F - 32)$.

   (a) What is the temperature on the Celsius scale when it is 113 degrees Fahrenheit?
   (b) What is the temperature on the Fahrenheit scale when it is 35 degrees Celsius?

3. A package delivery company added vans to its fleet at one of its centers between the years 1995 and 2002. The number of vans in the center's fleet is given by the equation $y = 10x + 90$, where y is the number of vans and x is the number of years after 1995.

   (a) How many vans were in the fleet for the years 1995, 1999, and 2002?
   (b) In what year did the center have 110 vans?

SOLUTIONS

1. (a) Her annual sales level was \$190,000. Let $x = 190{,}000$ in the equation $y = 0.08x + 15{,}000$ and compute y.

$$y = 0.08(190{,}000) + 15{,}000$$

$$y = 15{,}200 + 15{,}000 = 30{,}200$$

Her annual salary was \$30,200.

(b) Her annual salary was \$25,080. Let $y = 25{,}080$ in the equation and solve for x.

$$25{,}080 = 0.08x + 15{,}000$$

$$10{,}080 = 0.08x$$

$$\frac{10{,}080}{0.08} = x$$

$$126{,}000 = x$$

Her annual sales level was \$126,000.

2. (a) Substitute $F = 113$ in $C = \frac{5}{9}(F - 32)$.

$$C = \frac{5}{9}(113 - 32)$$

$$C = \frac{5}{9}(81) = 45$$

The temperature is 45 degrees Celsius.

(b) Substitute $C = 35$ in the equation and solve for F.

$$35 = \frac{5}{9}(F - 32)$$

$$9(35) = \left(9 \cdot \frac{5}{9}\right)(F - 32)$$

$$315 = 5(F - 32)$$

$$315 = 5F - 160$$

$$475 = 5F$$

$$95 = F$$

The temperature is 95 degrees Fahrenheit.

3. (a) The year 1995 is 0 years after 1995, so $x = 0$. Substitute $x = 0$ in the equation $y = 10x + 90$ and compute y.

$$y = 10(0) + 90$$

$$y = 90$$

The center had 90 vans in its fleet in the year 1995.

The year 1999 is 4 years after 1995, so $x = 4$.

$$y = 10(4) + 90$$
$$y = 40 + 90 = 130$$

The center had 130 vans in its fleet in the year 1999.
The year 2002 is 7 years after 1995, so $x = 7$.

$$y = 10(7) + 90$$
$$y = 70 + 90 = 160$$

The center had 160 vans in its fleet in the year 2002.

(b) Let $y = 110$ in the equation and solve for x.

$$110 = 10x + 90$$
$$20 = 10x$$
$$2 = x$$

There were 110 vans in the fleet when $x = 2$, that is, in the year $1995 + 2 = 1997$.

In the last problems in this section, we will be given enough information to find a linear equation. In the first problem set, we will be given enough information to find two points on the line. In the second problem set, we will be given enough information to find a point and the slope.

EXAMPLES

- A company pays its entry-level sales representatives a commission that is a percentage of their monthly sales plus a certain base salary. This month, the sales representative from City A earned $5000 from sales of $35,000. The sales representative from City B earned $5300 on sales of $37,500. What percentage of monthly sales does the company pay in commission? What is its base salary?

  Salaries that are based on commission (with or without a base salary) are based on a linear equation. If y is the amount paid, m is the commission percentage, and x is the sales level, then the equation is $y = mx$ (without a base salary) and $y = mx + b$ (with base salary b). What do the ordered pairs (x, y) mean for this problem? The x-coordinate is a sales representative's sales level, and the y-coordinate is his pay amount. With this in mind, we can view the sentence, "The sales representative from City A earned $5000 from sales of $35,000" as the ordered pair $(35{,}000, 5000)$ on the line $y = mx + b$. The other sales representative's pay amount of $5300 on sales of

\$37,500 becomes the point (37,500, 5300). Now that we have two points on a line, we can find the equation of the line containing the points.

$$m = \frac{5300 - 5000}{37{,}500 - 35{,}000} = \frac{300}{2500} = 0.12$$

Using $m = 0.12$ and (35,000, 5000) in $y - y_1 = m(x - x_1)$, we get

$$y - 5000 = 0.12(x - 35{,}000)$$

$$y - 5000 = 0.12x - 4200$$

$$y = 0.12x + 800.$$

The commission rate is 12% of sales and the monthly base rate is \$800.

- The manager of a grocery store notices that sales of bananas are proportionate to sales of milk. On one Friday, 400 pounds of bananas are sold and 1700 gallons of milk are sold. On the following Friday, 360 pounds of bananas are sold and 1540 gallons of milk are sold. Find an equation that gives the number of gallons of milk sold in terms of the number of pounds of bananas sold.

  We will find a linear equation in the form $y = mx + b$. For some problems, it does not matter which quantity x represents and which y represents. In this problem, it does matter because of the sentence, "Find an equation that gives the number of gallons of milk sold in terms of the number of pounds of bananas sold." The equation $y = mx + b$ gives y in terms of x. This means that y will need to represent the number of gallons of milk and x will represent the number of pounds of bananas. The ordered pairs will be (bananas, milk). Our points, then, are (360, 1540) and (400, 1700).

$$m = \frac{1700 - 1540}{400 - 360} = \frac{160}{40} = 4$$

Using $m = 4$ and (400, 1700) in $y - y_1 = m(x - x_1)$, we get

$$y - 1700 = 4(x - 400)$$

$$y - 1700 = 4x - 1600$$

$$y = 4x + 100$$

What does the slope mean in these two problems? In the first equation, $y = 0.12x + 800$, the slope tells us how a sales representative's pay increases for each one-dollar increase in sales.

$$m = \frac{\$0.12}{\$1.00} = \frac{\text{increase in pay}}{\text{increase in sales}}$$

In the second equation, $y = 4x + 100$, the slope tells us that each pound of bananas sold results in four gallons of milk sold.

$$m = 4 = \frac{4}{1} = \frac{4 \text{ gallons of milk}}{1 \text{ pound of bananas}}$$

PRACTICE

1. A marketing director notices that the sales level for a certain product and amount spent on television advertising are linearly related. When \$6000 is spent on television advertising, sales for the product are \$255,000, and when \$8000 is spent on television advertising, sales for the product are \$305,000. Find an equation that gives the sales level for the product in terms of the amount spent on television advertising.
2. Show that the formula $C = \frac{5}{9}(F - 32)$ gives the degrees Celsius in terms of degrees Fahrenheit. Use the fact that water freezes at 0°C and 32°F and boils at 100°C and 212°F.
3. A car rental company charges a daily fee plus a mileage fee. A businesswoman's bill for one day was \$42.55 after driving 55 miles. The bill for the next day was \$36.40 after driving 40 miles. How much did it cost for each mile? What was the daily fee?

SOLUTIONS

1. Because we want the sales level in terms of the amount spent on television advertising, we will let y represent the sales level and x represent the amount spent on advertising. Our points are (6000, 255,000) and (8000, 305,000).

$$m = \frac{305{,}000 - 255{,}000}{8000 - 6000} = \frac{50{,}000}{2000} = 25$$

$$y - 255{,}000 = 25(x - 6000)$$
$$y - 255{,}000 = 25x - 150{,}000$$
$$y = 25x + 105{,}000$$

(Every dollar in advertising results in $25 in sales.)

2. We will treat C like y and F like x. Our points will have the form (degrees Fahrenheit, degrees Celsius), that is, $(32, 0)$ and $(212, 100)$.

$$m = \frac{100 - 0}{212 - 32} = \frac{100}{180} = \frac{5}{9}$$
$$C - 0 = \frac{5}{9}(F - 32)$$
$$C = \frac{5}{9}(F - 32)$$

3. In the equation $y = mx + b$, we will let x represent the number of miles driven and y represent the daily cost. The ordered pair (x, y) is (miles, cost). The points are $(55, 42.55)$ and $(40, 36.40)$.

$$m = \frac{36.40 - 42.55}{40 - 55} = \frac{-6.15}{-15} = 0.41$$
$$y - 36.40 = 0.41(x - 40)$$
$$y - 36.40 = 0.41x - 16.40$$
$$y = 0.41x + 20$$

The daily fee is $20 and each mile costs $0.41.

In these last problems, we will be given one pair of numbers, which will be a point on the line, and information on the rate of change. The rate of change is the slope.

EXAMPLES

- A utilities company charges $4\frac{1}{2}$ cents per kilowatt hour for electricity plus a monthly base charge. Find an equation that gives the monthly costs in terms of the number of kilowatt hours of electricity used if the bill for one month for a certain family is $62.55 for 1050 kilowatt hours of electricity used.

  In the equation $y = mx + b$ (where y is the cost and x is the number of kilowatt hours used), the slope is the cost per kilowatt hour of electricity used. This means that $m = 0.045$, and a point is $(1050, 62.55)$. We could use either $y - y_1 = m(x - x_1)$ (as before) or $y = mx + b$. We will use $y = mx + b$ with $y = 62.55$, $m = 0.045$ and $x = 1050$.

$$62.55 = 0.045(1050) + b$$
$$15.30 = b$$

The equation is $y = 0.045x + 15.30$.

- A recipe calls for two cups of biscuit mix and $\frac{2}{3}$ cups of milk. Find a linear equation that gives the amount of milk in terms of the amount of biscuit mix.

  Because we need to give the milk in terms of the biscuit mix, we will let y represent the number of cups of milk and x represent the number of cups of biscuit mix. The ordered pair (x, y) is (biscuit mix, milk). Also, the slope is (change in y)/(change in x) which is (change in milk)/(change in mix). We will use the fact that if we increase the number of cups of biscuit mix by two cups, we need to increase the number of cups of milk by $\frac{2}{3}$, giving us a slope of

$$\frac{2/3}{2} = \frac{2}{3} \div 2 = \frac{2}{3} \cdot \frac{1}{2} = \frac{1}{3}.$$

  So far we have $y = \frac{1}{3}x + b$. We need more information to find b. Although another point is not explicitly given, we can figure one out—when no biscuit mix is used, no milk is used. In other words, $(0, 0)$ is a point on the line. This means that b, the y-intercept, is 0. The equation is $y = \frac{1}{3}x$.

- The dosage for a certain cattle drug is $4.5\,\text{cm}^3$ per 100 pounds of body weight. Find an equation that gives the amount of the drug in terms of a cow's weight.

  We want the amount of the drug in terms of a cow's weight, so we will let y represent the number of cubic centimeters of the drug and x represent a cow's weight in pounds. What is the slope of our line?

$$m = \frac{\text{change in drug amount}}{\text{change in weight}} = \frac{4.5}{100} = 0.045$$

  This means that $0.045\,\text{cm}^3$ of the drug is needed for each pound of a cow's weight. Again, $(0, 0)$ is a point on the line, so $b = 0$. The equation is $y = 0.045x$.

PRACTICE

1. Each unit of a product costs $1.75 to make. The total cost to produce 20,000 units one week was $41,000. Find an equation that gives the total cost in terms of the number of units produced.
2. The manager of a movie theater believes that for every 200 tickets sold, 15 buckets of popcorn are sold. Find an equation that gives the amount of buckets of popcorn sold in terms of the number of tickets sold.
3. An office manager notices that the office copier uses one container of toner for every 25 reams of paper. Find an equation that gives the amount of toner used in terms of the amount of paper used.
4. A garden hose is used to fill a tall rectangular tank. The water level rises six inches every 20 minutes. If the water level was already eight inches before the water was turned on, find an equation that gives the water level in terms of the time the hose is used.

SOLUTIONS

1. Let y represent the total cost and x the number of units produced. This means that $(20{,}000, 41{,}000)$ is a point on the line. Each unit costs $1.75 to produce, so the slope is 1.75. We will let $x = 20{,}000$, $y = 41{,}000$, and $m = 1.75$ in $y = mx + b$.

$$41{,}000 = 1.75(20{,}000) + b$$

$$6000 = b$$

The equation is $y = 1.75x + 6000$. ($6000 represents *fixed costs*, costs such as rent, loan payments, salaries, etc.)

2. Let y represent the number of buckets of popcorn sold and x represent the number of tickets sold. The slope is

$$m = \frac{\text{change in popcorn sales}}{\text{change in tickets sold}} = \frac{15}{200} = 0.075.$$

Because 0 buckets of popcorn are sold when 0 tickets are sold, $(0, 0)$ is on the line and b, the y-intercept, is 0. The equation is $y = 0.075x$.

3. Let y represent the number of toner containers used and x the number of reams of paper used. The slope is

$$m = \frac{\text{change in toner used}}{\text{change in paper used}} = \frac{1}{25}.$$

The point $(0, 0)$ is on the graph, so b, the y-intercept, is 0. The equation is $y = \frac{1}{25}x$.

4. Let y represent the water level in inches and x the time in minutes that the hose is used. When the time is 0 minutes, the water level is 8 inches, giving us the point $(0, 8)$. This means that b, the y-intercept, is 8. The slope is

$$m = \frac{6 \text{ inches}}{20 \text{ minutes}} = 0.3 \text{ inches per minute.}$$

The equation is $y = 0.3x + 8$.

Parabolas

The graph of any quadratic equation ($y = ax^2 + bx + c$) will look like one of the graphs below.

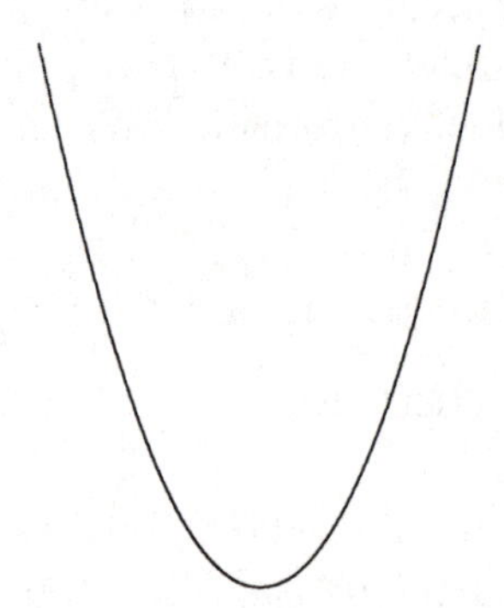

Fig. 4-38.

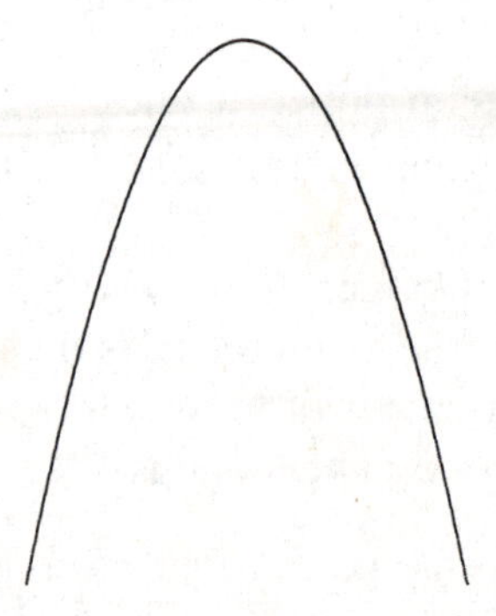

Fig. 4-39.

These graphs are called *parabolas*. Parabolas occur in nature. To see a parabola, toss a small object up and watch its path—it will be part of a parabola.

The graph of every parabola has a *vertex*, the point where the graph turns around. For a parabola that opens up, the vertex is the lowest point. The vertex is the highest point for a graph that opens down. When a quadratic equation is in the form $y = a(x - h)^2 + k$, the vertex is the point (h, k).

EXAMPLES

- $y = 2(x - 4)^2 + 5$
  $a = 2,\ h = 4,\ k = 5$. The vertex is $(4, 5)$.
- $y = (x - 2)^2 - 1$
  $a = 1,\ h = 2,\ k = -1$. The vertex is $(2, -1)$.
- $y = -3(x - 1)^2 + 2$
  $a = -3,\ h = 1,\ k = 2$. The vertex is $(1, 2)$.
- $y = \frac{1}{2}(x + 4)^2 + 1$
  $a = \frac{1}{2},\ h = -4,\ k = 1$. The vertex is $(-4, 1)$.
- $y = 2x^2 - 4 = 2(x - 0)^2 - 4$
  $a = 2,\ h = 0,\ k = -4$. The vertex is $(0, -4)$.
- $y = (x - 8)^2 = (x - 8)^2 + 0$
  $a = 1,\ h = 8,\ k = 0$. The vertex is $(8, 0)$.

PRACTICE

Identify a and the vertex.

1. $y = -(x - 2)^2 + 4$
2. $y = 10(x + 1)^2 - 2$
3. $y = \frac{1}{2}(x + 5)^2 + 4$
4. $y = (x + 6)^2$
5. $y = -x^2$

SOLUTIONS

1. $a = -1$. The vertex is $(2, 4)$.
2. $a = 10$. The vertex is $(-1, -2)$.
3. $a = \frac{1}{2}$. The vertex is $(-5, 4)$.
4. $a = 1$. The vertex is $(-6, 0)$.
5. $a = -1$. The vertex is $(0, 0)$.

When graphing parabolas, we will begin with the vertex. We will graph two points to the left and to the right of the vertex. One pair of points should be fairly close to the vertex to show the curving around the vertex. Another pair should be further away to show how steep the ends are. What do "fairly close" and "a little further away" mean? There is no standard answer. For some parabolas, one unit is "close" but for others, one unit is "far away." It all depends on a. A good rule of thumb is to plot two points a units to the left and to the right of the vertex and two other points that are $2a$ units to the left and right of the vertex. The sign on a is also important. When a is positive, the parabola opens up (see Fig. 4-40). When a is negative, the parabola opens down (see Fig. 4-41).

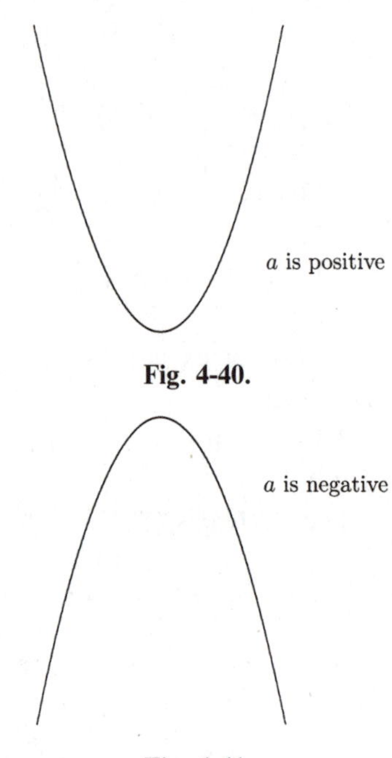

Fig. 4-40.

Fig. 4-41.

We will start the T-table (Table 4-1) with five x-values. Because parabolas are symmetric (the left half is a mirror-image of the right half), the y-values

Table 4-1

x	y
$h-2a$	
$h-a$	
h	k vertex
$h+a$	
$h+2a$	

for $h - a$ and $h + a$ will be the same; and the y-values for $h - 2a$ and $h + 2a$ will be the same. This can save some computation.

EXAMPLES

- $y = 2(x - 4)^2 - 8$
  $a = 2,\ h = 4,\ k = -8$ (Table 4-2).

Table 4-2

x	y
0	
2	
4	-8 vertex
6	
8	

$$h - 2a = 4 - 2(2) = 0$$
$$h - a = 4 - 2 = 2$$
$$h + a = 4 + 2 = 6$$
$$h + 2a = 4 + 2(2) = 8$$

Compute the y-values (Table 4-3).

Table 4-3

x	y
0	24
2	0
4	-8
6	0
8	24

$$x = 0 \qquad y = 2(0 - 4)^2 - 8 = 24$$
$$x = 2 \qquad y = 2(2 - 4)^2 - 8 = 0$$
$$x = 6 \qquad y = 2(6 - 4)^2 - 8 = 0$$
$$x = 8 \qquad y = 2(8 - 4)^2 - 8 = 24$$

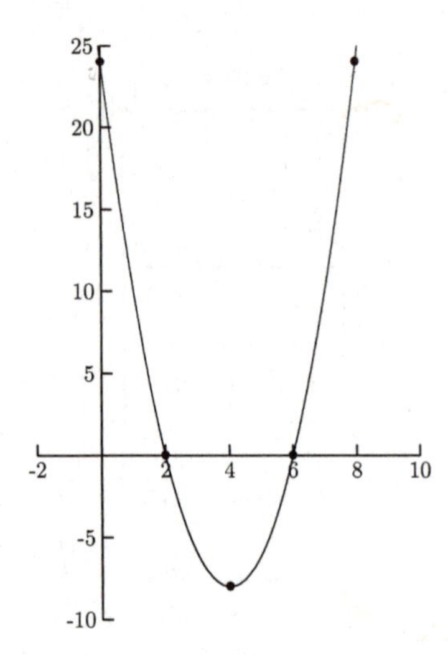

Fig. 4-42.

- $-2(x-1)^2 + 5$

  $a = -2, \; h = 1, \; k = 5$

$$h - 2a = 1 - 2(-2) = 5$$
$$h - a = 1 - (-2) = 3$$
$$h + a = 1 + (-2) = -1$$
$$h + 2a = 1 + 2(-2) = -3$$

Compute the y-values (Table 4-4).

$$x = 5 \qquad y = -2(5 - 1)^2 + 5 = -27$$
$$x = 3 \qquad y = -2(3 - 1)^2 + 5 = -3$$
$$x = -1 \qquad y = -2(-1 - 1)^2 + 5 = -3$$
$$x = -3 \qquad y = -2(-3 - 1)^2 + 5 = -27$$

Table 4-4

x	y
5	−27
3	−3
1	5
−1	−3
−3	−27

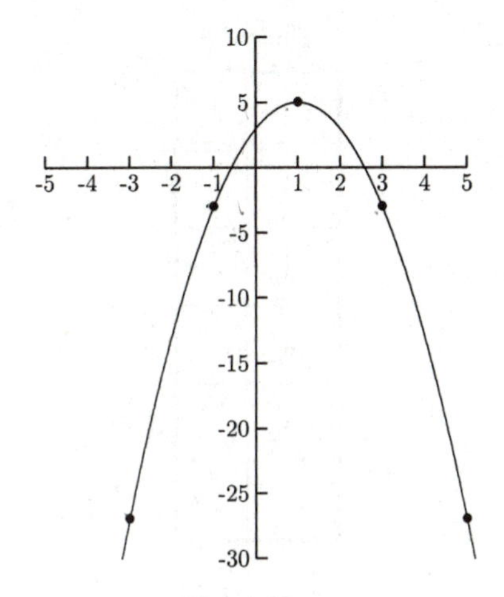

Fig. 4-43.

- $y = \frac{1}{2}(x+4)^2 + 1$

  $a = \frac{1}{2},\ h = -4,\ k = 1$

$$h - 2a = -4 - 2\left(\frac{1}{2}\right) = -5$$

$$h - a = -4 - \frac{1}{2} = -4\frac{1}{2}$$

$$h + a = -4 + \frac{1}{2} = -3\frac{1}{2}$$

$$h + 2a = -4 + 2\left(\frac{1}{2}\right) = -3$$

Compute the y-values (Table 4-5).

$$x = -5 \qquad y = \frac{1}{2}(-5+4)^2 + 1 = 1\frac{1}{2}$$

$$x = -4\frac{1}{2} \qquad y = \frac{1}{2}\left(-4\frac{1}{2}+4\right)^2 + 1 = 1\frac{1}{8}$$

$$x = -3\frac{1}{2} \qquad y = \frac{1}{2}\left(-3\frac{1}{2}+4\right)^2 + 1 = 1\frac{1}{8}$$

$$x = -3 \qquad y = \frac{1}{2}(-3+4)^2 + 1 = 1\frac{1}{2}$$

Table 4-5

x	y
-5	$1\frac{1}{2}$
$-4\frac{1}{2}$	$1\frac{1}{8}$
-4	1
$-3\frac{1}{2}$	$1\frac{1}{8}$
-3	$1\frac{1}{8}$

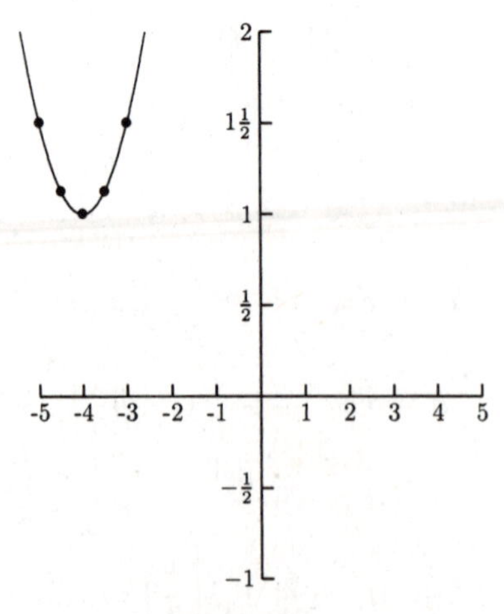

Fig. 4-44.

PRACTICE

Graph the quadratic equations.

1. $y = -(x-2)^2 + 4$
2. $y = (x-3)^2 - 1$
3. $y = \frac{1}{2}(x+5)^2 + 4$
4. $y = (x+6)^2$
5. $y = -x^2$ (Square x then take the negative. For example $-3^2 = -9$.)

SOLUTIONS

1. $a = -1,\ h = 2,\ k = 4$ (Table 4-6).

Table 4-6

x	y
4	0
3	3
2	4
1	3
0	0

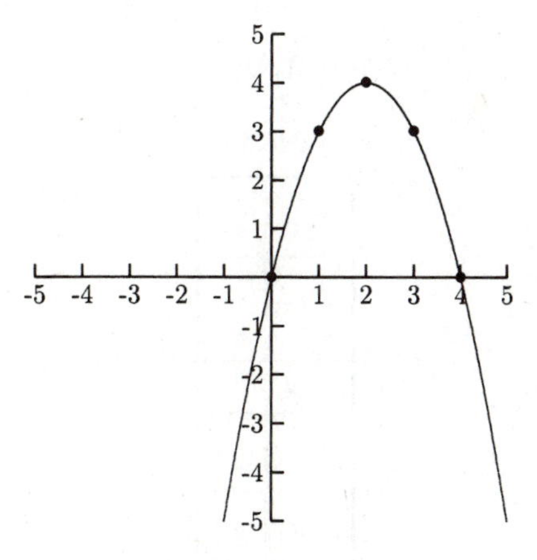

Fig. 4-45.

2. $a = 1$, $h = 3$, $k = -1$ (Table 4-7).

Table 4-7

x	y
1	3
2	0
3	−1
4	0
5	3

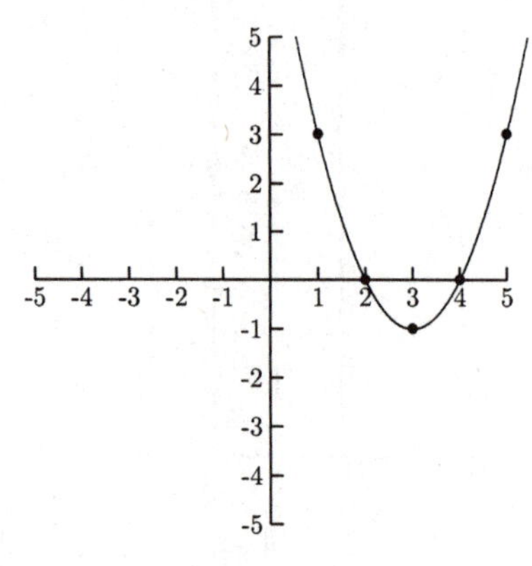

Fig. 4-46.

3. $a = \frac{1}{2}$, $h = -5$, $k = 4$ (Table 4-8).

Table 4-8

x	y
-6	$4\frac{1}{2}$
$-5\frac{1}{2}$	$4\frac{1}{8}$
-5	4
$-4\frac{1}{2}$	$4\frac{1}{8}$
-4	$4\frac{1}{2}$

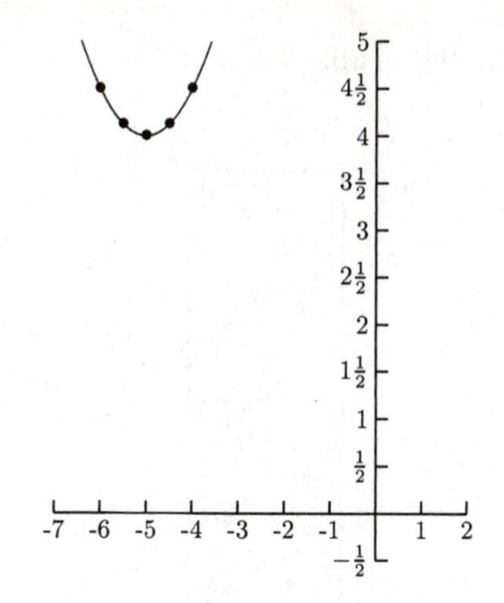

Fig. 4-47.

4. $a = 1,\ h = -6,\ k = 0$ (Table 4-9).

Table 4-9

x	y
−8	4
−7	1
−6	0
−5	1
−4	4

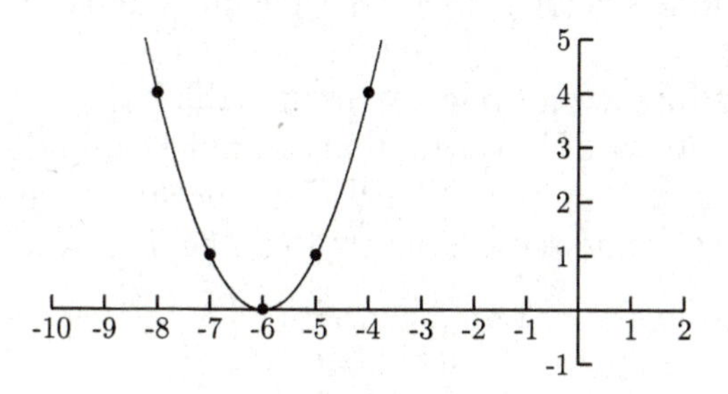

Fig. 4-48.

5. $a = -1$, $h = 0$, $k = 0$ (Table 4-10).

Table 4-10

x	y
2	-4
1	-1
0	0
-1	-1
-2	-4

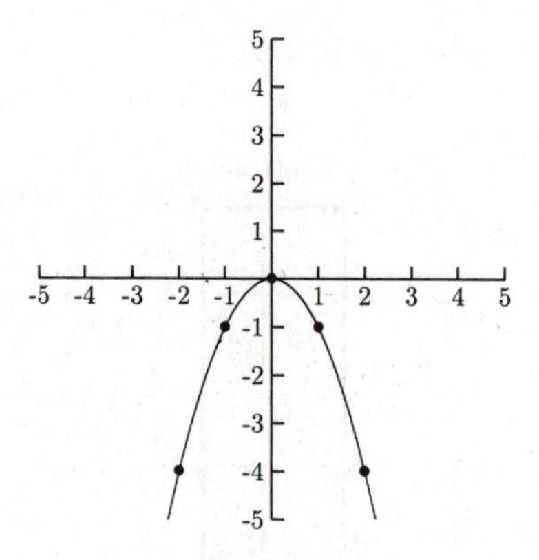

Fig. 4-49.

It might seem that $y = a(x - h)^2 + k$ should either be $y = a(x + h)^2 + k$ or $y = a(x - h)^2 - k$. The reason that the signs in front of h and k are different is that k (the y-coordinate of the vertex) is on the same side as x. If k were on the same side as y, then the signs on h and k would be the same: $y - k = a(x - h)^2$.

Quadratic equations are normally written in the form $y = ax^2 + bx + c$, not $y = a(x - h)^2 + k$. As with circle equations, to put an equation in the first form into one that is in the second form we will need to complete the square. Completing the square on these equations can be more of a problem.

1. Factor a from the x^2 and x terms.
2. Complete the square on the x^2 and x terms.

3. Compute the constant that needs to be added to c. This step is the difficult part.
4. Write the expression in the parentheses as a perfect square.

EXAMPLES

- $y = x^2 - 6x + 1$
  Because $a = 1$, Step 1 is not necessary. Complete the square on $x^2 - 6x$.

$$y = (x^2 - 6x + 9) + 1 + ?$$

We need to balance the equation so that it is the same as the original equation. To balance "+9", add "−9" to 1.

$$y = (x^2 - 6x + 9) + 1 + (-9)$$
$$y = (x^2 - 6x + 9) - 8$$
$$y = (x - 3)^2 - 8$$

- $y = x^2 + 8x + 5$
  Because $a = 1$, Step 1 is not necessary. Complete the square on $x^2 + 8x$.

$$y = (x^2 + 8x + 16) + 5 + ?$$

To balance the "+16" in the parentheses, add "−16" to 5.

$$y = (x^2 + 8x + 16) + 5 + (-16)$$
$$y = (x^2 + 8x + 16) - 11$$
$$y = (x + 4)^2 - 11$$

- $y = 2x^2 + 16x - 1$
  Factor 2 from $2x^2 + 16x$. We must do this step *before* completing the square.

$$y = 2(x^2 + 8x) - 1$$
$$y = 2(x^2 + 8x + 16) - 1 + ?$$

Adding "−16" to −1 might seem to be the next step. This would not get an equivalent equation. We will simplify $y = 2(x^2 + 8x + 16) - 1$ to see what effect adding 16 in the parentheses has on the equation.

$$y = 2(x^2 + 8x + 16) - 1 = 2x^2 + 16x + 32 - 1$$

By putting "+16" in the parentheses, we are really adding $2(16) = 32$. To balance the equation, we need to add "-32" to -1.

$$y = 2(x^2 + 8x + 16) - 1 - 32$$

$$y = 2(x^2 + 8x + 16) - 33$$

$$y = 2(x + 4)^2 - 33$$

When factoring fractions (or other unusual quantities) from algebraic expressions, write each term to be factored as the numerator of a fraction. The quantity to be factored will be the denominator of the fraction. Simplify this fraction. The simplified fraction is what goes inside the parentheses.

EXAMPLES

- Factor $\frac{1}{4}$ from $2x - 3$.

$$\frac{2x}{1/4} = 2x \div \frac{1}{4} = 2x \cdot 4 = 8x$$

$$\frac{-3}{1/4} = -3 \div \frac{1}{4} = -3 \cdot 4 = -12$$

$$2x - 3 = \frac{1}{4}(8x - 12)$$

- Factor $-\frac{5}{3}$ from $6x^2 - 2x + 1$.

$$\frac{6x^2}{-5/3} = 6x^2 \div -\frac{5}{3} = 6x^2 \cdot -\frac{3}{5} = -\frac{18x^2}{5} = -\frac{18}{5}x^2$$

$$\frac{-2x}{-5/3} = -2x \div -\frac{5}{3} = -2x \cdot -\frac{3}{5} = \frac{6x}{5} = \frac{6}{5}x$$

$$\frac{1}{-5/3} = 1 \div -\frac{5}{3} = 1 \cdot -\frac{3}{5} = -\frac{3}{5}$$

$$6x^2 - 2x + 1 = -\frac{5}{3}\left(-\frac{18}{5}x^2 + \frac{6}{5}x - \frac{3}{5}\right)$$

Now we can complete the square on $y = ax^2 + bx + c$ when a is a fraction.

EXAMPLES

Put the following equations in the form $y = a(x - h)^2 + k$.

- $y = \frac{1}{2}x^2 + 3x - 4$

First we will factor $\frac{1}{2}$ from $\frac{1}{2}x^2 + 3x$.

$$\frac{(1/2)x^2}{1/2} = x^2 \qquad \frac{1}{2} \text{ cancels}$$

$$\frac{3x}{1/2} = 3x \div \frac{1}{2} = 3x \cdot 2 = 6x$$

$$y = \frac{1}{2}(x^2 + 6x) - 4$$

Now we can complete the square.

$$y = \frac{1}{2}(x^2 + 6x + 9) - 4 + ?$$

By putting "+9" in the parentheses, we are really adding $\frac{1}{2}(9) = \frac{9}{2}$. To balance the equation, add $-\frac{9}{2}$ to -4.

$$y = \frac{1}{2}(x^2 + 6x + 9) - 4 + \left(-\frac{9}{2}\right)$$

$$y = \frac{1}{2}(x^2 + 6x + 9) - \frac{17}{2}$$

$$y = \frac{1}{2}(x + 3)^2 - \frac{17}{2}$$

- $y = -3x^2 + 12x + 2$

  First we will factor -3 from $-3x^2 + 12x$.

$$\frac{-3x^2}{-3} = x^2$$

$$\frac{12x}{-3} = -4x$$

$$y = -3(x^2 - 4x) + 2$$

We are ready to complete the square.

$$y = -3(x^2 - 4x + 4) + 2 + ?$$

By putting "+4" in the parentheses, we are really adding $-3(4) = -12$. Balance this by adding 12 to 2.

$$y = -3(x^2 - 4x + 4) + 2 + 12$$

$$y = -3(x^2 - 4x + 4) + 14$$

$$y = -3(x - 2)^2 + 14$$

PRACTICE

Put the quadratic equations in the form $y = a(x - h)^2 + k$.

1. $y = x^2 - 10x + 6$
2. $y = -x^2 - 4x + 3$
3. $y = 5x^2 + 10x + 6$
4. $y = 3x^2 + 12x - 4$
5. $y = -2x^2 + 4x + 3$
6. $y = \frac{2}{3}x^2 + 4x - 2$

SOLUTIONS

1.

$$y = (x^2 - 10x + 25) + 6 + (-25)$$

$$y = (x^2 - 10x + 25) - 19$$

$$y = (x - 5)^2 - 19$$

2.

$$y = -x^2 - 4x + 3$$

$$y = -(x^2 + 4x) + 3$$

$$y = -(x^2 + 4x + 4) + 3 + ?$$

By putting "+4" in the parentheses, we are adding $-(4) = -4$. We will balance this by adding 4 to 3.

$$y = -(x^2 + 4x + 4) + 3 + 4$$

$$y = -(x^2 + 4x + 4) + 7$$

$$y = -(x + 2)^2 + 7$$

3.

$$y = 5x^2 + 10x + 6$$

$$y = 5(x^2 + 2x) + 6$$

$$y = 5(x^2 + 2x + 1) + 6 + ?$$

By putting "+1" in the parentheses, we are adding $5(1) = 5$. Balance this by adding -5 to 6.

$$y = 5(x^2 + 2x + 1) + 6 + (-5)$$

$$y = 5(x^2 + 2x + 1) + 1$$

$$y = 5(x + 1)^2 + 1$$

4.

$$y = 3x^2 + 12x - 4$$

$$y = 3(x^2 + 4x) - 4$$

$$y = 3(x^2 + 4x + 4) - 4 + ?$$

By putting "+4" inside the parentheses, we are adding $3(4) = 12$. Balance this by adding -12 to -4.

$$y = 3(x^2 + 4x + 4) - 4 + (-12)$$

$$y = 3(x^2 + 4x + 4) - 16$$

$$y = 3(x + 2)^2 - 16$$

5.

$$y = -2x^2 + 4x + 3$$

$$y = -2(x^2 - 2x) + 3$$

$$y = -2(x^2 - 2x + 1) + 3 + ?$$

By putting "+1" inside the parentheses, we are adding $-2(1) = -2$. Balance this by adding 2 to 3.

$$y = -2(x^2 - 2x + 1) + 3 + 2$$

$$y = -2(x^2 - 2x + 1) + 5$$

$$y = -2(x - 1)^2 + 5$$

6.

$$y = \frac{2}{3}x^2 + 4x - 2$$

Factor $\frac{2}{3}$ from $\frac{2}{3}x^2 + 4x$.

$$\frac{(2/3)x^2}{2/3} = x^2$$

$$\frac{4x}{2/3} = 4x \div \frac{2}{3} = 4x \cdot \frac{3}{2} = 6x$$

$$y = \frac{2}{3}(x^2 + 6x) - 2$$

$$y = \frac{2}{3}(x^2 + 6x + 9) - 2 + ?$$

By putting "+9" in the parentheses, we are adding $\frac{2}{3}(9) = 6$. Balance this by adding -6 to -2.

$$y = \frac{2}{3}(x^2 + 6x + 9) - 2 + (-6)$$

$$y = \frac{2}{3}(x^2 + 6x + 9) - 8$$

$$y = \frac{2}{3}(x + 3)^2 - 8$$

There is a shortcut for finding the vertex of a parabola without having to put the equation in the form $y = a(x - h)^2 + k$. The shortcut involves a formula for h. We can compute k by putting $x = h$ in the equation. The shortcut for h comes from completing the square on $y = ax^2 + bx + c$.

$$y = ax^2 + bx + c$$

Factor a from $ax^2 + bx$.

$$\frac{ax^2}{a} = x^2$$

$$\frac{bx}{a} = \frac{b}{a}x$$

$$y = a\left(x^2 + \frac{b}{a}x\right) + c$$

Complete the square by adding $(1/2 \cdot b/a)^2 = b^2/4a^2$. Putting this in the parentheses is adding $a(b^2/4a^2) = b^2/4a$. Subtract this from c.

$$y = a\left(x^2 + \frac{b}{a}x + \frac{b^2}{4a^2}\right) + c - \frac{b^2}{4a}$$

$$y = a\left(x + \frac{b}{2a}\right)^2 + c - \frac{b^2}{4a}$$

$$h = -\frac{b}{2a} \qquad\qquad k = c - \frac{b^2}{4a}$$

It is usually easier to compute k by letting $x = h$ in the equation rather than using the above formula for k.

EXAMPLES

Find the vertex using $h = -b/2a$.

- $y = x^2 + 6x + 4$

$$a = 1 \qquad b = 6 \qquad h = \frac{-b}{2a} = \frac{-6}{2(1)} = -3$$

$$k = (-3)^2 + 6(-3) + 4 = -5$$

The vertex is $(-3, -5)$.

- $y = 2x^2 - 12x - 7$

$$a = 2 \qquad b = -12 \qquad h = \frac{-b}{2a} = \frac{-(-12)}{2(2)} = 3$$

$$k = 2(3)^2 - 12(3) - 7 = -25$$

The vertex is $(3, -25)$.

- $y = -x^2 + 2x - 4$

$$a = -1, \qquad b = 2 \qquad h = \frac{-b}{2a} = \frac{-2}{2(-1)} = 1$$

$$k = -(1)^2 + 2(1) - 4 = -3$$

The vertex is $(1, -3)$.

PRACTICE

Find the vertex using $h = -b/2a$.

1. $y = x^2 + 6x + 5$
2. $y = \frac{1}{2}x^2 - 3x + 4$
3. $y = 4x^2 - 6x + 8$
4. $y = -x^2 - 5x + 3$

SOLUTIONS

1. $a = 1$, $b = 6$

$$h = \frac{-b}{2a} = \frac{-6}{2(1)} = -3$$

$$k = (-3)^2 + 6(-3) + 5 = -4$$

The vertex is $(-3, -4)$.

2. $a = \frac{1}{2}$, $b = -3$

$$h = \frac{-b}{2a} = \frac{-(-3)}{2 \cdot 1/2} = \frac{3}{1} = 3$$

$$k = \frac{1}{2}(3)^2 - 3(3) + 4 = -\frac{1}{2}$$

The vertex is $(3, -\frac{1}{2})$.

3. $a = 4$, $b = -6$

$$h = \frac{-b}{2a} = \frac{-(-6)}{2(4)} = \frac{3}{4}$$

$$k = 4\left(\frac{3}{4}\right)^2 - 6\left(\frac{3}{4}\right) + 8$$

$$= 4\left(\frac{9}{16}\right) - \frac{9}{2} + 8 = \frac{23}{4}$$

The vertex is $(\frac{3}{4}, \frac{23}{4})$.

4. $a = -1$, $b = -5$

$$h = \frac{-b}{2a} = \frac{-(-5)}{2(-1)} = -\frac{5}{2}$$

$$k = -\left(-\frac{5}{2}\right)^2 - 5\left(-\frac{5}{2}\right) + 3$$

$$= -\left(\frac{25}{4}\right) + \frac{25}{2} + 3 = \frac{37}{4}$$

The vertex is $(-\frac{5}{2}, \frac{37}{4})$.

Chapter 4 Review

1. Find the slope of the line containing the points $(-1, 4)$ and $(3, 2)$.
   a) -2 b) 2 c) $-\frac{1}{2}$ d) $\frac{1}{2}$

2. What is the vertex for $y = 2(x-1)^2 + 3$?
   a) $(2, 3)$ b) $(-2, 3)$ c) $(1, 3)$ d) $(-1, 3)$

3. What is the slope and y-intercept for the line $y = -\frac{2}{5}x + 4$?
   a) The slope is $\frac{2}{5}$, and the y-intercept is 4.
   b) The slope is $-\frac{2}{5}$, and the y-intercept is 4.
   c) The slope is 4, and the y-intercept is $\frac{2}{5}$.
   d) The slope is 4, and the y-intercept is $-\frac{2}{5}$.

4. Rewrite $y = 2x^2 - 8x + 5$ in the form $y = a(x-h)^2 + k$.
   a) $y = 2(x-4)^2 - 11$ b) $y = 2(x-2)^2 + 9$
   c) $y = 2(x-2)^2 - 3$ d) $y = 2(x-4)^2 + 21$

5. Are the lines $2x - 3y = 9$ and $2x + 3y = -6$ parallel, perpendicular, or neither?
   a) Parallel b) Perpendicular c) Neither d) Cannot be determined

6. What are the intercepts for $y = x^2 - 2x - 8$?
   a) The x-intercept is -8, there is no y-intercept.
   b) The x-intercept is 4, and the y-intercept is -8.
   c) The x-intercepts are -4 and 2, and the y-intercept is -8.
   d) The x-intercepts are 4 and -2, and the y-intercept is -8.

7. For $y = \frac{3}{4}(x^2 - 8x + \_) + 3 + \_$, what numbers should be put in the blanks to write the equivalent equation in the form $y = a(x-h)^2 + k$?
   a) Put 16 in the first blank and -12 in the second blank.
   b) Put 16 in the first blank and 12 in the second blank.
   c) Put 16 in the first blank and 16 in the second blank.
   d) Put 16 in the first blank and -16 in the second blank.

8. Suppose the slope of a line is $\frac{2}{3}$. What is the slope of a line that is perpendicular to it?
   a) $\frac{3}{2}$ b) $-\frac{3}{2}$ c) $\frac{2}{3}$ d) $-\frac{2}{3}$

9. Find an equation of the line containing the points (5, 1) and (6, 8).
 a) $7x - y = 34$ b) $7x + y = 36$ c) $x - 7y = -2$ d) $x - 7y = -50$

10. What is the vertex for $y = 3x^2 + 12x - 4$?
 a) $(-2, -16)$ b) $(2, 32)$ c) $(-4, -4)$ d) $(4, 60)$

11. The equation $y = 0.05x + 10$ is a formula for a power company's monthly charge for its service, where y is the monthly bill and x is the number of kilowatt hours of electricity used. How many kilowatts of electricity are used for a monthly bill to be \$47.50?
 a) 700 b) 750 c) 800 d) 850

12. The parabola in Fig. 4-50 is the graph of which equation?
 a) $y = (x - 1)^2 + 2$ b) $y = (x - 1)^2 - 2$
 c) $y = (x + 1)^2 + 2$ d) $y = (x + 1)^2 - 2$

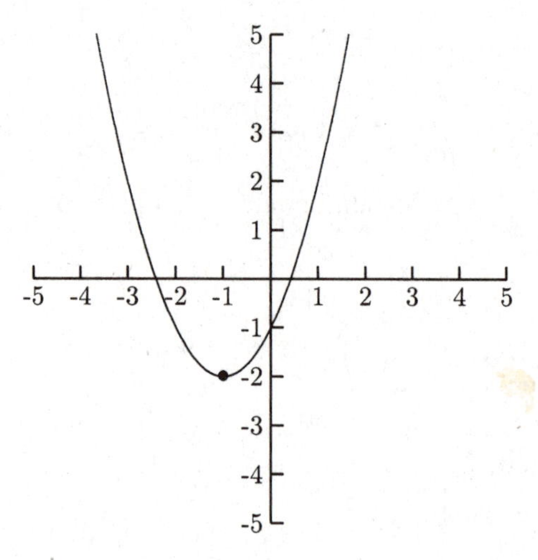

Fig. 4-50.

13. Which of the following lines is perpendicular to the line $x = -2$?
 a) $x = \frac{1}{2}$ b) $y = -2$ c) $x = 2$ d) $y = \frac{1}{2}x$

14. A gravy mix calls for two cups of water for each $\frac{3}{4}$ cups of mix. Find an equation that gives the amount of water in terms of the amount of mix.
 a) $y = \frac{8}{3}x$ b) $y = \frac{3}{8}x$ c) $y = \frac{3}{2}x$ d) $y = \frac{2}{3}x$

15. What is the equation of the line in Fig. 4-51?
a) $y = \frac{1}{2}x + 2$ b) $y = \frac{2}{3}x + 2$ c) $y = -\frac{1}{2}x + 2$
d) $y = -\frac{2}{3}x + 2$

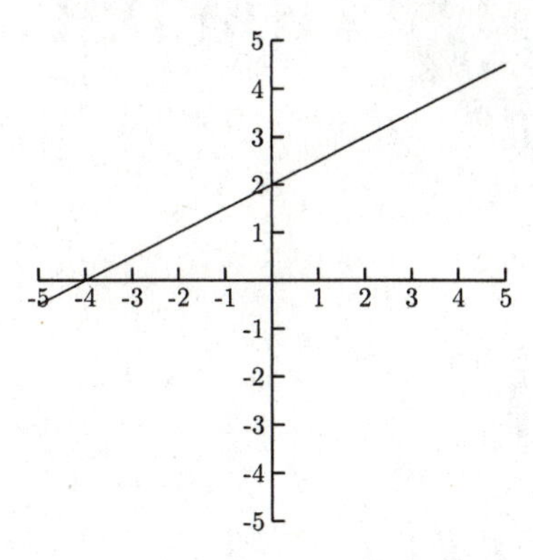

Fig. 4-51.

SOLUTIONS

1. c) 2. c) 3. b) 4. c) 5. c) 6. d) 7. a)
8. b) 9. a) 10. a) 11. b) 12. d) 13. b)
14. a) 15. a)

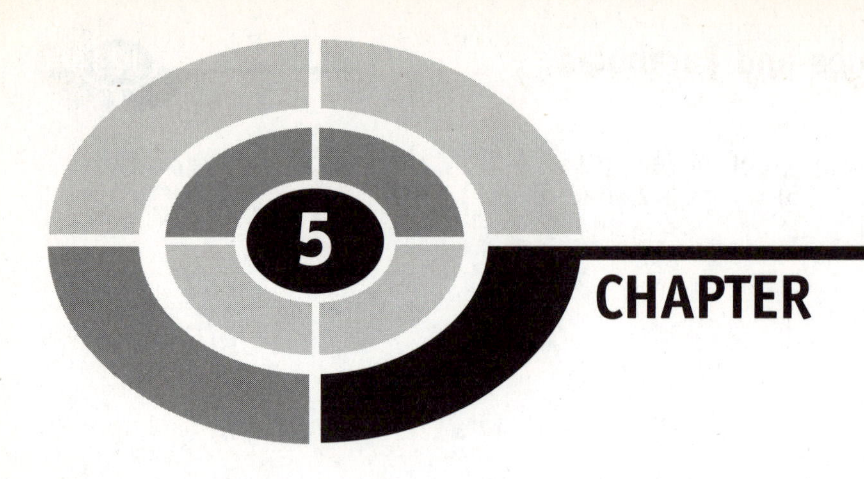

Nonlinear Inequalities

There are times in algebra and other mathematics courses where we might need to know where a graph is above and/or below the x-axis.

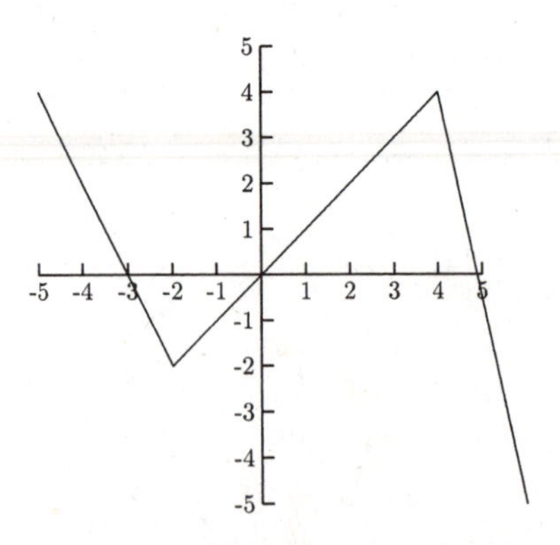

Fig. 5-1.

For what x-values is this graph above the x-axis? Below the x-axis?

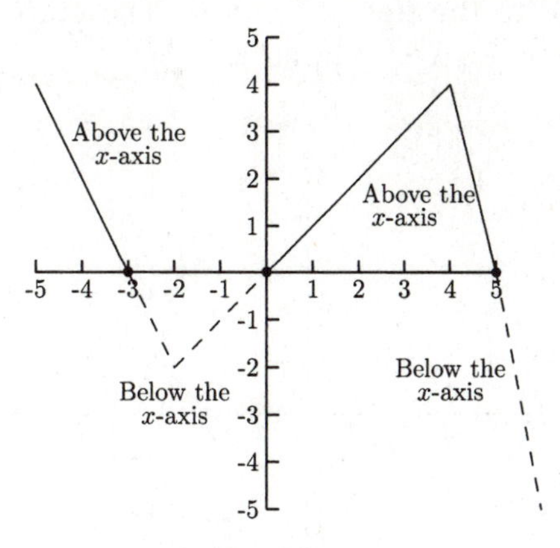

Fig. 5-2.

The graph is above the x-axis to the left of $x = -3$ and between $x = 0$ and $x = 5$ (see the solid part of the graph). The graph is below the x-axis between $x = -3$ and $x = 0$ and to the right of $x = 5$ (see the dashed part of the graph).

When answering questions about graphs, we usually need to answer the question in interval notation. For example, to represent "to the left of $x = -3$," we write $(-\infty, -3)$. To represent "between $x = 0$ and $x = 5$," we write $(0, 5)$.

EXAMPLES

Determine where the following graphs are above the x-axis and where they are below the x-axis.

•

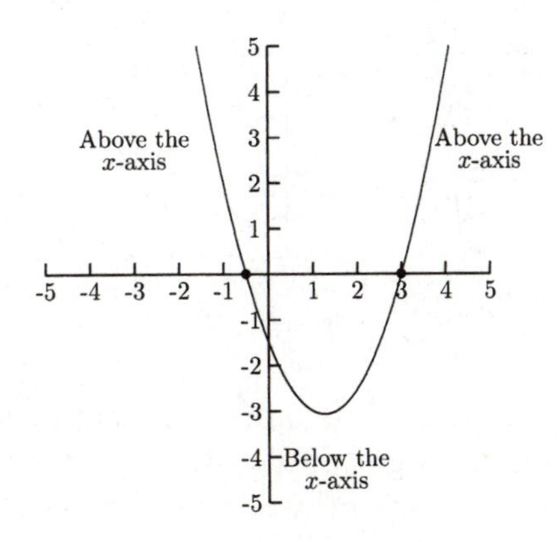

Fig. 5-3.

The graph is above the x-axis on $(-\infty, -\frac{1}{2})$ (to the left of $x = -\frac{1}{2}$) and on $(3, \infty)$ (to the right of $x = 3$). The graph is below the x-axis on $(-\frac{1}{2}, 3)$ (between $x = -\frac{1}{2}$ and $x = 3$).

-

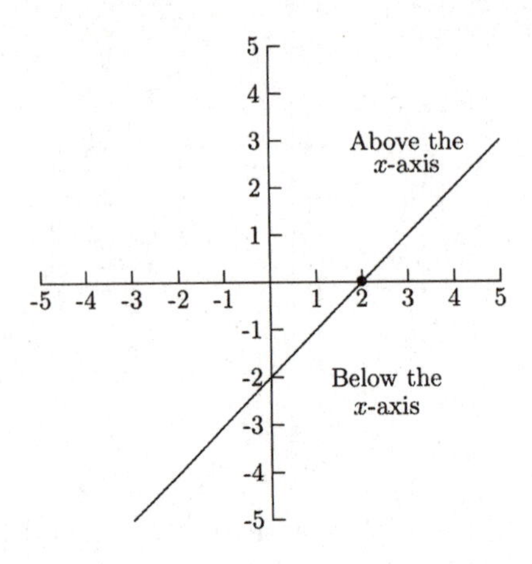

Fig. 5-4.

The graph is above the x-axis on $(2, \infty)$ (to the right of $x = 2$) and below the x-axis on $(-\infty, 2)$ (to the left of $x = 2$).

-

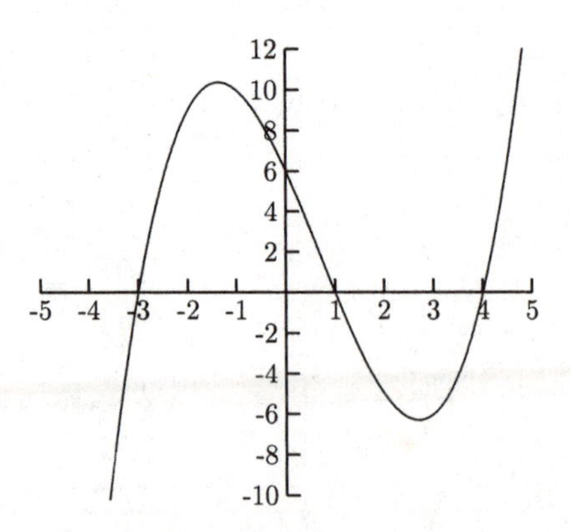

Fig. 5-5.

The graph is above the x-axis on $(-3, 1)$ (between $x = -3$ and $x = 1$) and on $(4, \infty)$ (to the right of $x = 4$). The graph is below the x-axis on $(-\infty, -3)$ (to the left of $x = -3$) and on $(1, 4)$ (between $x = 1$ and $x = 4$).

•

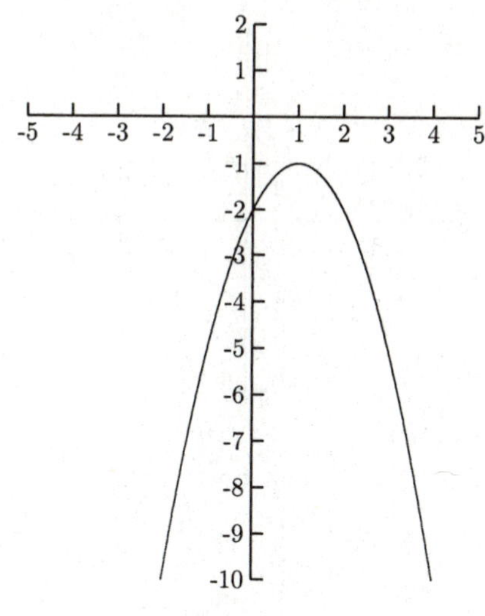

Fig. 5-6.

The graph is never above the x-axis. The graph is below the x-axis on $(-\infty, \infty)$ (this is interval notation for "all real numbers").

PRACTICE

Give the intervals of x where the graph is above the x-axis and below the x-axis.

1.

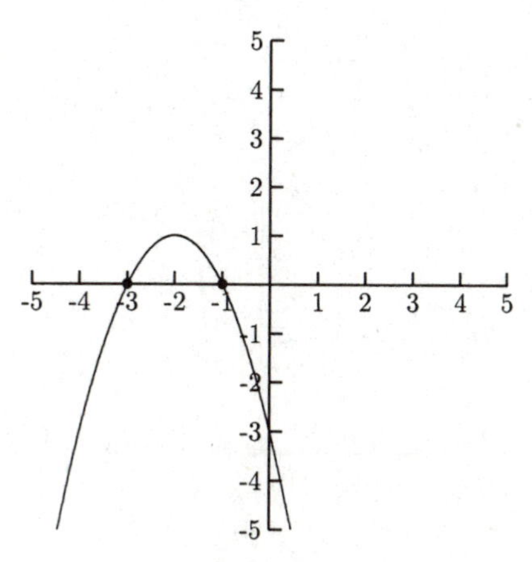

Fig. 5-7.

2.

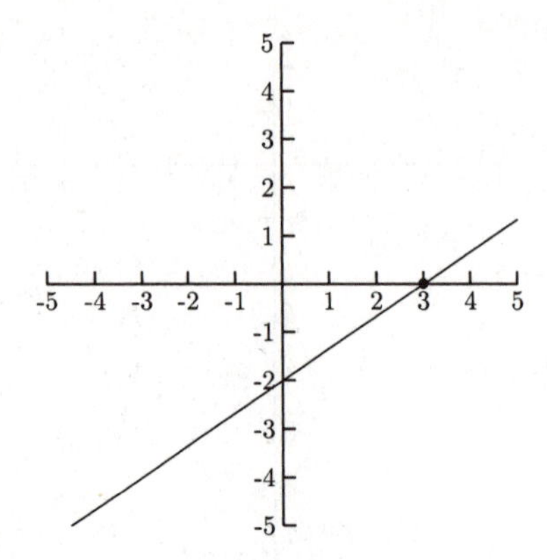

Fig. 5-8.

3.

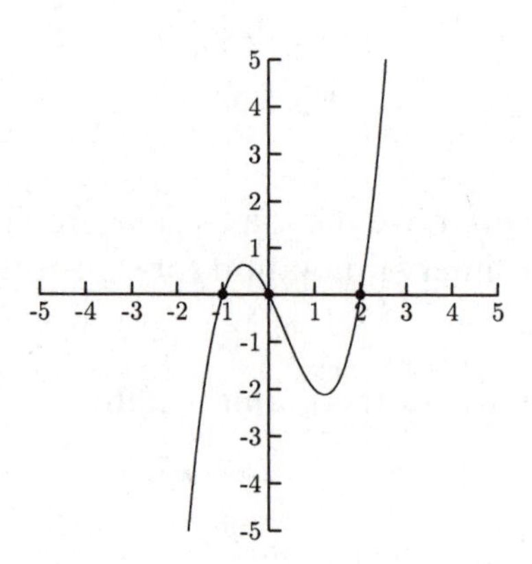

Fig. 5-9.

4.

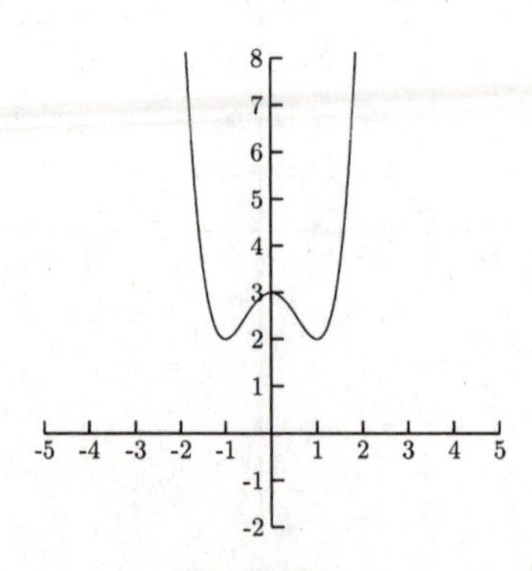

Fig. 5-10.

SOLUTIONS

1. The graph is above the x-axis on $(-3, -1)$. The graph is below the x-axis on $(-\infty, -3)$ and $(-1, \infty)$.
2. The graph is above the x-axis on $(3, \infty)$ and below the x-axis on $(-\infty, 3)$.
3. The graph is above the x-axis on $(-1, 0)$ and $(2, \infty)$. The graph is below the x-axis on $(-\infty, -1)$ and $(0, 2)$.
4. The graph is above the x-axis on $(-\infty, \infty)$ (everywhere). The graph is never below the x-axis.

Solving Nonlinear Inequalities

Solving linear inequalities is much like solving linear equations (except when multiplying or dividing by a negative number, we need to reverse the inequality symbol). Nonlinear inequalities are solved with a different method. In order for this method to make sense, we need to look at the graphs of some nonlinear equations. Let us look at Fig. 5-5 again. The graph is above the x-axis when the y-values are positive. The graph is below the x-axis when the y-values are negative. Between any two consecutive x-intercepts (where the graph touches the x-axis) y-values are either *all* positive or they are *all* negative. To the left of the smallest x-intercept, either all the y-values are positive or they are all negative. To the right of the largest x-intercept, the y-values are either all positive or all negative. We will use these facts to solve nonlinear inequalities.

For the graph in Fig. 5-5, the y-values are all negative to the left of $x = -3$, the smallest x-intercept. The y-values are all positive between x-intercepts -3 and 1. The y-values are negative between x-intercepts 1 and 4 and are again positive to the right of $x = 4$, the largest x-intercept.

Here is an example of a nonlinear inequality.

$$x^2 - 2x - 3 > 0$$

This inequality is asking the question, "For what values of x are the y-values for $y = x^2 - 2x - 3$ positive?" The graph of $y = x^2 - 2x - 3$ is shown in Fig. 5-11.

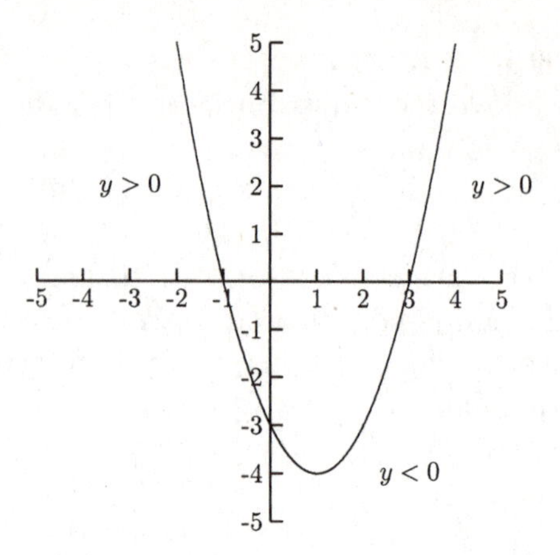

Fig. 5-11.

We can see from the graph that $x^2 - 2x - 3$ is positive for $(-\infty, -1)$ (to the left of $x = -1$) and for $(3, \infty)$ (to the right of $x = 3$). According to the graph, the solution to $x^2 - 2x - 3 > 0$ is $(-\infty, -1) \cup (3, \infty)$.

Graphs are not usually used to solve nonlinear inequalities—algebra is. To use algebra to solve nonlinear inequalities, we first need to find the x-intercepts. Then we need to see if the y-values are positive or negative around the x-intercepts. We need to test one y-value to the left of the smallest x-intercept, one y-value between each pair of consecutive x-intercepts, and one y-value to the right of the largest x-intercept.

We will find the x-intercepts for $y = x^2 - 2x - 3$ by setting $y = 0$ and solving for x.

$$0 = x^2 - 2x - 3$$

$$0 = (x - 3)(x + 1)$$

$$x - 3 = 0 \qquad x + 1 = 0$$

$$x = 3 \qquad x = -1$$

The two x-intercepts are -1 and 3. Are the y-values to the left of $x = -1$ positive or negative? We can answer this question by choosing *any* x-value smaller than -1. We will use $x = -2$ here. Is the y-value for $x = -2$ positive or negative? If it is positive, all y-values to the left of $x = -1$ are positive. If it is negative, all y-values to the left of $x = -1$ are negative.

$$y = (-2)^2 - 2(-2) - 3 = +5$$

This y-value is positive. All y-values to the left of $x = -1$ are positive.

Are the y-values positive or negative between $x = -1$ and $x = 3$? We only need to check one y-value. We can choose any x-value between $x = -1$ and $x = 3$. We will use $x = 0$ here.

$$y = 0^2 - 2(0) - 3 = -3$$

The y-value for $x = 0$ is negative. All the y-values between $x = -1$ and $x = 3$ are negative.

Are the y-values positive or negative to the right of $x = 3$? We can choose any x-value larger than 3. We will use $x = 4$ here.

$$y = 4^2 - 2(4) - 3 = +5$$

The y-value for $x = 4$ is positive. All y-values to the right of $x = 3$ are positive.

Values of x which have positive y-values are either smaller than -1 or larger than 3. The solution for the inequality $x^2 - 2x - 3 > 0$ is $(-\infty, -1) \cup (3, \infty)$.

Using a *sign graph* will help us to keep track of x-intervals having positive y-values and those having negative y-values. First we will draw the number line. Next we will compute the x-intercepts then put the x-intercepts on the sign graph. Then we will write a plus sign over the interval(s) having positive y-values and a minus sign over the interval(s) having negative y-values. The sign graph for the inequality $x^2 - 2x - 3 > 0$ looks like this.

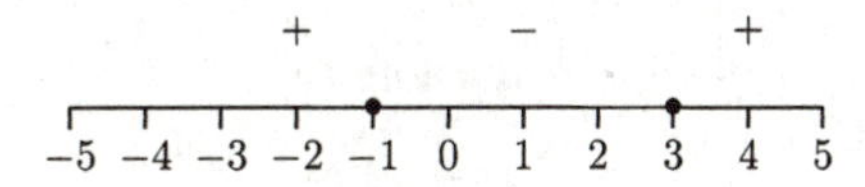

Fig. 5-12.

Here are the steps for solving these kinds of nonlinear inequalities.

1. Get a zero on one side of the inequality.
2. Find the x-intercepts.
3. Mark the x-intercepts on a sign graph.
4. Choose an x-value in each interval to test whether the y-value is positive or negative.
5. Mark each interval with a plus sign or a minus sign, depending on whether the y-value for the interval is positive or negative.

6. Look at the inequality to decide if the solution is the plus interval(s) or the minus interval(s).
7. Write the solution in interval notation.

EXAMPLES

- $x^2 + 2x - 8 < 0$

  Step 1 is not necessary because one side of the inequality is already 0. We will find the x-intercept(s) for $y = x^2 + 2x - 8$ by setting y equal to 0 and solving for x.

$$0 = x^2 + 2x - 8$$

$$0 = (x + 4)(x - 2)$$

$$x + 4 = 0 \qquad x - 2 = 0$$

$$x = -4 \qquad x = 2$$

Next we will put the x-intercepts on the sign graph.

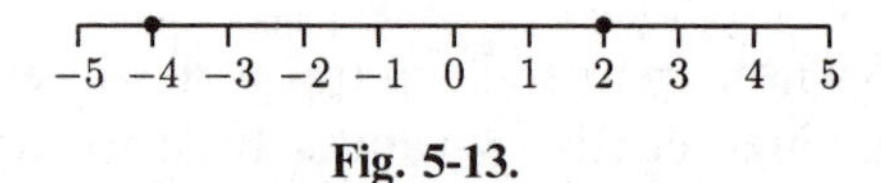

Fig. 5-13.

We will use $x = -5$ for the point to the left of $x = -4$; $x = 0$ for the point between $x = -4$ and $x = 2$; and $x = 3$ for the point the right of $x = 2$.

For $x = -5$, $y = (-5)^2 + 2(-5) - 8 = +7$. We need to put a plus sign on the sign graph to the left of −4.

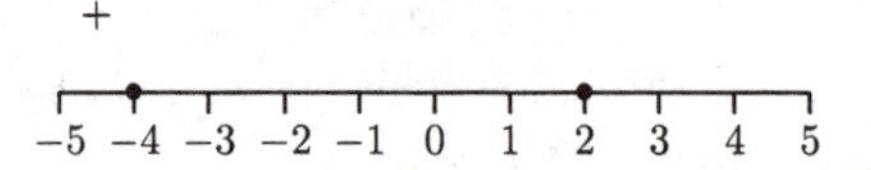

Fig. 5-14.

For $x = 0$, $y = 0^2 + 2(0) - 8 = -8$. We need to put a minus sign on the sign graph between $x = -4$ and $x = 2$.

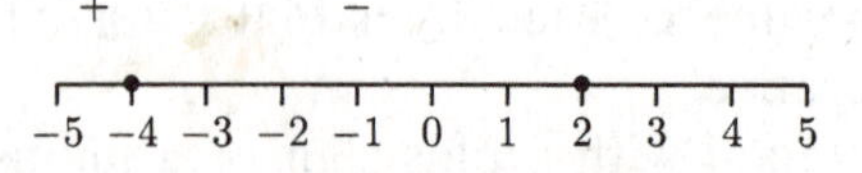

Fig. 5-15.

For $x = 3$, $y = 3^2 + 2(3) - 8 = +7$. We need to put a plus sign on the sign graph to the right of $x = 2$.

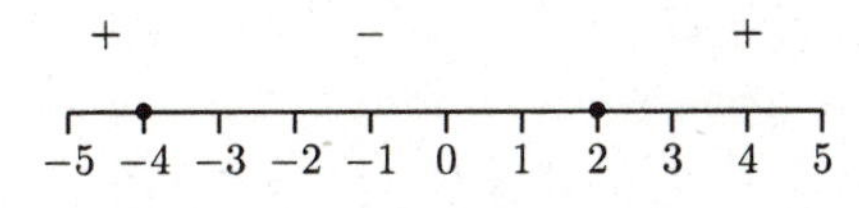

Fig. 5-16.

The inequality reads “< 0” which means we want the negative y-values. The solution is the interval of numbers between $x = -4$ and $x = 2$: $(-4, 2)$.

- $x^3 + x^2 - 2x \geq 0$

We will find the x-intercepts by factoring $y = x^3 + x^2 - 2x$ and setting each factor equal to 0.

$$x^3 + x^2 - 2x = x(x^2 + x - 2)$$

$$= x(x + 2)(x - 1)$$

$$x = 0 \qquad x + 2 = 0 \qquad x - 1 = 0$$

$$x = -2 \qquad x = 1$$

Now we can put the x-intercepts on the graph.

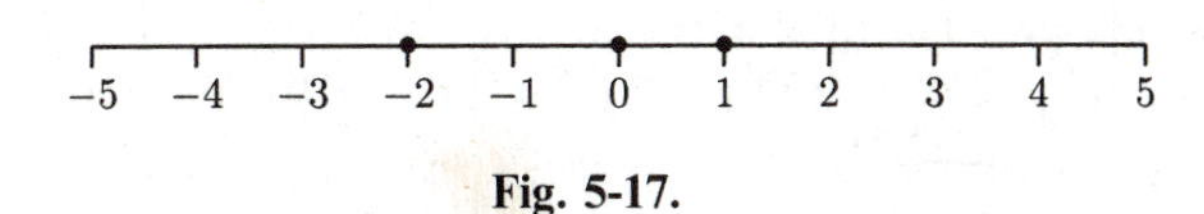

Fig. 5-17.

We will use $x = -3$ for the point to the left of $x = -2$; $x = -1$ for the point between $x = -2$ and $x = 0$; $x = 0.5$ for the point between $x = 0$ and $x = 1$; and $x = 2$ for the point to the right of $x = 1$.

For $x = -3$, $y = (-3)^3 + (-3)^2 - 2(-3) = -12$. A minus sign goes to the left of $x = -2$.

For $x = -1$, $y = (-1)^3 + (-1)^2 - 2(-1) = +2$. A plus sign goes between $x = -2$ and $x = 0$.

For $x = 0.5$, $y = (0.5)^3 + (0.5)^2 - 2(0.5) = -0.625$. A minus sign goes between $x = 0$ and $x = 1$.
For $x = 2$, $y = 2^3 + 2^2 - 2(2) = +8$. A plus sign goes to the right of $x = 1$.

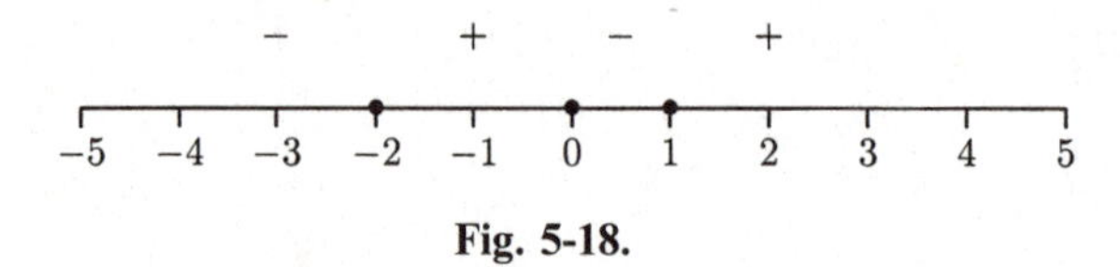

Fig. 5-18.

The inequality is "≥ 0" which means we want the positive intervals. The solution is $[-2, 0] \cup [1, \infty)$.

It seems that the signs always alternate between plus and minus signs. Signs on the sign graphs do not always alternate.

- $(x - 3)^2(x + 2)(x + 1) < 0$

$$(x-3)^2 = 0 \qquad x + 2 = 0 \qquad x + 1 = 0$$
$$x - 3 = 0 \qquad x = -2 \qquad x = -1$$
$$x = 3$$

For $x = -3$, $y = (-3 - 3)^2(-3 + 2)(-3 + 1) = +72$. We will put a plus sign to the left of $x = -2$.
For $x = -1.5$, $y = (-1.5 - 3)^2(-1.5 + 2)(-1.5 + 1) = -5.0625$. We will put a minus sign between $x = -1$ and $x = -2$.
For $x = 0$, $y = (0 - 3)^2(0 + 2)(0 + 1) = +18$. We will put a plus sign between $x = -1$ and $x = 3$.
For $x = 4$, $y = (4 - 3)^2(4 + 2)(4 + 1) = +30$. We will put a plus sign to the right of $x = 3$.

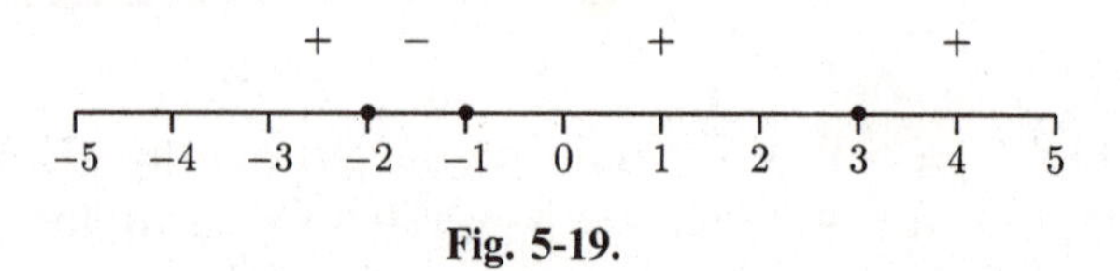

Fig. 5-19.

The inequality is "<0" which means we want the minus interval. The solution is $(-2, -1)$.

- $x^2 + 5x + 9 < 3$
  We need to subtract 3 from each side of the inequality so that 0 is on one side.

$$x^2 + 5x + 6 < 0$$

$$x^2 + 5x + 6 = (x + 2)(x + 3)$$

$$x + 2 = 0 \qquad x + 3 = 0$$

$$x = -2 \qquad x = -3$$

  For $x = -4$, $y = (-4 + 2)(-4 + 3) = +2$.
  For $x = -2.5$, $y = (-2.5 + 2)(-2.5 + 3) = -0.25$.
  For $x = 0$, $y = (0 + 2)(0 + 3) = +6$.

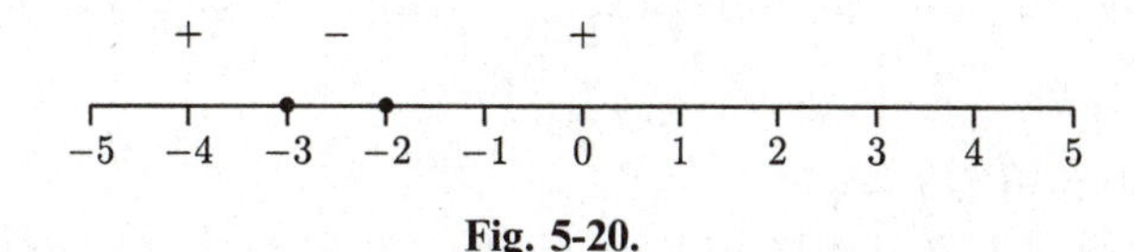

Fig. 5-20.

  The solution is $(-3, -2)$.
- $x^2 + 1 > 0$
  The equality $x^2 + 1 = 0$ has no solution, so the graph of $y = x^2 + 1$ has no x-intercept. This means that either all y-values are positive or they are all negative. We need to check only one y-value. Choose any x-value. We will use $x = 0$.
  For $x = 0$, $y = 0^2 + 1 = +1$.
  Because this y-value is positive, *all* y-values are positive. The solution is $(-\infty, \infty)$.

PRACTICE

Solve the inequalities. Give solutions in interval notation.

1. $x^2 + 3x - 4 < 0$
2. $-x^2 - x + 6 \geq -14$
3. $x^3 - 2x^2 - 4x + 8 > 0$. Hint: factor by grouping.
4. $x^4 - 13x^2 + 36 \geq 0$. Hint: $x^4 - 13x^2 + 36 = (x^2 - 4)(x^2 - 9) = (x - 2)(x + 2)(x - 3)(x + 3)$
5. $x^2 + 9 \leq 0$

SOLUTIONS

1. $x^2 + 3x - 4 = (x + 4)(x - 1)$

$$x + 4 = 0 \qquad x - 1 = 0$$

$$x = -4 \qquad x = 1$$

For $x = -5$, $y = (-5 + 4)(-5 - 1) = 6$.
For $x = 0$, $y = (0 + 4)(0 - 1) = -4$.
For $x = 2$, $y = (2 + 4)(2 - 1) = 6$.

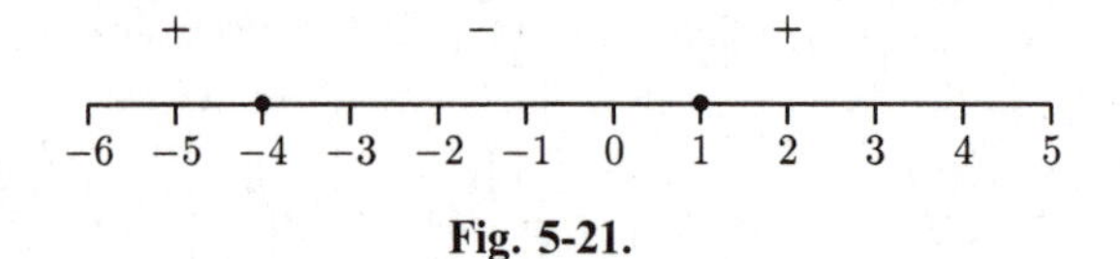

Fig. 5-21.

The solution is $(-4, 1)$.

2. First we need to add 14 to both sides of the inequality to get $-x^2 - x + 20 \geq 0$. Then $-x^2 - x + 20 = -(x^2 + x - 20) = -(x + 5)(x - 4)$.

$$-(x + 5) = 0 \qquad x - 4 = 0$$

$$-x - 5 = 0 \qquad x = 4$$

$$-x = 5$$

$$x = -5$$

For $x = -6$, $y = -(-6 + 5)(-6 - 4) = -10$.
For $x = 0$, $y = -(0 + 5)(0 - 4) = 20$.
For $x = 5$, $y = -(5 + 5)(5 - 4) = -10$.

– + –

–7 –6 –5 –4 –3 –2 –1 0 1 2 3 4 5 6 7

Fig. 5-22.

The solution is $[-5, 4]$.

3.

$$x^3 - 2x^2 - 4x + 8 = x^2(x-2) - 4(x-2)$$
$$= (x^2-4)(x-2)$$
$$= (x-2)(x+2)(x-2)$$
$$= (x-2)^2(x+2)$$

$$(x-2)^2 = 0 \qquad x+2=0$$
$$x-2=0 \qquad x=-2$$
$$x=2$$

For $x=-3$, $y=(-3-2)^2(-3+2)=-25$.
For $x=0$, $y=(0-2)^2(0+2)=+8$.
For $x=3$, $y=(3-2)^2(3+2)=+5$.

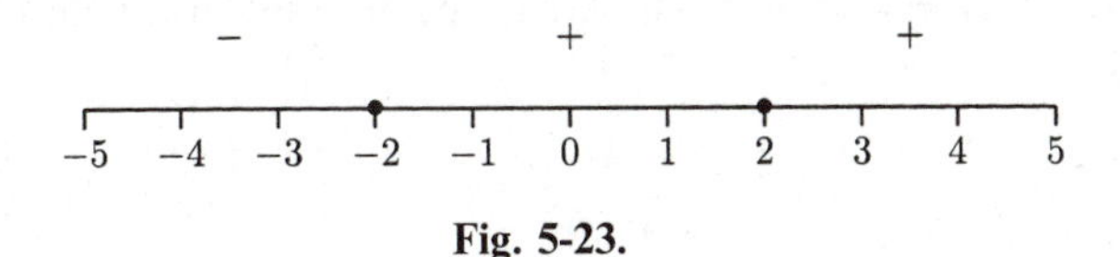

Fig. 5-23.

The solution is $(-2,2)\cup(2,\infty)$. The solution is not $(-2,\infty)$ because that would imply $x=2$ is part of the solution ($2^3-2(2)^2-4(2)+8=0$, not ">0"). Because the inequality is strict (not allowing equality), $x=2$ is not part of the solution. If the inequality had been "≥ 0", then $x=2$ would be part of the solution and the solution would be $[-2,\infty)$.

4. $x^4-13x^2+36=(x^2-4)(x^2-9)=(x-2)(x+2)(x-3)(x+3)$

$$x-2=0 \qquad x+2=0 \qquad x-3=0 \qquad x+3=0$$
$$x=2 \qquad x=-2 \qquad x=3 \qquad x=-3$$

For $x=-4$, $y=(-4-2)(-4+2)(-4-3)(-4+3)=+84$.
For $x=-2.5$, $y=(-2.5-2)(-2.5+2)(-2.5-3)(-2.5+3)=-6.1875$.

For $x = 0$, $y = (0 - 2)(0 + 2)(0 - 3)(0 + 3) = +36$.
For $x = 2.5$, $y = (2.5 - 2)(2.5 + 2)(2.5 - 3)(2.5 + 3) = -6.1875$.
For $x = 4$, $y = (4 - 2)(4 + 2)(4 - 3)(4 + 3) = +84$.

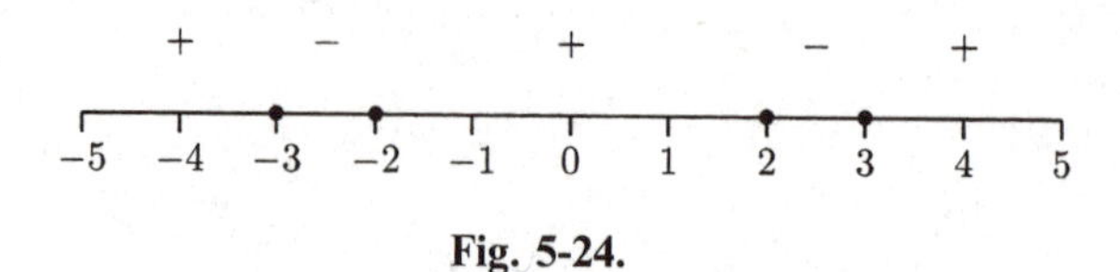

Fig. 5-24.

The solution is $(-\infty, -3] \cup [-2, 2] \cup [3, \infty)$.

5. The equation $x^2 + 9 = 0$ has no solution. This means that the graph of $y = x^2 + 9$ has no x-intercepts, so either all y-values are positive or they are all negative. We need to check only one y-value. Let $x = 0$.

$$0^2 + 9 = 9$$

This y-value is positive, so all y-values are positive. Because the inequality is "≤ 0", we want negative y-values. There is no solution.

Graphs of equations that have variables in denominators usually have separate parts.

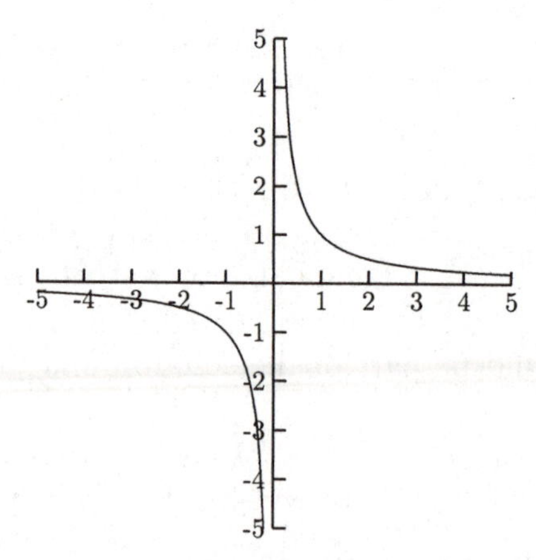

Fig. 5-25.

The graph of $y = 1/x$ comes in two parts, one to the left of the y-axis and one to the right of the y-axis.

For every x-value that makes the denominator zero, the graph will have a break. In $y = 1/x$, if we let $x = 0$, there is a zero in the denominator. To the right of the break at $x = 0$, the y-values are positive and to the left, they are negative. A break in the graph acts like an x-intercept—the y-values can change from positive to negative (or from negative to positive).

EXAMPLE
The graph shown in Fig. 5-26 is the graph of the equation

$$y = \frac{x+2}{x-1}.$$

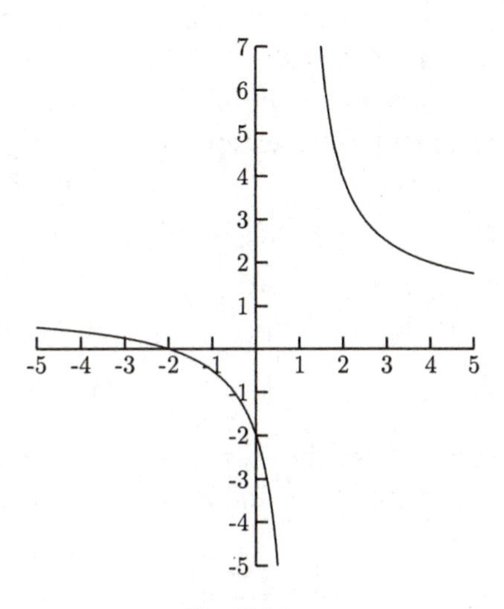

Fig. 5-26.

This graph has both an x-intercept (at $x = -2$) and a break (at $x = 1$). The y-values are positive to the left of the x-intercept and to the right of the break. The y-values are negative between the x-intercept and the break.

Solving inequalities with variables in a denominator is much like solving earlier inequalities.

1. Get zero on one side of the inequality. (Sometimes this step is not necessary.)
2. Rewrite the nonzero side of the inequality as one fraction. (Sometimes this step is not necessary.)
3. Find the x-intercept(s) by setting the numerator equal to zero and solving for x. If the numerator does not have a variable, there will be no x-intercept. Put any x-intercepts on the sign graph.

4. Find the break(s) in the graph by setting the denominator equal to zero and solving for x. Put this x-value or values on the sign graph. If there is no x-value that makes the denominator equal to zero, then there is no break in the graph.
5. Test an x-value in each interval on the sign graph. If the y-value is positive, put a plus sign over the interval. If the y-value is negative, put a minus sign over the interval.
6. Look at the inequality to decide if you want the "plus" interval(s) or the "minus" interval(s). Be careful to exclude from the solution any x-value that causes a zero in a denominator.

EXAMPLES

- $(2x - 6)/(x + 4) > 0$

  Step 1 and Step 2 are not necessary. For Step 3, set the numerator equal to zero to find the x-intercept(s).

$$2x - 6 = 0$$
$$2x = 6$$
$$x = 3$$

We need to put $x = 3$ on the sign graph. For Step 4, we will set the denominator equal to zero to find any break in the graph.

$$x + 4 = 0$$
$$x = -4$$

We will put $x = -4$ on the sign graph. For Step 5, test an x-value smaller than -4 (we will use $x = -5$), between -4 and 3 (we will use $x = 0$), and larger than 3 (we will use $x = 4$).

$$x = -5 \qquad y = \frac{2(-5) - 6}{-5 + 4} = 16$$

$$x = 0 \qquad y = \frac{2(0) - 6}{0 + 4} = -\frac{3}{2}$$

$$x = 4 \qquad y = \frac{2(4) - 6}{4 + 4} = \frac{1}{4}$$

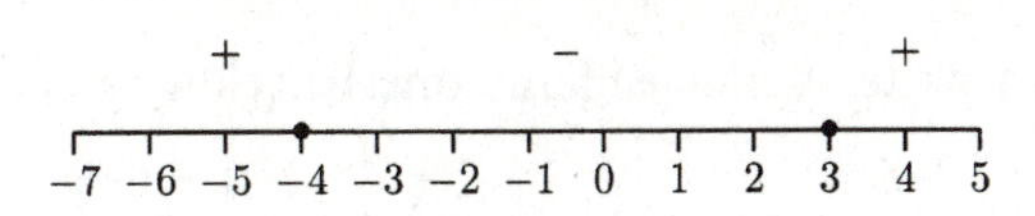

Fig. 5-27.

The solution is $(-\infty, -4) \cup (3, \infty)$.

- $(5x+8)/(x^2+1) < 0$

  We will set the numerator equal to zero to find the x-intercepts.

$$5x + 8 = 0$$
$$5x = -8$$
$$x = -\frac{8}{5}$$

We need to put $x = -\frac{8}{5}$ on the sign graph. We will set the denominator equal to zero to find the breaks in the graph.

$$x^2 + 1 = 0$$

This equation has no solution, so there are no breaks in the graph. We will test $x = -2$ for the point to the left of $x = -\frac{8}{5}$ and $x = 0$ for the point to the right of $x = -\frac{8}{5}$.

$$x = -2 \qquad y = \frac{5(-2)+8}{(-2)^2+1} = -\frac{2}{5}$$

$$x = 0 \qquad y = \frac{5(0)+8}{0^2+1} = 8$$

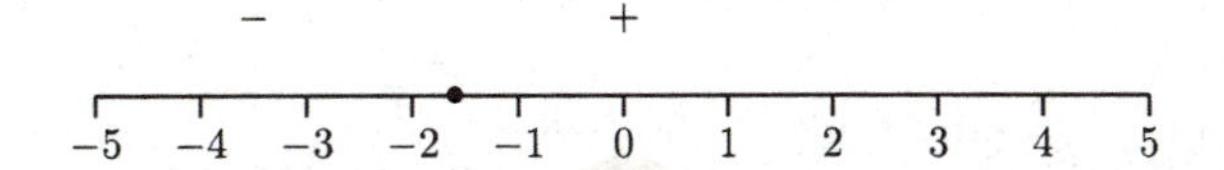

Fig. 5-28.

The solution is $(-\infty, -\frac{8}{5})$.

- $(3-2x)/(x+4) > 2$

  One side of the inequality needs to be zero. We need to subtract 2 from each side.

$$\frac{3-2x}{x+4} - 2 > 0$$

Now we can write the left side as one fraction.

$$\frac{3-2x}{x+4}-2=\frac{3-2x}{x+4}-2\,\frac{x+4}{x+4}$$
$$=\frac{3-2x}{x+4}-\frac{2(x+4)}{x+4}$$
$$=\frac{3-2x-2(x+4)}{x+4}$$
$$=\frac{3-2x-2x-8}{x+4}$$
$$=\frac{-4x-5}{x+4}$$

The inequality can be rewritten.

$$\frac{-4x-5}{x+4}>0$$

$$-4x-5=0 \qquad x+4=0$$
$$-4x=5 \qquad x=-4$$
$$x=-\frac{5}{4}$$

Put $x=-\frac{5}{4}$ and $x=-4$ on the sign graph. We will test $x=-5$ for the point to the left of $x=-4$, $x=-2$ for the point between $x=-4$ and $x=-\frac{5}{4}$, and $x=0$ for the point to the right of $-\frac{5}{4}$.

$$x=-5 \qquad y=\frac{-4(-5)-5}{-5+4}=-15$$
$$x=-2 \qquad y=\frac{-4(-2)-5}{-2+4}=\frac{3}{2}$$
$$x=0 \qquad y=\frac{-4(0)-5}{0+4}=-\frac{5}{4}$$

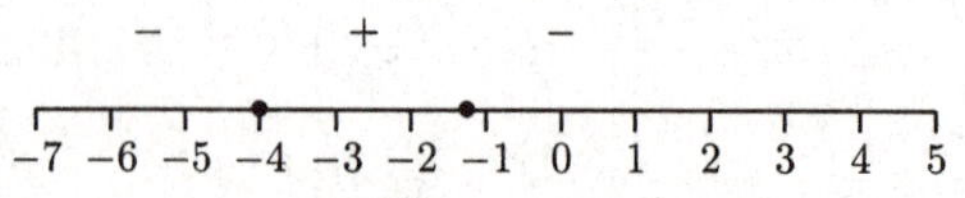

Fig. 5-29.

The solution is $(-4, -1\frac{1}{4})$.

- $(x^2 + 6x + 8)/(x - 3) \le 0$

$$x^2 + 6x + 8 = 0$$
$$(x + 2)(x + 4) = 0$$

$$x + 2 = 0 \qquad x + 4 = 0$$
$$x = -2 \qquad x = -4$$

$$x - 3 = 0$$
$$x = 3$$

Put $x = -2$, $x = -4$, and $x = 3$ on the sign graph. Test an x-value smaller than -4 (we will use $x = -5$), between -2 and -4 (we will use $x = -3$), between -2 and 3 (we will use $x = 0$), and larger than 3 (we will use $x = 4$).

$$x = -5 \qquad y = \frac{(-5)^2 + 6(-5) + 8}{-5 - 3} = -\frac{3}{8}$$

$$x = -3 \qquad y = \frac{(-3)^2 + 6(-3) + 8}{-3 - 3} = \frac{1}{6}$$

$$x = 0 \qquad y = \frac{0^2 + 6(0) + 8}{0 - 3} = -\frac{8}{3}$$

$$x = 4 \qquad y = \frac{4^2 + 6(4) + 8}{4 - 3} = 48$$

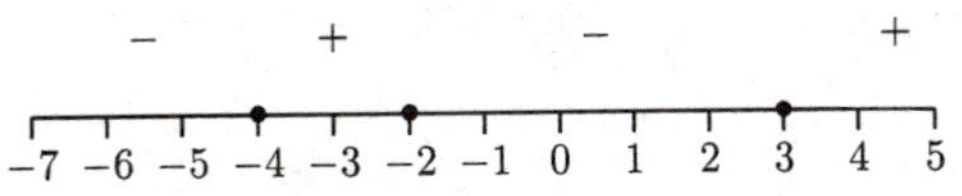

– + – +

–7 –6 –5 –4 –3 –2 –1 0 1 2 3 4 5

Fig. 5-30.

The solution is *not* $(-\infty, -4] \cup [-2, 3]$. By using a square bracket around 3, we are implying that $x = 3$ is a solution, but $x = 3$ leads to a zero in the denominator. Use a parenthesis around 3 to indicate that $x = 3$ is not part of the solution. The solution is $(-\infty, -4] \cup [-2, 3)$.

PRACTICE

Solve the inequalities, giving the solutions in interval notation.

1. $\frac{x+5}{x-2} > 0$
2. $\frac{2x-8}{x^2+3x+2} < 0$
3. $\frac{3x+2}{x-4} \leq 1$

SOLUTIONS

1.

$$x+5=0 \qquad x-2=0$$

$$x=-5 \qquad x=2$$

$$x=-6 \qquad y=\frac{-6+5}{-6-2}=\frac{1}{8}$$

$$x=0 \qquad y=\frac{0+5}{0-2}=-\frac{5}{2}$$

$$x=3 \qquad y=\frac{3+5}{3-2}=8$$

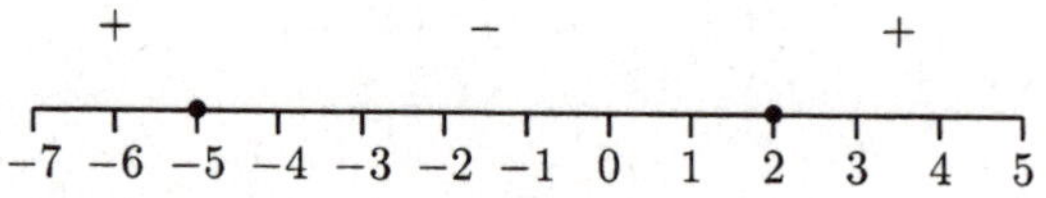

+ − +

−7 −6 −5 −4 −3 −2 −1 0 1 2 3 4 5

Fig. 5-31.

The solution is $(-\infty, -5) \cup (2, \infty)$.

2.

$$2x-8=0$$
$$2x=8$$
$$x=4$$

$$x^2 + 3x + 2 = 0$$
$$(x + 2)(x + 1) = 0$$
$$x + 2 = 0 \qquad x + 1 = 0$$
$$x = -2 \qquad x = -1$$

$x = -3$ $\qquad y = \dfrac{2(-3) - 8}{(-3)^2 + 3(-3) + 2} = -7$

$x = -1.5$ $\qquad y = \dfrac{2(-1.5) - 8}{(-1.5)^2 + 3(-1.5) + 2} = 44$

$x = 0$ $\qquad y = \dfrac{2(0) - 8}{0^2 + 3(0) + 2} = -4$

$x = 5$ $\qquad y = \dfrac{2(5) - 8}{5^2 + 3(5) + 2} = \dfrac{1}{21}$

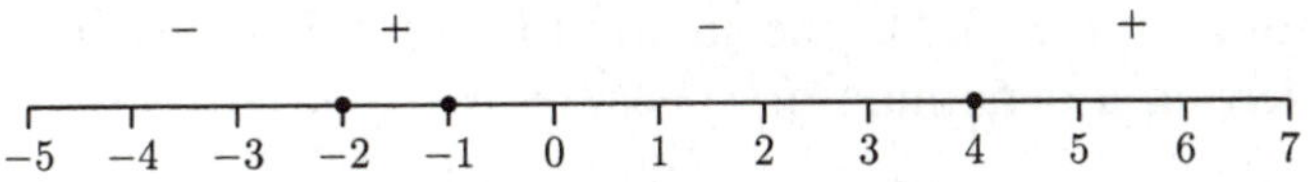

Fig. 5-32.

The solution is $(-\infty, -2) \cup (-1, 4)$.

3.

$$\frac{3x + 2}{x - 4} \le 1$$
$$\frac{3x + 2}{x - 4} - 1 \le 0$$
$$\frac{3x + 2}{x - 4} - 1 \cdot \frac{x - 4}{x - 4} \le 0$$
$$\frac{3x + 2 - (x - 4)}{x - 4} \le 0$$
$$\frac{3x + 2 - x + 4}{x - 4} \le 0$$
$$\frac{2x + 6}{x - 4} \le 0$$

$$2x + 6 = 0 \qquad x - 4 = 0$$

$$x = -3 \qquad x = 4$$

$$x = -4 \qquad \frac{2(-4) + 6}{-4 - 4} = \frac{1}{4}$$

$$x = 0 \qquad \frac{2(0) + 6}{0 - 4} = -\frac{3}{2}$$

$$x = 5 \qquad \frac{2(5) + 6}{5 - 4} = 16$$

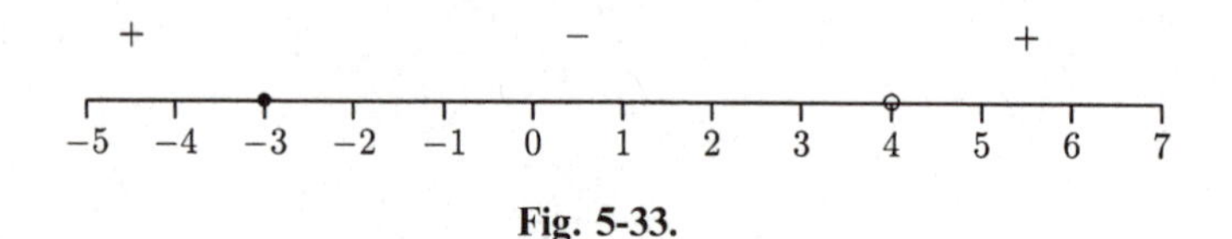

Fig. 5-33.

The solution is $[-3, 4)$. The solution is *not* $[-3, 4]$ because $x = 4$ leads to a zero in a denominator.

Chapter 5 Review

Solve the inequalities.

1. $x^2 - 5x + 6 \leq 0$
   a) $(2, 3)$ b) $[2, 3]$ c) $(-\infty, 2) \cup (-\infty, 3)$ d) $(-\infty, 2] \cup (-\infty, 3]$
2. $-x^2 - 3x + 18 > 0$
   a) $(-6, 3)$ b) $(-\infty, -6) \cup (3, \infty)$
   c) $(-6, \infty) \cup (3, \infty)$ d) $(-\infty, -6) \cup (-\infty, 3)$
3. $x^2 + 1 > 0$
   a) $(-\infty, -1) \cup (1, \infty)$ b) $(-\infty, \infty)$ c) $(-1,1)$ d) No solution
4. $(x + 4)^2(x - 1) < 0$
   a) $(-\infty, -4) \cup (1, \infty)$ b) $(-\infty, -1)$
   c) $(-\infty, -4) \cup (-4, 1)$ d) $(-\infty, -4)$
5. $(x - 3)/(x + 2) < 0$
   a) $(-2, 3)$ b) $(-\infty, -2) \cup (-\infty, 3)$ c) $(-3, 2)$ d) No solution

6. $(x - 6)/(x^2 + 6x + 8) \geq 0$

   a) $[-4, -2] \cup [6, \infty)$ b) $(-4, -2) \cup (6, \infty)$

   c) $(-4, -2) \cup [6, \infty)$ d) $[-4, \infty) \cup [6, \infty)$

7. $(2x - 6)/(x + 1) < 2$

   a) $(-1, 3)$ b) $(-1, 1)$ c) $(-1, \infty)$ d) $(-\infty, -1) \cup (-\infty, 3)$

SOLUTIONS

1. b) 2. a) 3. b) 4. c) 5. a) 6. c) 7. c)

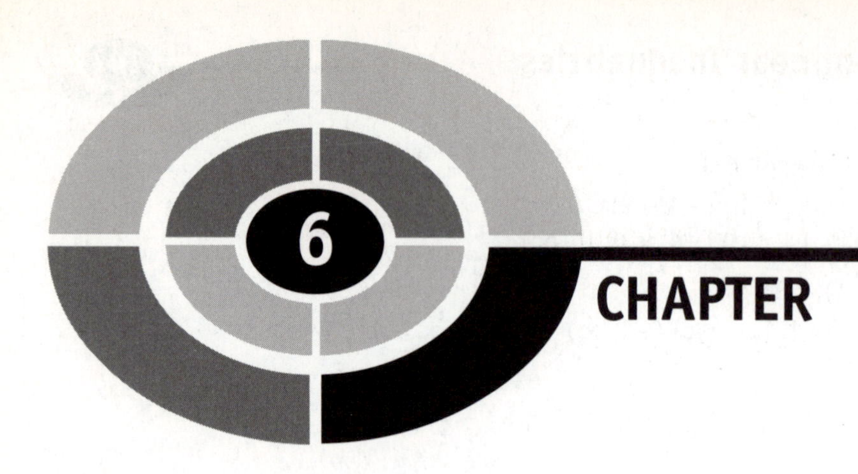

Functions

A function is a special type of relationship where the value of one variable *depends* on the value of one or more other variables. Functions occur all around us. For example, a person's weight *depends* on many variables—age, sex, height, food intake, activity level, and so on. An hourly worker's pay *depends* on the number of hours worked. College algebra students are mostly concerned with one variable depending on another. Normally, y depends on x. An example of an equation where y depends on x is the linear equation $y = mx + b$. For quadratic equations of the form $y = ax^2 + bx + c$, y is also a function of x. We call x the *independent* variable and y the *dependent* variable.

Technically, a function is a relation between two sets, A and B, where every element in A is assigned exactly one element in B. What this means for x and y is that for *every* x-value, there is *exactly* one y-value. In the function $y = x^2 + 2x + 3$, once an x-value is chosen, exactly one y-value follows. If $x = 2$, then $y = 2^2 + 2(2) + 3 = 11$. No matter what we put in for x, there is exactly one y-value for that particular value of x.

What kind of equations are not functions? Equations that have at least one x-value that has more than one y-value. For example, in the equation

$x^2 + y^2 = 9$, y is not a function of x. If we let $x = 0$, then we get $y^2 = 9$, so $y = 3$ or $y = -3$. So $x = 0$ has *two* y-values, 3 and -3.

When asked to determine whether or not an equation "gives y as a function of x," solve the equation for y. Then decide if there can be any x-value that has more than one y-value.

EXAMPLES

Determine if y is a function of x.

- $y^3 + x^2 - 3x = 7$

  Solve for y.

  $$y^3 = -x^2 + 3x + 7$$

  $$y = \sqrt[3]{-x^2 + 3x + 7} \qquad \text{(We only need } \pm \text{ for even roots.)}$$

  Each x-value has only one y-value, so y is a function of x.

- $(x+1)^2 + (y-8)^2 = 9$

  We solve this equation for y.

  $$(x+1)^2 + (y-8)^2 = 9$$

  $$(y-8)^2 = 9 - (x+1)^2$$

  $$y - 8 = \pm\sqrt{9 - (x+1)^2}$$

  $$y = 8 \pm \sqrt{9 - (x+1)^2}$$

  Most x-values have *two* y-values, $y = 8 + \sqrt{9 - (x+1)^2}$ and $y = 8 - \sqrt{9 - (x+1)^2}$. This means that y is not a function of x.

PRACTICE

Determine if y is a function of x.

1. $x^2 + (y-3)^2 = 16$
2. $x^2 - 2y = 4$
3. $|y| = x$

SOLUTIONS

1.

$$x^2 + (y-3)^2 = 16$$

$$(y-3)^2 = 16 - x^2$$

$$y - 3 = \pm\sqrt{16 - x^2}$$

$$y = 3 \pm \sqrt{16 - x^2}$$

Therefore y is not a function of x. To see this, let $x = 0$: $y = 3 \pm \sqrt{16 - 0^2} = 3 \pm \sqrt{16}$ or $y = 3 + 4 = 7$, and $y = 3 - 4 = -1$.

2.

$$x^2 - 2y = 4$$

$$-2y = 4 - x^2$$

$$y = \frac{4 - x^2}{-2}$$

Therefore y is a function of x.

3. In the equation $|y| = x$, y is not a function of x because every positive x-value has two y-values. For example, if $x = 3$, $|y| = 3$ has the solutions $y = 3$ and $y = -3$.

Domain and Range

When y is a function of x, the domain of a function is the collection of all possible values for x. The range is the collection of all y-values. When asked to find the domain of a function, think in terms of what can and cannot be done. For now, keep in mind that we cannot divide by 0 and we cannot take an even root of a negative number. For example, in the function $y = 1/x$ we cannot let $x = 0$, so 0 is not in the domain of this function. The domain is the set of all nonzero real numbers. We might also say that the domain is $x \neq 0$.

When asked to find the domain of a function that has x in one or more denominators, we need to set each denominator (that has x in it) equal to zero and solve for x. The domain will not include these numbers.

EXAMPLES

Find the domain for the following functions. Give the domain in interval notation.

- $y = 2x/(x - 4)$
  Because the denominator has an x in it, we will set it equal to zero and solve for x. The solution to $x - 4 = 0$ is $x = 4$. The domain is all real numbers except 4. The interval notation is $(-\infty, 4) \cup (4, \infty)$.

- $y = (2x+5)/(x^2 - x - 6) = (2x+5)/(x-3)(x+2)$

$$x - 3 = 0 \qquad x + 2 = 0$$
$$x = 3 \qquad x = -2$$

The domain is all real numbers except -2 and 3. The interval notation for $x \neq -2, 3$ is $(-\infty, -2) \cup (-2, 3) \cup (3, \infty)$.

- $y = (x^2 + x - 8)/(x^2 + 1)$
Because $x^2 + 1 = 0$ has no real solution, we can let x be any real number. This means that the domain is all real numbers. The interval notation for *all real numbers* is $(-\infty, \infty)$.

- $y = (3/(x^2 - 4)) + 5$

$$x^2 - 4 = 0$$
$$x^2 = 4$$
$$x = \pm 2$$

The domain is all real numbers except 2 and -2: $(-\infty, -2) \cup (-2, 2) \cup (2, \infty)$.

PRACTICE
Find the domain for the following functions. Give your solutions in interval notation.

1. $y = \dfrac{x^2 - 3x + 5}{x + 6}$
2. $y = \dfrac{4}{x^2 + 2x - 8} + 12x$
3. $y = \dfrac{6x}{4x^2 + 1}$
4. $y = 3x - 6 + \dfrac{1}{x} + \dfrac{x}{x + 5}$

SOLUTIONS

1. The solution to $x + 6 = 0$ is $x = -6$. The domain is all real numbers except -6: $(-\infty, -6) \cup (-6, \infty)$.

2. Solve $x^2 + 2x - 8 = 0$

$$x^2 + 2x - 8 = (x + 4)(x - 2)$$
$$x + 4 = 0 \qquad x - 2 = 0$$
$$x = -4 \qquad x = 2$$

The domain is all real numbers except -4 and 2: $(-\infty, -4) \cup (-4, 2) \cup (2, \infty)$.

3 Solve $4x^2 + 1 = 0$

$$4x^2 + 1 = 0$$
$$x^2 = -\frac{1}{4}$$
$$x = \pm\sqrt{-\frac{1}{4}} \quad \left(\sqrt{-\frac{1}{4}} \text{ is not a real number}\right)$$

There are no real solutions to $4x^2 + 1 = 0$, so the domain is all real numbers: $(-\infty, \infty)$.

4 Solve $x = 0$ and $x + 5 = 0$

$$x = 0 \qquad x + 5 = 0$$
$$x = -5$$

The domain is all real numbers except 0 and -5: $(-\infty, -5) \cup (-5, 0) \cup (0, \infty)$.

Functions that have a variable under an even root also might have limited domains. We can find the domain of these functions by setting the expression under the root sign greater than or equal to zero and solving the inequality.

EXAMPLES

Find the domain. Give your answers in interval notation.

- $y = \sqrt{x - 6}$

$$x - 6 \geq 0$$
$$x \geq 6$$

The domain is $x \geq 6$: $[6, \infty)$.

- $y = \sqrt{16 - 4x}$

$$16 - 4x \geq 0$$
$$-4x \geq -16$$
$$x \leq 4$$

The domain is $x \leq 4$: $(-\infty, 4]$.

- $y = \sqrt[4]{x^2 - 3x - 4}$

$$x^2 - 3x - 4 \geq 0 \quad \text{becomes} \quad (x-4)(x+1) \geq 0$$
$$x - 4 = 0 \qquad x + 1 = 0$$
$$x = 4 \qquad x = -1$$

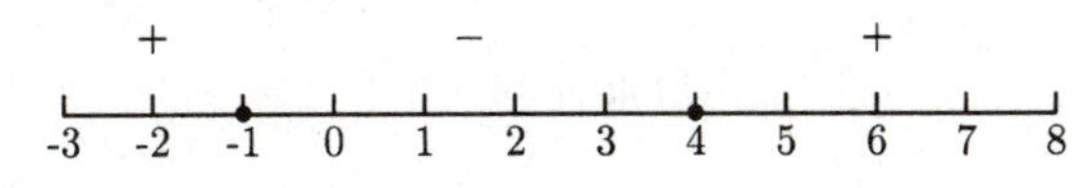

Fig. 6-1.

Because the inequality is "≥," we want the + intervals. The domain is "$x \leq -1$ or $x \geq 4$": $(-\infty, -1] \cup [4, \infty)$.

- $y = \sqrt[4]{x^2 + 3}$

$$x^2 + 3 \geq 0$$

The inequality $x^2 + 3 \geq 0$ is true for all real numbers (since $x^2 \geq 0$ for all x), making the domain all real numbers: $(-\infty, \infty)$.

- $y = \sqrt[3]{5x - 4}$
  Because we can take odd roots of negative numbers, the domain for this function is all real numbers: $(-\infty, \infty)$.

PRACTICE

Find the domain, expressing the answer in interval notation.

1. $y = \sqrt{6x - 8}$
2. $y = \sqrt[5]{4x + 9}$
3. $y = \sqrt{x^2 + 5x + 6}$
4. $y = \sqrt[6]{x^2 + 1}$

SOLUTIONS

1.

$$6x - 8 \geq 0$$
$$x \geq \frac{8}{6} = \frac{4}{3}$$

The domain is $x \geq \frac{4}{3}$: $[\frac{4}{3}, \infty)$.

2. Because the fifth root is odd, the domain is all real numbers: $(-\infty, \infty)$.
3. $x^2 + 5x + 6 \geq 0$ becomes $(x+2)(x+3) \geq 0$

$$x + 2 = 0 \qquad x + 3 = 0$$
$$x = -2 \qquad x = -3$$

The sign graph is shown in Fig. 6-2.

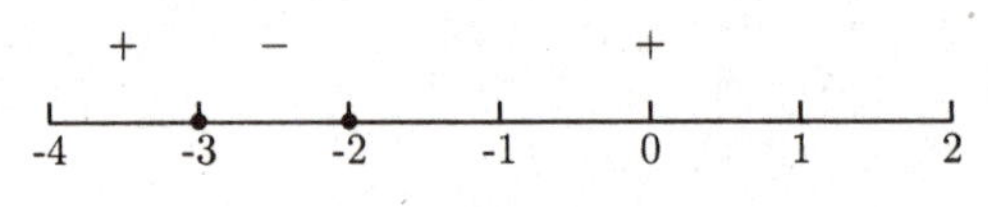

Fig. 6-2.

The domain is $(-\infty, -3] \cup [-2, \infty)$.

4. Because $x^2 + 1 \geq 0$ is true for all real numbers, the domain is all real numbers: $(-\infty, \infty)$.

Some functions are combinations of different kinds of functions. The domain of a combination of two or more kinds of functions is the set of all x-values that are possible for each part. The function

$$y = \frac{\sqrt{x+4}}{x+3}$$

is made up of the parts $\sqrt{x+4}$ and $1/(x+3)$. For $\sqrt{x+4}$, we need $x \geq -4$. For $1/(x+3)$, we need $x \neq -3$.

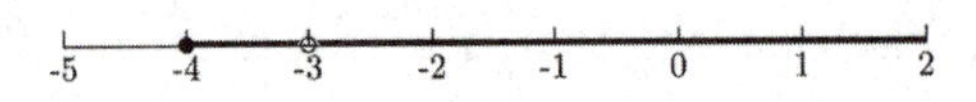

Fig. 6-3.

As we can see from the shaded region in Fig. 6-3, the domain for the function is $[-4, -3) \cup (-3, \infty)$.

EXAMPLE

Find the domain for the function.

- $y = x/\sqrt[4]{x-7}$

  Because $\sqrt[4]{x-7}$ is in the denominator, $\sqrt[4]{x-7}$ cannot be zero. Because $x - 7$ is under an even root, it cannot be negative. Putting these two together means that $x - 7 > 0$ (instead of $x - 7 \geq 0$). The domain of this function is $x > 7$: $(7, \infty)$.

PRACTICE

Find the domain for the functions. Give your solutions in interval notation.

1. $y = \dfrac{\sqrt{2x+5}}{x-6}$

2. $y = \dfrac{1}{\sqrt{3 - 2x}}$

3. $y = \dfrac{\sqrt[4]{x - 2}}{x^2 - x - 12}$

4. $y = \sqrt{x^2 + 3x - 18} + \dfrac{1}{x - 5}$

SOLUTIONS

1.

$$2x + 5 \geq 0 \quad \text{and} \quad x - 6 \neq 0$$
$$x \geq -\frac{5}{2} \qquad\qquad x \neq 6$$

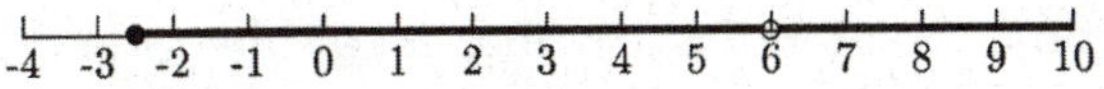

-4 -3 -2 -1 0 1 2 3 4 5 6 7 8 9 10

Fig. 6-4.

The domain is $[-\frac{5}{2}, 6) \cup (6, \infty)$.

2.

$$3 - 2x > 0$$
$$x < \frac{3}{2}$$

The domain is $(-\infty, \frac{3}{2})$.

3.

$$x - 2 \geq 0 \quad \text{and} \quad x^2 - x - 12 \neq 0$$
$$x \geq 2 \qquad (x - 4)(x + 3) \neq 0$$
$$x - 4 \neq 0 \quad x + 3 \neq 0$$
$$x \neq 4 \quad \text{and} \quad x \neq -3$$

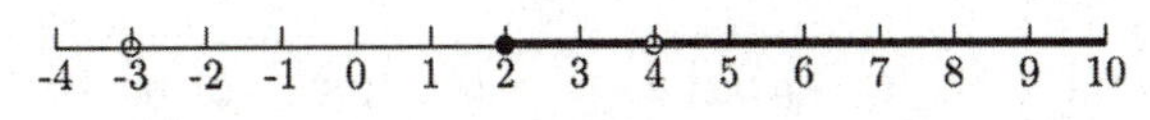

Fig. 6-5.

The domain is $[2, 4) \cup (4, \infty)$. (Because $x \geq 2$, $x = -3$ is not in the domain, anyway.)

4.

$$x^2 + 3x - 18 \geq 0 \quad \text{and} \quad x - 5 \neq 0$$
$$(x + 6)(x - 3) \geq 0 \qquad x \neq 5$$

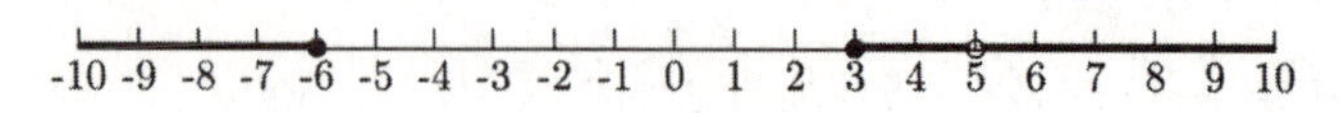

Fig. 6-6.

The domain is $(-\infty, -6] \cup [3, 5) \cup (5, \infty)$.

Evaluating Functions

Functions are often given letter names. The most common name is "$f(x)$." Instead of writing $y = 2x + 1$, we write $f(x) = 2x + 1$. Usually y and $f(x)$ are the same. The notation "$f(x)$" means "the function f evaluated at x."

Evaluating a function at a quantity means to substitute the quantity for x. If the function is $f(x) = 2x + 1$, then to evaluate the function at 3 means to find the y-value for $x = 3$. Sometimes you might be asked to "find f at 3," "evaluate $f(3)$,"or "let $x = 3$ in the equation."

EXAMPLES

Evaluate the functions at the given values.

- Find $f(1)$, $f(-2)$, and $f(0)$ for $f(x) = 3x^2 + 4$.
  We need to substitute the number in the parentheses for x.

$$f(1) = 3(1)^2 + 4 = 3(1) + 4 = 7$$

$$f(-2) = 3(-2)^2 + 4 = 3(4) + 4 = 16$$

$$f(0) = 3(0)^2 + 4 = 3(0) + 4 = 4$$

- Find $f(-4)$, $f(10)$, and $f(1)$ for $f(x) = (6x + 5)/(x^2 + 2)$.

$$f(-4) = \frac{6(-4) + 5}{(-4)^2 + 2} = \frac{-24 + 5}{16 + 2} = \frac{-19}{18}$$

$$f(10) = \frac{6(10) + 5}{10^2 + 2} = \frac{60 + 5}{100 + 2} = \frac{65}{102}$$

$$f(1) = \frac{6(1) + 5}{1^2 + 2} = \frac{6 + 5}{1 + 2} = \frac{11}{3}$$

- Find $g(0)$, $g(1)$, and $g(-6)$ for $g(t) = \sqrt{3-t}$. (Treat the variable t like the variable x.)

$$g(0) = \sqrt{3-0} = \sqrt{3}$$

$$g(1) = \sqrt{3-1} = \sqrt{2}$$

$$g(-6) = \sqrt{3-(-6)} = \sqrt{9} = 3$$

Functions that have no variable (other than $y, f(x)$, or $g(t)$, etc.) are called *constant functions*. The y-values do not change. No matter what x is, the y-value (or functional value) stays the same.

EXAMPLE

- Evaluate $f(x) = 10$ at $x = 3$, $x = -8$, and $x = \pi$.
  No matter what x is, $f(x) = 10$.

$$f(3) = 10 \qquad f(-8) = 10 \qquad f(\pi) = 10$$

PRACTICE

1. Find $f(4)$, $f(-6)$, and $f(0)$ for $f(x) = 3x - 2$.
2. Find $f(-10)$, $f(6)$, and $f(\pi^{23})$ for $f(x) = \sqrt{17}$.
3. Find $h(0)$, $h(5)$, and $h(-2)$ for $h(t) = (2t+4)/(t^2-7)$.
4. Find $f(0)$ and $f(-\frac{1}{2})$ for $f(x) = \sqrt{2x+1}$.

SOLUTIONS

1.

$$f(4) = 3(4) - 2 = 10$$

$$f(-6) = 3(-6) - 2 = -20$$

$$f(0) = 3(0) - 2 = -2$$

2.

$$f(-10) = \sqrt{17}$$

$$f(6) = \sqrt{17}$$

$$f(\pi^{23}) = \sqrt{17}$$

3.

$$h(0) = \frac{2(0)+4}{0^2-7} = -\frac{4}{7}$$
$$h(5) = \frac{2(5)+4}{5^2-7} = \frac{14}{18} = \frac{7}{9}$$
$$h(-2) = \frac{2(-2)+4}{(-2)^2-7} = \frac{0}{-3} = 0$$

4.

$$f(0) = \sqrt{2(0)+1} = \sqrt{1} = 1$$
$$f\left(-\frac{1}{2}\right) = \sqrt{2\left(-\frac{1}{2}\right)+1} = \sqrt{0} = 0$$

Piecewise Functions

Piecewise-defined functions come in two parts. One part is an interval for x, the other part is the formula for computing y.

EXAMPLES

- $f(x) = \begin{cases} x-4, & \text{if } x < 1; \\ 2x+5 & \text{if } x \geq 1 \end{cases}$

  This function is telling us that for x-values smaller than 1, the y-values are computed using $x-4$. For x-values greater than or equal to 1, the y-values are computed using $2x+5$. When asked to evaluate $f(\textit{number})$, we first need to decide what interval of x the number is in, then compute the y-value using the formula to the left of the interval. We will evaluate this function at $x=6$, $x=0$, $x=-2$, and $x=10$.

  $f(6)$: Does $x=6$ belong to the interval $x<1$ or to $x \geq 1$? Since $6 \geq 1$, we will use $2x+5$ to compute y.

$$f(6) = 2(6)+5 = 17$$

  $f(0)$: Does $x=0$ belong to the interval $x<1$ or to $x \geq 1$? Since $0<1$, we will use $x-4$ to compute y.

$$f(0) = 0-4 = -4$$

$f(-2)$: Does $x = -2$ belong to the interval $x < 1$ or to $x \geq 1$? Since $-2 < 1$, we will use $x - 4$ to compute y.

$$f(-2) = -2 - 4 = -6$$

$f(10)$: Does $x = 10$ belong to the interval $x < 1$ or to $x \geq 1$? Since $10 \geq 1$, we will use $2x + 5$ to compute y.

$$f(10) = 2(10) + 5 = 25$$

- Evaluate $f(-4)$, $f(0)$, $f(6)$, and $f(-2)$.

$$f(x) = \begin{cases} x^2 & \text{if } x < -2; \\ 4x + 8 & \text{if } -2 \leq x < 3; \\ 16 & \text{if } x \geq 3 \end{cases}$$

$f(-4)$: Since $x = -4$ belongs to the interval $x < -2$, we will use x^2 to compute y.

$$f(-4) = (-4)^2 = 16$$

$f(0)$: Since $x = 0$ belongs to the interval $-2 \leq x < 3$, we will use $4x + 8$ to compute y.

$$f(0) = 4(0) + 8 = 8$$

$f(6)$: Since $x = 6$ belongs to the interval $x \geq 3$, the y-value is 16.

$$f(6) = 16$$

$f(-2)$: Since $x = -2$ belongs to the interval $-2 \leq x < 3$, we will use $4x + 8$ to compute y.

$$f(-2) = 4(-2) + 8 = 0$$

PRACTICE

1. Find $f(0)$, $f(-1)$, $f(7)$, and $f(-6)$.

$$f(x) = \begin{cases} x^2 - 2x & \text{if } x < -4; \\ x + 5 & \text{if } x \geq -4 \end{cases}$$

2. Find $f(-4)$, $f(3)$, $f(-1)$, and $f(1)$.

$$f(x) = \begin{cases} 3x^2 + 2x & \text{if } x \leq -1; \\ x + 4 & \text{if } -1 < x \leq 1; \\ 6x & \text{if } x > 1 \end{cases}$$

3. Find $f(3)$, $f(-2)$, and $f(0)$.

$$f(x) = \begin{cases} 0 & \text{if } x < 0; \\ 1 & \text{if } x \geq 0 \end{cases}$$

4. Find $f(3)$, $f(0)$, $f(-4)$, $f(2)$, and $f(1)$.

$$f(x) = \begin{cases} -2 & \text{if } x < 1; \\ 3x - 4 & \text{if } 1 \leq x < 3; \\ x^2 - 2x + 2 & \text{if } x \geq 3 \end{cases}$$

SOLUTIONS

1.

$$f(0) = 0 + 5 = 5 \qquad f(-1) = -1 + 5 = 4$$
$$f(7) = 7 + 5 = 12 \qquad f(-6) = (-6)^2 - 2(-6) = 48$$

2.

$$f(-4) = 3(-4)^2 + 2(-4) = 40 \qquad f(3) = 6(3) = 18$$
$$f(-1) = 3(-1)^2 + 2(-1) = 1 \qquad f(1) = 1 + 4 = 5$$

3.

$$f(3) = 1 \qquad f(-2) = 0 \qquad f(0) = 1$$

4.

$$f(3) = 3^2 - 2(3) + 2 = 5 \qquad f(0) = -2$$
$$f(-4) = -2 \qquad f(2) = 3(2) - 4 = 2$$
$$f(1) = 3(1) - 4 = -1$$

More Evaluating Functions

Functions can be evaluated at quantities other than numbers, even at other functions. Keep in mind that evaluating a function means to substitute whatever is in the parentheses for the variable, even if what is in the parentheses is another variable. The function $f(x) = 2x + 1$ says, "Double the quantity in

the parentheses, then add 1." Suppose we are asked to find $f(b)$. We need to substitute b for x in the equation, that is, double b then add 1.

$$f(b) = 2b + 1$$

Similarly $f(v^2) = 2v^2 + 1$ and $f(a+b) = 2(a+b) + 1 = 2a + 2b + 1$.

EXAMPLES

- Find $f(a)$, $f(2a)$, and $f(a+1)$ for $f(x) = x^2 + 3x + 2$.

$$f(a) = a^2 + 3a + 2$$

$$f(2a) = (2a)^2 + 3(2a) + 2 = 4a^2 + 6a + 2$$

$$f(a+1) = (a+1)^2 + 3(a+1) + 2 = (a+1)(a+1) + 3(a+1) + 2$$

$$= a^2 + 2a + 1 + 3a + 3 + 2 = a^2 + 5a + 6$$

- Find $f(u)$, $f(3u)$, $f(u-v)$, and $f(u^2)$ for $f(x) = (6x-1)/(x^2+3)$.

$$f(u) = \frac{6u-1}{u^2+3}$$

$$f(3u) = \frac{6(3u)-1}{(3u)^2+3} = \frac{18u-1}{9u^2+3}$$

$$f(u-v) = \frac{6(u-v)-1}{(u-v)^2+3} = \frac{6u-6v-1}{(u-v)(u-v)+3} = \frac{6u-6v-1}{u^2-2uv+v^2+3}$$

$$f(u^2) = \frac{6u^2-1}{(u^2)^2+3} = \frac{6u^2-1}{u^4+3}$$

- Find $f(a)$, $f(a+h)$, $f(1/a)$, and $f(-x)$ for $f(x) = 1/(x-1)$

$$f(a) = \frac{1}{a-1}$$

$$f(a+h) = \frac{1}{(a+h)-1} = \frac{1}{a+h-1}$$

$$f\left(\frac{1}{a}\right) = \frac{1}{(1/a)-1} = \frac{1}{(1/a)-(a/a)} = \frac{1}{(1-a)/a}$$

$$= 1 \div \frac{1-a}{a} = 1 \cdot \frac{a}{1-a} = \frac{a}{1-a}$$

$$f(-x) = \frac{1}{-x-1} \quad \text{or} \quad \frac{1}{-(x+1)} = -\frac{1}{x+1}$$

- Find $g(u)$, $g(u^2+v)$, and $g(3u-1)$ for $g(t)=12$. Because g is a constant function, $g(t)=12$ no matter what is in the parentheses.

$$g(u) = 12$$
$$g(u^2+v) = 12$$
$$g(3u-1) = 12$$

PRACTICE

1. Find $f(a)$, $f(2a)$, $f(a^2)$, and $f(a+h)$ for $f(x)=3x-8$.
2. Find $g(a)$, $g(a+1)$, $g(-a)$, and $g(a+h)$ for $g(t)=7$.
3. Find $f(u)$, $f(uv)$, $f(a)$, and $f(a+h)$ for $f(x)=2x^2-x+1$.
4. Find $f(-t)$, $f(a)$, $f(a+h)$, and $f(1/a)$ for $f(t)=(3-t)/t$.
5. Find $f(-x)$, $f(1/x)$, $f(a)$, and $f(a+h)$ for $f(x)=(4-x)/(1-2x)$.

SOLUTIONS

1.

$$f(a) = 3a-8$$
$$f(2a) = 3(2a)-8 = 6a-8$$
$$f(a^2) = 3a^2-8$$
$$f(a+h) = 3(a+h)-8 = 3a+3h-8$$

2.

$$g(a) = 7$$
$$g(a+1) = 7$$
$$g(-a) = 7$$
$$g(a+h) = 7$$

3.

$$f(u) = 2u^2-u+1$$
$$f(uv) = 2(uv)^2-uv+1 = 2u^2v^2-uv+1$$
$$f(a) = 2a^2-a+1$$
$$\begin{aligned} f(a+h) &= 2(a+h)^2-(a+h)+1 \\ &= 2(a+h)(a+h)-(a+h)+1 \\ &= 2(a^2+2ah+h^2)-a-h+1 \\ &= 2a^2+4ah+2h^2-a-h+1 \end{aligned}$$

4.

$$f(-t) = \frac{3-(-t)}{-t} = \frac{3+t}{-t} \quad \text{or} \quad -\frac{3+t}{t}$$

$$f(a) = \frac{3-a}{a}$$

$$f(a+h) = \frac{3-(a+h)}{a+h} = \frac{3-a-h}{a+h}$$

$$f\left(\frac{1}{a}\right) = \frac{3-1/a}{1/a} = \frac{(3a/a)-(1/a)}{1/a}$$

$$= \frac{(3a-1)/a}{1/a} = \frac{3a-1}{a} \div \frac{1}{a} = \frac{3a-1}{a} \cdot \frac{a}{1} = 3a-1$$

5.

$$f(-x) = \frac{4-(-x)}{1-2(-x)} = \frac{4+x}{1+2x}$$

$$f\left(\frac{1}{x}\right) = \frac{4-(1/x)}{1-2(1/x)} = \frac{(4x/x)-(1/x)}{1-(2/x)} = \frac{(4x-1)/x}{(x/x)-(2/x)}$$

$$= \frac{(4x-1)/x}{(x-2)/x} = \frac{4x-1}{x} \div \frac{x-2}{x} = \frac{4x-1}{x} \cdot \frac{x}{x-2} = \frac{4x-1}{x-2}$$

$$f(a) = \frac{4-a}{1-2a}$$

$$f(a+h) = \frac{4-(a+h)}{1-2(a+h)} = \frac{4-a-h}{1-2a-2h}$$

Newton's Quotient

A very important expression in mathematics is *Newton's quotient*, sometimes written as

$$\frac{f(a+h)-f(a)}{h},$$

where f is some function. In fact, Newton's quotient is the basis for differential calculus. Algebra students work with Newton's quotient so that when (and if) they study calculus, they do not struggle with complicated algebra.

Evaluating Newton's quotient is really not much more than function evaluation. First, we need to find $f(a)$ and $f(a+h)$ for the function given to us. Second, we need to perform the subtraction $f(a+h)-f(a)$ and simplify. Third, we need to divide this by h and simplify. The previous practice problems gave us experience in evaluating $f(a)$ and $f(a+h)$. Now we will practice finding $f(a+h)-f(a)$.

EXAMPLES

Find $f(a+h)-f(a)$ for the functions.

- $f(x)=3x+5$.

$$f(a)=3a+5 \quad \text{and} \quad f(a+h)=3(a+h)+5=3a+3h+5$$

$$\begin{aligned} f(a+h)-f(a) &= 3a+3h+5-(3a+5) \\ &= 3a+3h+5-3a-5 \\ &= 3h \end{aligned}$$

- $f(t)=t^2+1$

$$f(a)=a^2+1 \quad \text{and} \quad f(a+h)=(a+h)^2+1=a^2+2ah+h^2+1$$

$$\begin{aligned} f(a+h)-f(a) &= a^2+2ah+h^2+1-(a^2+1) \\ &= a^2+2ah+h^2+1-a^2-1 \\ &= 2ah+h^2 \end{aligned}$$

- $f(x)=6$

$$f(a)=6 \quad \text{and} \quad f(a+h)=6$$

$$f(a+h)-f(a)=6-6=0$$

- $f(t)=\dfrac{1}{t+3}$

$$f(a)=\frac{1}{a+3} \quad \text{and} \quad f(a+h)=\frac{1}{a+h+3}$$

$$f(a+h)-f(a)=\frac{1}{a+h+3}-\frac{1}{a+3}$$

$$=\frac{a+3}{a+3}\cdot\frac{1}{a+h+3}-\frac{a+h+3}{a+h+3}\cdot\frac{1}{a+3}$$

$$=\frac{a+3-(a+h+3)}{(a+h+3)(a+3)}$$

$$=\frac{a+3-a-h-3}{(a+h+3)(a+3)}=\frac{-h}{(a+h+3)(a+3)}$$

PRACTICE

Find $f(a+h)-f(a)$ for the functions.

1. $f(x)=3x-4$
2. $f(x)=x^2+5$
3. $f(x)=x^2-3x-6$
4. $f(t)=-19$
5. $f(t)=1/t$

SOLUTIONS

1. $f(a+h)=3(a+h)-4$ and $f(a)=3a-4$

$$f(a+h)-f(a)=3(a+h)-4-(3a-4)$$

$$=3a+3h-4-3a+4$$

$$=3h$$

2. $f(a+h)=(a+h)^2+5$ and $f(a)=a^2+5$

$$f(a+h)-f(a)=(a+h)^2+5-(a^2+5)$$

$$=a^2+2ah+h^2+5-a^2-5$$

$$=2ah+h^2$$

3. $f(a+h)=(a+h)^2-3(a+h)-6$ and $f(a)=a^2-3a-6$

$$f(a+h)-f(a)=(a+h)^2-3(a+h)-6-(a^2-3a-6)$$

$$=a^2+2ah+h^2-3a-3h-6-a^2+3a+6$$

$$=2ah+h^2-3h$$

4. $f(a+h) = -19$ and $f(a) = -19$

$$f(a+h) - f(a) = -19 - (-19) = -19 + 19 = 0$$

5. $f(a+h) = \frac{1}{a+h}$ and $f(a) = \frac{1}{a}$

$$\begin{aligned} f(a+h) - f(a) &= \frac{1}{a+h} - \frac{1}{a} \\ &= \frac{a}{a} \cdot \frac{1}{a+h} - \frac{a+h}{a+h} \cdot \frac{1}{a} \\ &= \frac{a}{a(a+h)} - \frac{a+h}{a(a+h)} = \frac{a-(a+h)}{a(a+h)} \\ &= \frac{a-a-h}{a(a+h)} = \frac{-h}{a(a+h)} \end{aligned}$$

The only steps remaining in evaluating Newton's quotient is to divide the difference $f(a+h) - f(a)$ by h. The following examples and practice problems are from the previous section.

EXAMPLES

Evaluate $(f(a+h) - f(a))/h$ for the functions.

- $f(x) = 3x + 5$.
  We found that $f(a+h) - f(a) = 3h$.

$$\frac{f(a+h) - f(a)}{h} = \frac{3h}{h} = 3$$

- $f(t) = t^2 + 1$.
  We found that $f(a+h) - f(a) = 2ah + h^2$.

$$\frac{f(a+h) - f(a)}{h} = \frac{2ah + h^2}{h} = 2a + h$$

- $f(x) = 6$.
  We found that $f(a+h) - f(a) = 0$.

$$\frac{f(a+h) - f(a)}{h} = \frac{0}{h} = 0$$

- $f(t) = 1/(t+3)$

  We found that $f(a+h) - f(a) = -h/((a+h+3)(a+3))$.

$$\frac{f(a+h)-f(a)}{h} = \frac{-h/((a+h+3)(a+3))}{h}$$
$$= \frac{-h}{(a+h+3)(a+3)} \div h = \frac{-h}{(a+h+3)(a+3)} \cdot \frac{1}{h}$$
$$= \frac{-1}{(a+h+3)(a+3)}$$

PRACTICE

Evaluate $(f(a+h) - f(a))/h$ for the functions. The first five functions are the same as in the previous practice problems.

1. $f(x) = 3x - 4$
2. $f(x) = x^2 + 5$
3. $f(x) = x^2 - 3x - 6$
4. $f(t) = -19$
5. $f(t) = 1/t$
6. $f(x) = 3x^2 - 5x + 2$

SOLUTIONS

1. We found that $f(a+h) - f(a) = 3h$.

$$\frac{f(a+h)-f(a)}{h} = \frac{3h}{h} = 3$$

2. We found that $f(a+h) - f(a) = 2ah + h^2$.

$$\frac{f(a+h)-f(a)}{h} = \frac{2ah+h^2}{h} = \frac{h(2a+h)}{h} = 2a+h$$

3. We found that $f(a+h) - f(a) = 2ah + h^2 - 3h$.

$$\frac{f(a+h)-f(a)}{h} = \frac{2ah+h^2-3h}{h} = \frac{h(2a+h-3)}{h} = 2a+h-3$$

4. We found that $f(a+h) - f(a) = 0$.

$$\frac{f(a+h)-f(a)}{h} = \frac{0}{h} = 0$$

5. We found that $f(a+h)-f(a) = -h/(a(a+h))$

$$\frac{f(a+h)-f(a)}{h} = \frac{-h/(a(a+h))}{h} = \frac{-h}{a(a+h)} \div h$$
$$= \frac{-h}{a(a+h)} \cdot \frac{1}{h} = \frac{-1}{a(a+h)}$$

6.

$$\frac{f(a+h)-f(a)}{h} = \frac{3(a+h)^2 - 5(a+h) + 2 - (3a^2 - 5a + 2)}{h}$$
$$= \frac{3a^2 + 6ah + 3h^2 - 5a - 5h + 2 - 3a^2 + 5a - 2}{h}$$
$$= \frac{6ah + 3h^2 - 5h}{h} = \frac{h(6a + 3h - 5)}{h}$$
$$= 6a + 3h - 5$$

Newton's quotient is really nothing more than the slope of the line containing the two points $(a, f(a))$ and $(a+h, f(a+h))$. Remember the slope formula for the line containing the points (x_1, y_1) and (x_2, y_2) is

$$m = \frac{y_2 - y_1}{x_2 - x_1}.$$

In Newton's quotient, $x_1 = a$, $y_1 = f(a)$, $x_2 = a+h$, $y_2 = f(a+h)$.

$$m = \frac{y_2 - y_1}{x_2 - x_1} = \frac{f(a+h)-f(a)}{a+h-a} = \frac{f(a+h)-f(a)}{h}$$

Functions and Their Graphs

Reading graphs, sketching graphs by hand, and sketching graphs using graphing calculators are all important in today's algebra courses. We will concentrate on reading graphs in this section.

A graph can give us a great deal of information about its equation. First, it can tell us if the graph is the graph of a function. Remember, if y is a function of x, then each x-value has exactly one y-value. What if we have a graph where an x-value has two or more y-values? A vertical line through that particular x-value would touch the graph in more than one point.

For example, a vertical line would touch the graph at both $(4, 2)$ and $(4, -2)$ in Fig. 6-7.

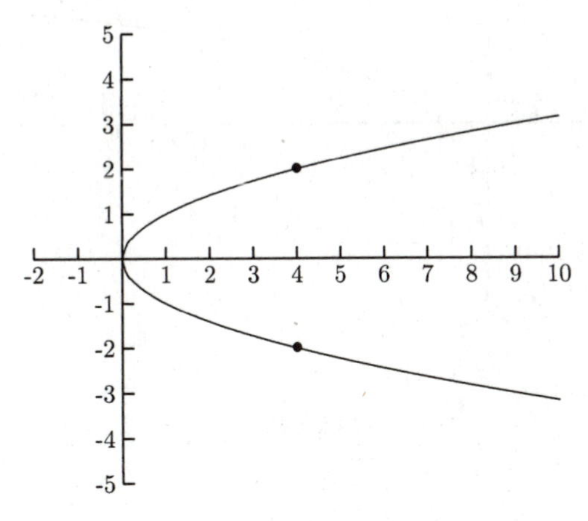

Fig. 6-7.

We can tell whether or not a graph is the graph of a function if any vertical line touches the graph in more than one point. If a vertical line touches the graph in more than one point, then the graph is not the graph of a function. If every vertical line touches the graph in one point or not at all, then the graph is the graph of a function. This is called the *vertical line test*. The graphs in Figs. 6-8 and 6-9 are graphs of functions.

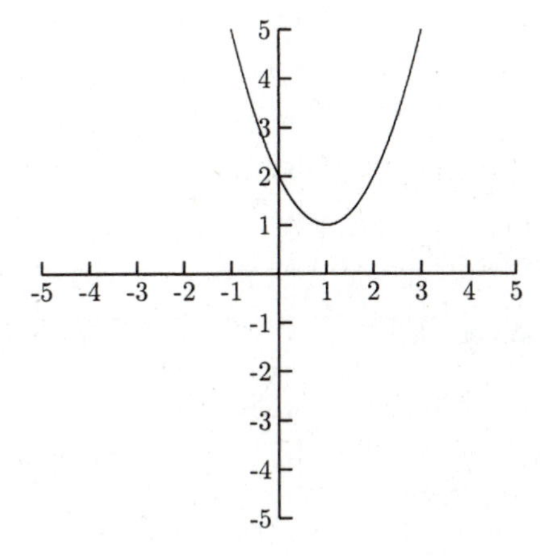

Fig. 6-8.

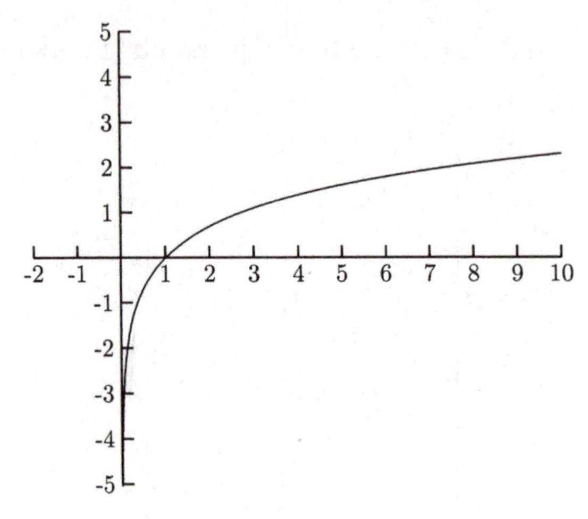

Fig. 6-9.

PRACTICE

Use the vertical line test to determine which of the graphs below are graphs of functions.

1.

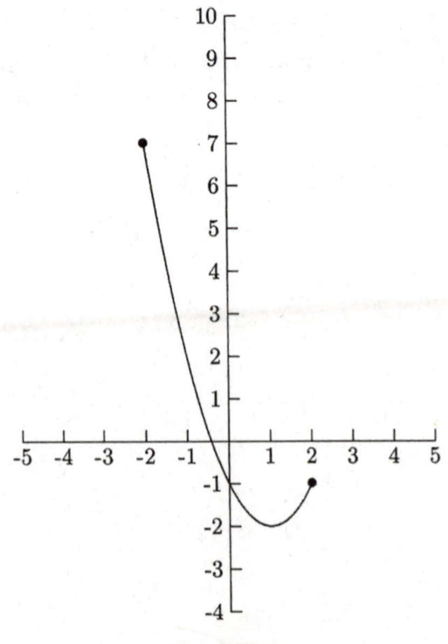

Fig. 6-10.

2.

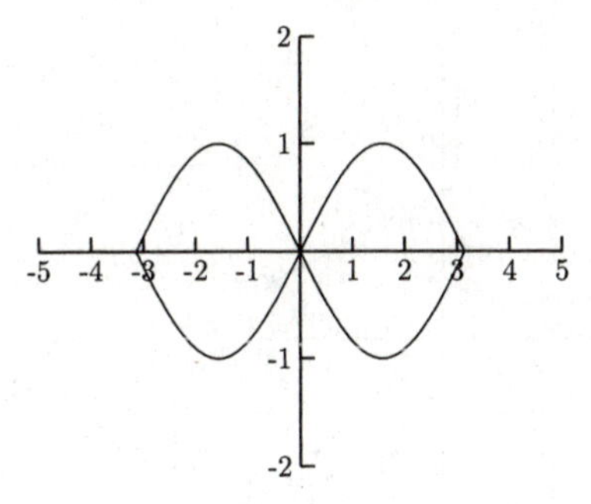

Fig. 6-11.

3.

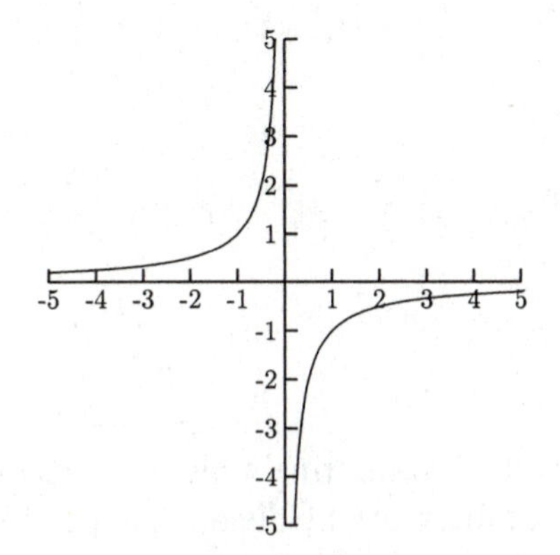

Fig. 6-12.

4.

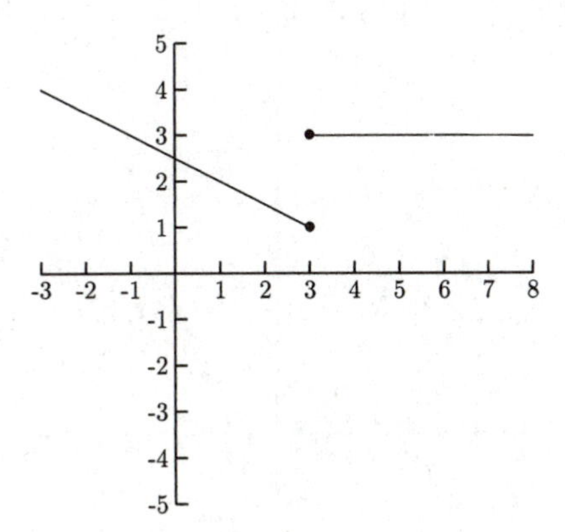

Fig. 6-13.

5.

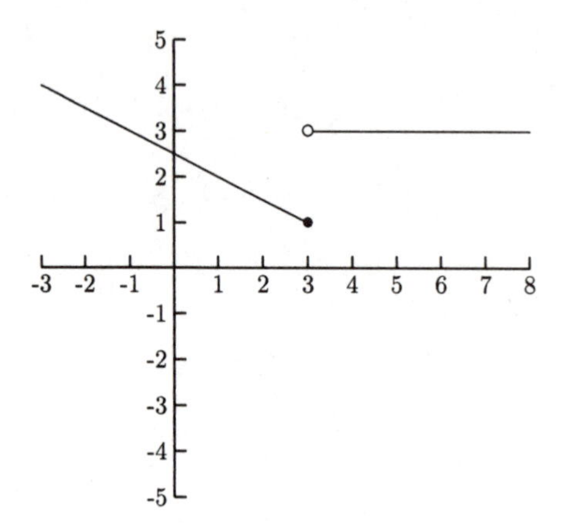

Fig. 6-14.

SOLUTIONS

1. Yes
2. No
3. Yes
4. No
5. Yes

Graphs are also useful in evaluating functions. Remember that points on the graph are pairs of numbers, x (the distance left or right of the origin), and y (the distance above or below the origin). Normally, y and $f(x)$ are the same. The point $(-1, 1)$ on the graph in Fig. 6-15 means that $f(-1) = 1$. The point $(2, 4)$ on the graph means that $f(2) = 4$. What is $f(0)$? In other words, when $x = 0$, what is y? Because the point $(0, 0)$ is on the graph, $f(0) = 0$.

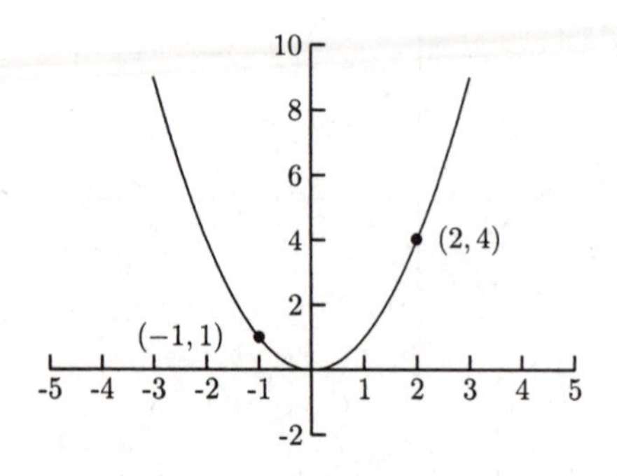

Fig. 6-15.

EXAMPLES

Refer to Fig. 6-16 for the following.

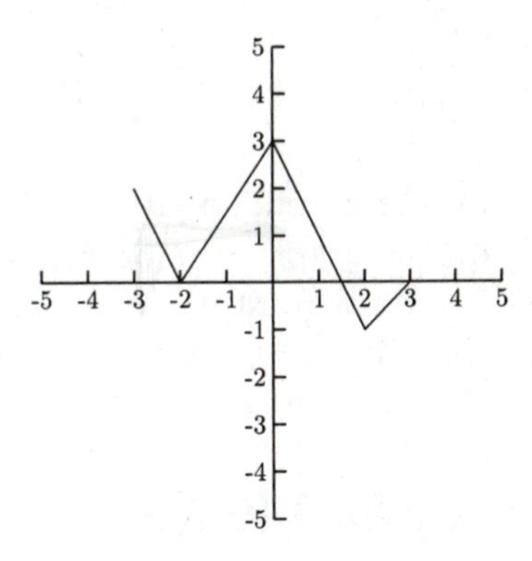

Fig. 6-16.

- Find $f(-3)$.
  Another way of saying, "Find $f(-3)$" is saying, "What is the y-value for the point on the graph for $x = -3$?" The point $(-3, 2)$ is on the graph, so $f(-3) = 2$.
- Find $f(-2)$.
  We need to look for the point on the graph where $x = -2$. The point $(-2, 0)$ is on the graph, so $f(-2) = 0$.
- Find $f(0)$.
  We need to look for the point on the graph where $x = 0$. The point $(0, 3)$ is on the graph, so $f(0) = 3$.

PRACTICE

Refer to Fig. 6-17 for the following.

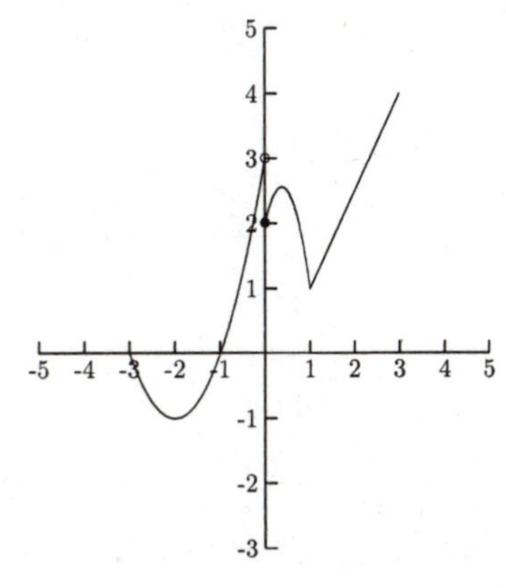

Fig. 6-17.

1. Find $f(0)$
2. Find $f(3)$
3. Find $f(-2)$
4. Find $f(-3)$

SOLUTIONS

1. The point $(0, 2)$ is on the graph, so $f(0) = 2$.
2. The point $(3, 4)$ is on the graph, so $f(3) = 4$.
3. The point $(-2, -1)$ is on the graph, so $f(-2) = -1$.
4. The point $(-3, 0)$ is on the graph, so $f(-3) = 0$.

FINDING THE DOMAIN AND RANGE

The graph of a function can tell us what its domain and range are. Remember that the domain of a function is the set of x-values that can be used in the function. We can find the domain by seeing how far left and right the graph goes. The range of a function is the set of y-values. We can find the range by seeing how far up and down the graph goes.

EXAMPLES

Find the domain and range. Give your answers in interval notation.

•

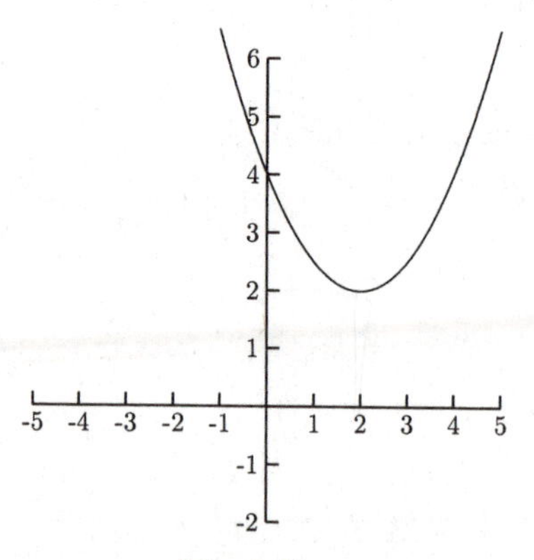

Fig. 6-18.

The domain is $(-\infty, \infty)$. The range is $[2, \infty)$ because the smallest y-value is 2 and there is no largest y-value.

•

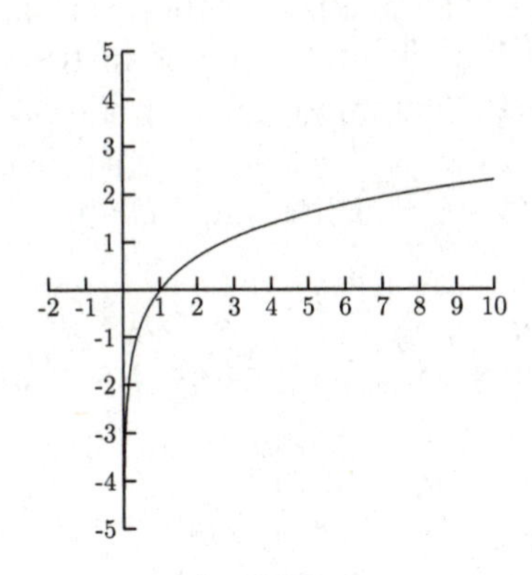

Fig. 6-19.

Because the entire graph is to the right of the y-axis (where $x = 0$), the domain is $(0, \infty)$. The range is $(-\infty, \infty)$.

•

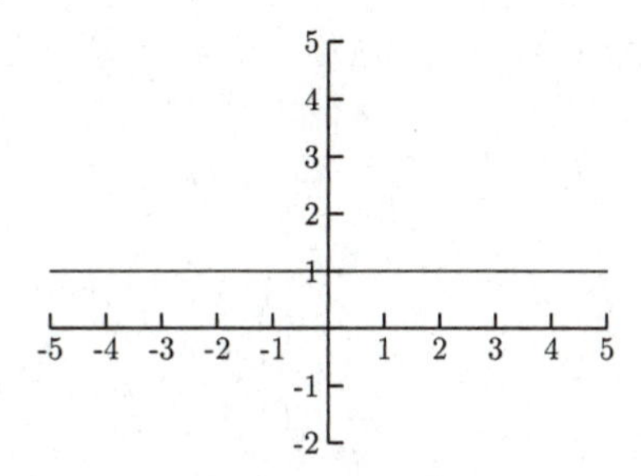

Fig. 6-20.

The domain is $(-\infty, \infty)$. The range consists of one number, $y = 1$. The range is $\{1\}$.

•

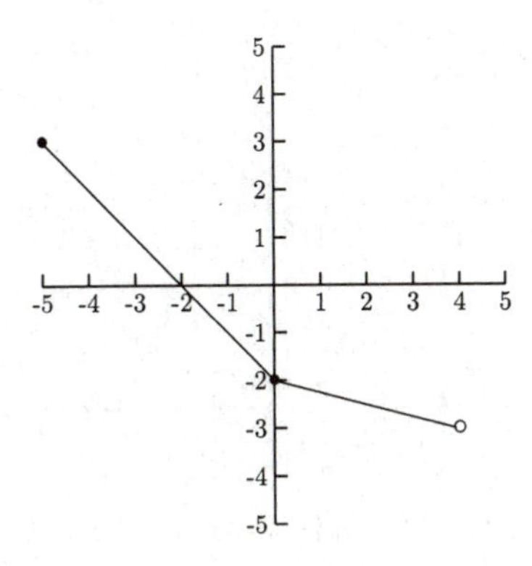

Fig. 6-21.

The solid dot at $(-5, 3)$ means that this point is part of the function, so $x = -5$ is in the domain and $y = 3$ is in the range. The open dot at $(4, -3)$ means that the domain goes up to $x = 4$ but does not include it, and that the range goes down to $y = -3$ but does not include it. The domain is $[-5, 4)$, and the range is $(-3, 3]$.

PRACTICE

Find the domain and range. Give your answers in interval notation.

1.

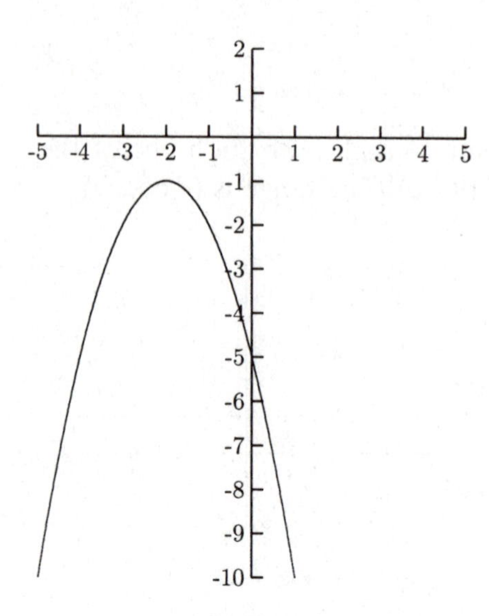

Fig. 6-22.

2.

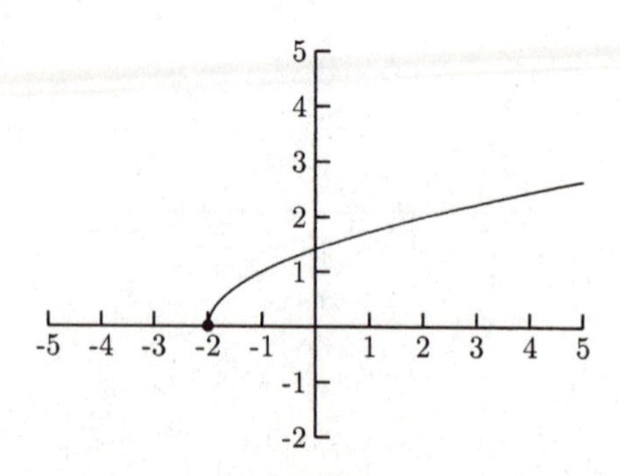

Fig. 6-23.

3.

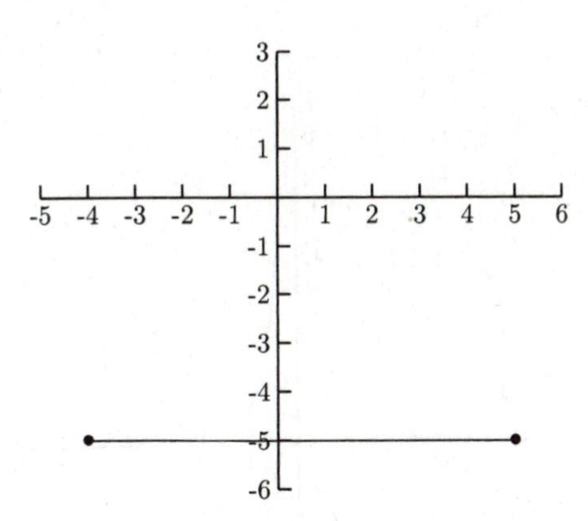

Fig. 6-24.

4.

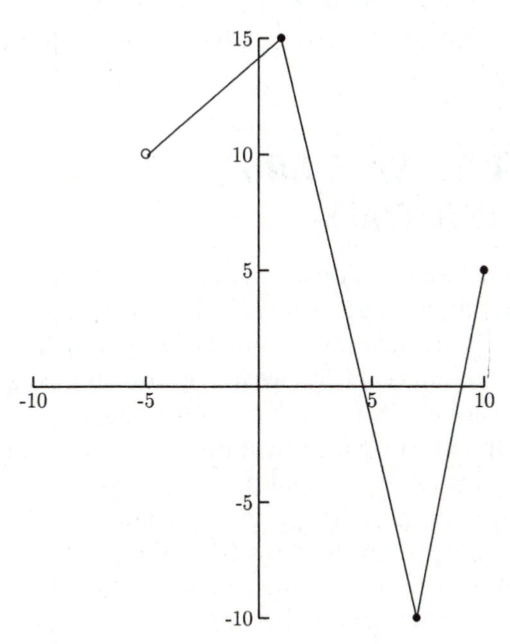

Fig. 6-25.

5.

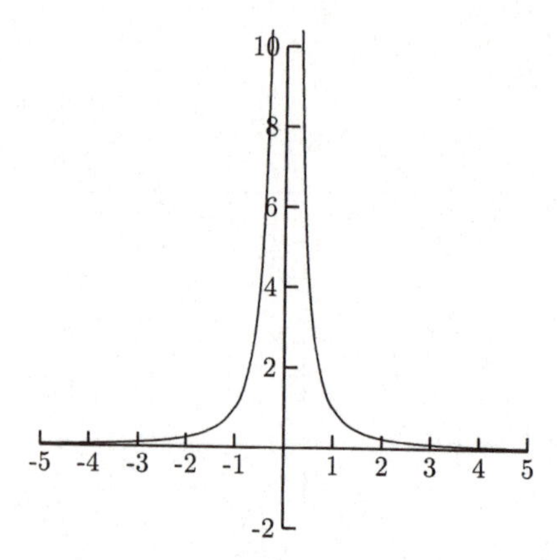

Fig. 6-26.

SOLUTIONS

1. The domain is $(-\infty, \infty)$. The range is $(-\infty, -1]$.
2. The domain is $[-2, \infty)$. The range is $[0, \infty)$.
3. The domain is $[-4, 5]$. The range is $\{-5\}$.
4. The domain is $(-5, 10]$. The range is $[-10, 15]$.
5. The domain is $(-\infty, 0) \cup (0, \infty)$. The range is $(0, \infty)$.

INCREASING INTERVALS AND DECREASING INTERVALS

Graphs can tell us where functions are going up (if anywhere) and where they are going down (if anywhere). Many functions go up in some places and down in others. A few functions do not go up or down. A function is said to be *increasing* on an interval if, as we move from left to right in the interval, the y-values are going up. A function is said to be *decreasing* on an interval if, as we move from left to right in the interval, the y-values are going down. A function is constant on an interval if, as we move from left to right in the interval, the y-values do not change. As an example, consider the graph in Fig. 6-27. If we are anywhere to the left of $x = -1$ and move to the right, the graph is going down. We say the function is decreasing on the interval $(-\infty, -1)$. If we are anywhere between $x = -1$ and $x = 0$ and move to the right, the graph is going up. We say the function is increasing on

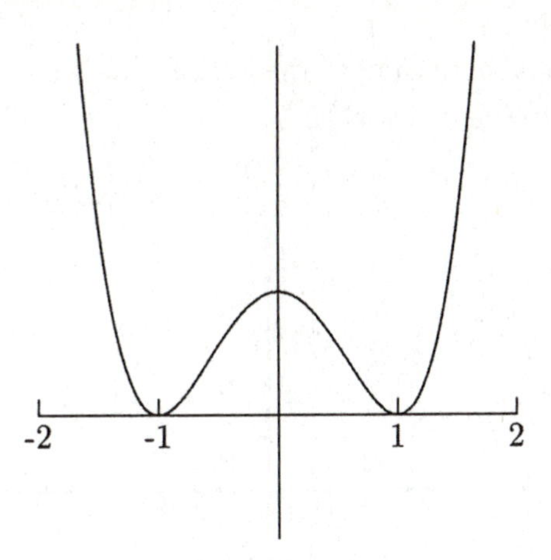

Fig. 6-27.

the interval $(-1, 0)$. If we are anywhere between $x = 0$ and $x = 1$ and move to the right, the graph is going back down. We say the function is decreasing on the interval $(0, 1)$. Finally, if we are anywhere to the right of $x = 1$ and move to the right, the function is going back up. We say the function is increasing on the interval $(1, \infty)$.

EXAMPLES

Determine where the functions are increasing, decreasing, or constant.

•

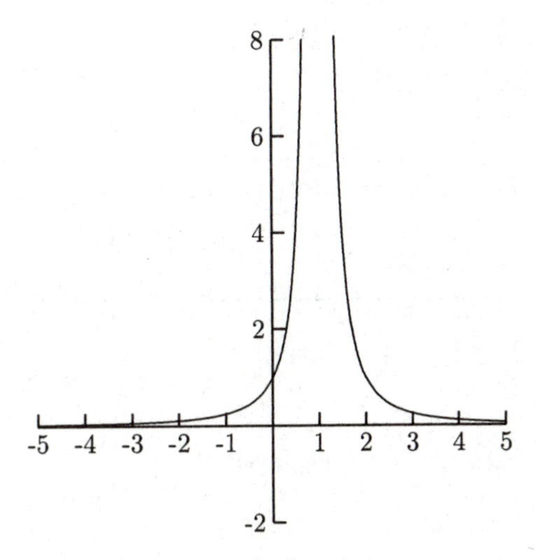

Fig. 6-28.

This function is increasing to the left of $x = 1$, $(-\infty, 1)$. It is decreasing to the right of $x = 1$, $(1, \infty)$.

•

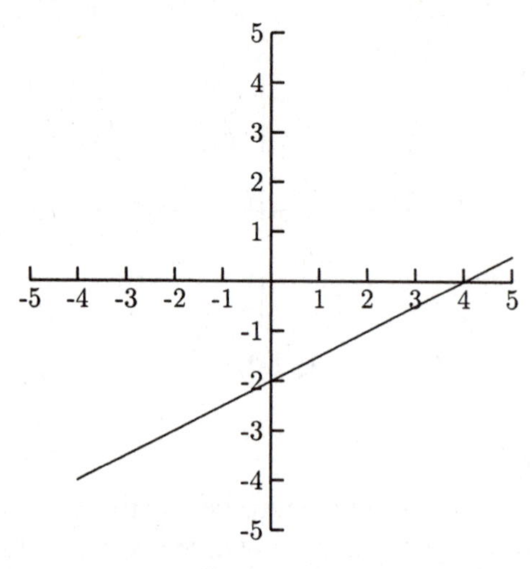

Fig. 6-29.

No matter where we are on this graph, as we move to the right, the graph is going up, so this function is increasing everywhere, $(-\infty, \infty)$.

•

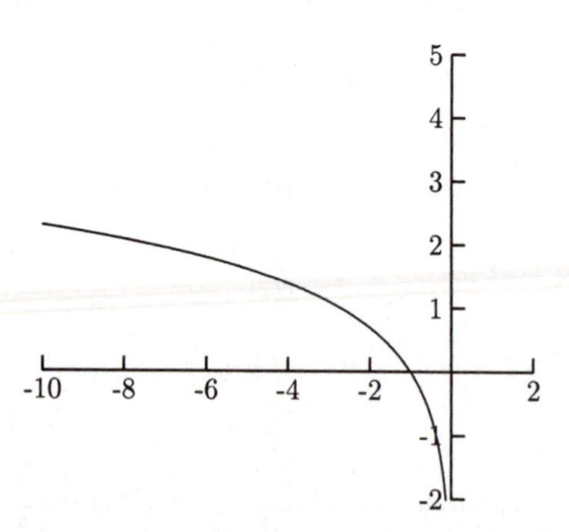

Fig. 6-30.

This function is decreasing on all of its domain, $(-\infty, 0)$.

•

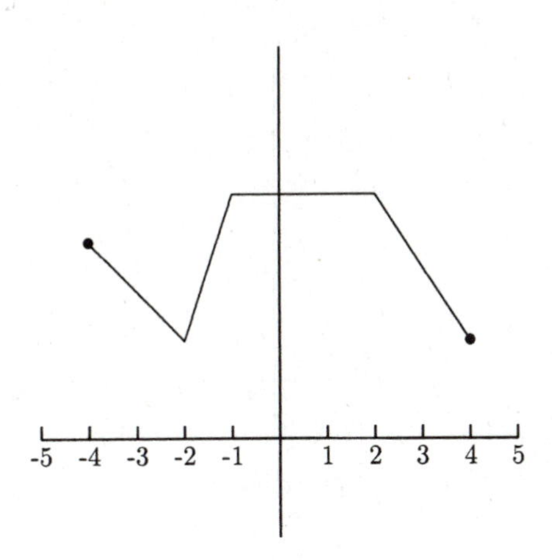

Fig. 6-31.

This function is decreasing on $(-4, -2)$ (between $x = -4$ and $x = -2$), increasing on $(-2, -1)$ (between $x = -2$ and $x = -1$), constant on $(-1, 2)$ (between $x = -1$ and $x = 2$), and decreasing on $(2, 4)$ (between $x = 2$ and $x = 4$).

PRACTICE

Determine where the functions are increasing, decreasing, or constant.

1.

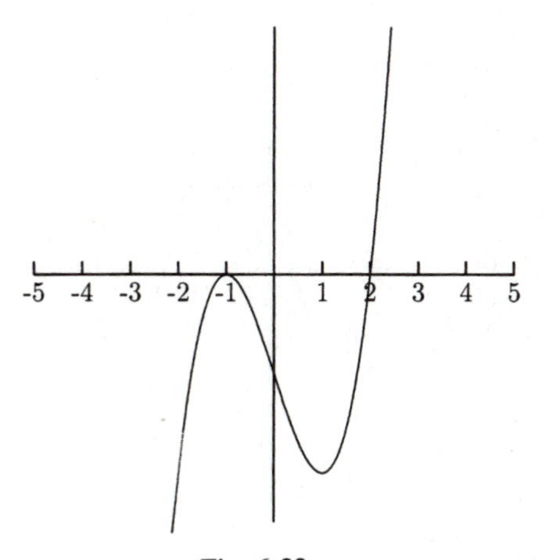

Fig. 6-32.

2.

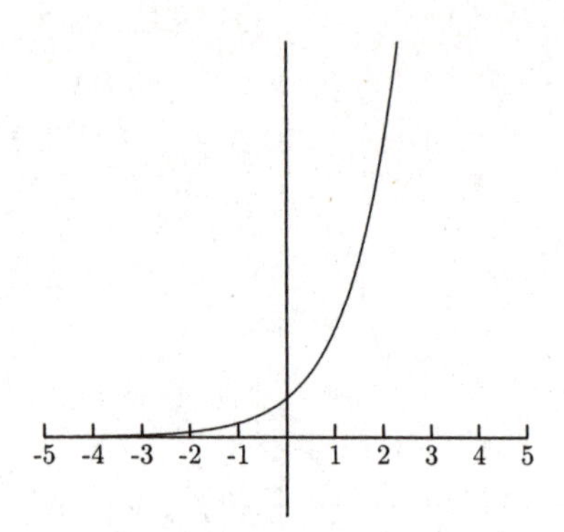

Fig. 6-33.

3.

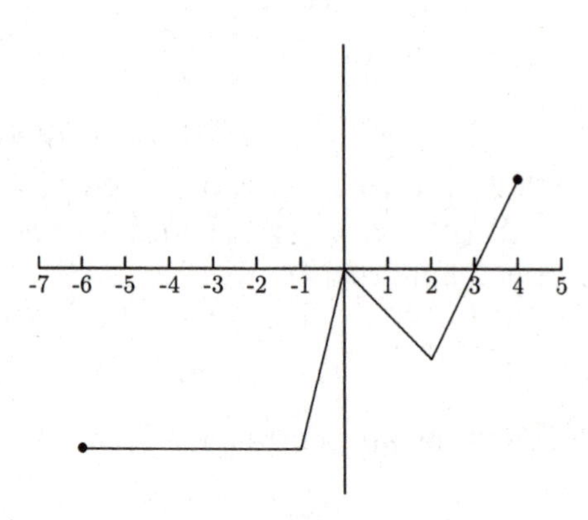

Fig. 6-34.

4.

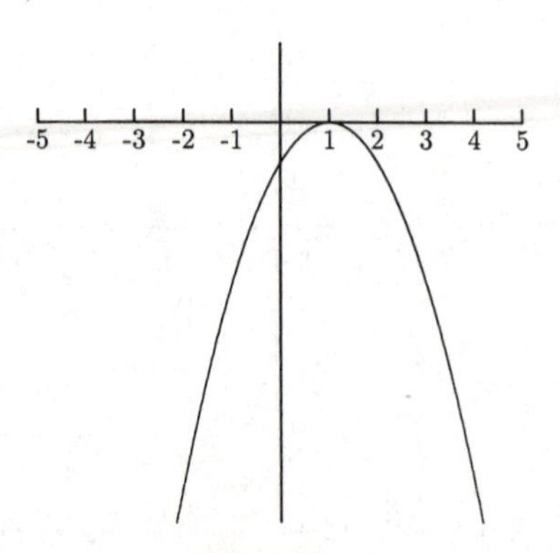

Fig. 6-35.

SOLUTIONS

1. The increasing intervals are $(-\infty, -1)$ and $(1, \infty)$. The decreasing interval is $(-1, 1)$.
2. The function is increasing everywhere, $(-\infty, \infty)$.
3. The constant interval is $(-6, -1)$. The increasing intervals are $(-1, 0)$ and $(2, 4)$. The decreasing interval is $(0, 2)$.
4. The increasing interval is $(-\infty, 1)$. The decreasing interval is $(1, \infty)$.

PIECEWISE FUNCTIONS

The graph of a piecewise-defined function comes in pieces. A piece might be part of a line, parabola, or some other shape. The important point is in determining which piece of which function is needed. For example, the following function comes in two pieces. The first piece is part of the line $y = x + 1$, and the second piece is part of the line $y = 2x$.

$$f(x) = \begin{cases} x + 1 & \text{if } x < 0 \\ 2x & \text{if } x \geq 0 \end{cases}$$

Because "$x < 0$" is written to the right of "$x + 1$", the part of the line $y = x + 1$ we need is to the left of $x = 0$.

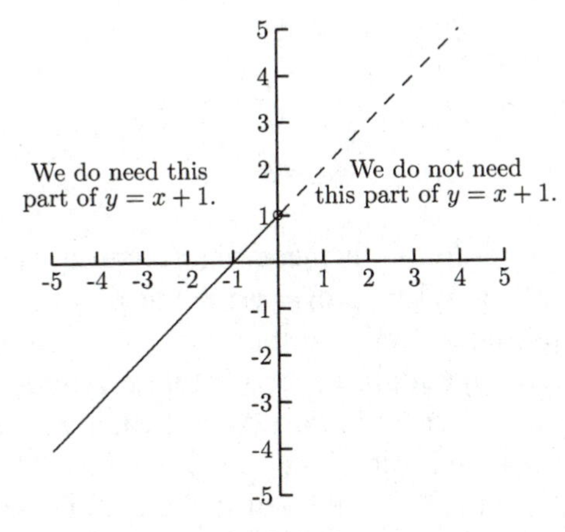

Fig. 6-36.

Because "$x \geq 0$" is written to the right of "$2x$," the part of the line $y = 2x$ we need is to the right of $x = 0$.

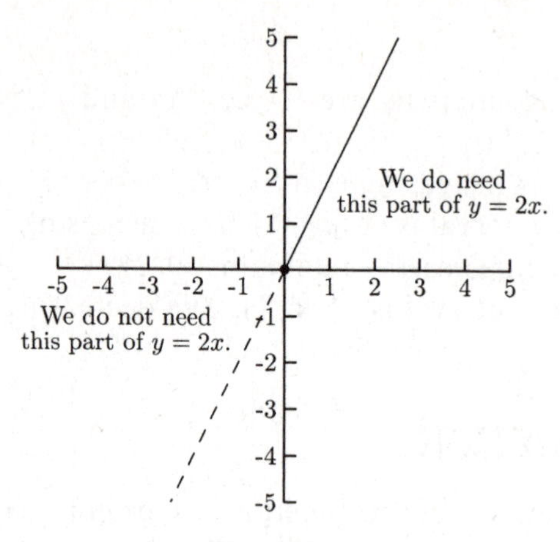

Fig. 6-37.

The graph of the function is shown in Fig. 6-38.

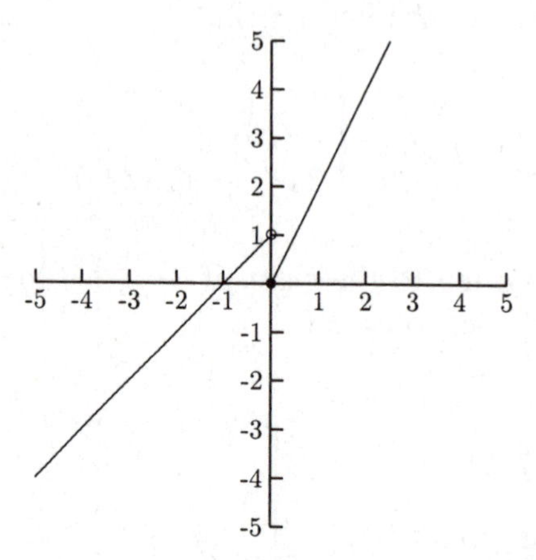

Fig. 6-38.

When sketching the graph of a piecewise function, the endpoint of each interval must be graphed. When plotting the endpoints, we will use an open dot "∘" for the inequalities "$x <$ number" and "$x >$ number." We will use a closed dot "•" for the inequalities "$x \leq$ number" and "$x \geq$ number."

In the above example, there are two pieces, each with one endpoint. For the piece $y = x + 1$, the endpoint is $x = 0$. Let $x = 0$ to get $y = 0 + 1 = 1$. Even though the point $(0, 1)$ is not part of the graph (because $x < 0$), we need to represent this point on the graph with an open dot to show that the graph goes all the way up to that point. For the piece $y = 2x$, the endpoint is also $x = 0$. We need to represent this point, $(0, 0)$, with a closed dot to show that this point *does* belong to the graph (because $x \geq 0$).

EXAMPLES

Sketch the graph of the piecewise functions.

- $f(x) = \begin{cases} 2x - 3 & \text{if } x \le 1 \\ -2 & \text{if } x > 1 \end{cases}$

  This graph comes in two pieces. One piece is part of the line $y = 2x - 3$, and the other piece is part of the horizontal line $y = -2$. We will start by making a table of values (Table 6-1), part of the table for $y = 2x - 3$ and the other part for $y = -2$. Because each piece is a line, we only need to plot two points for each piece. One of these points must be $x = 1$, the endpoint for each piece. The other x-value for the piece $y = 2x - 3$ can be anything to the left of $x = 1$. We will use $x = -1$. The other x-value for the piece $y = -2$ can be anything to the right of $x = 1$. We will use $x = 3$.

Table 6-1

x	$f(x)$	
1	-1	$y = 2(1) - 3 = -1$
-1	-5	$y = 2(-1) - 3 = -5$
1	-2	$y = -2$
3	-2	$y = -2$

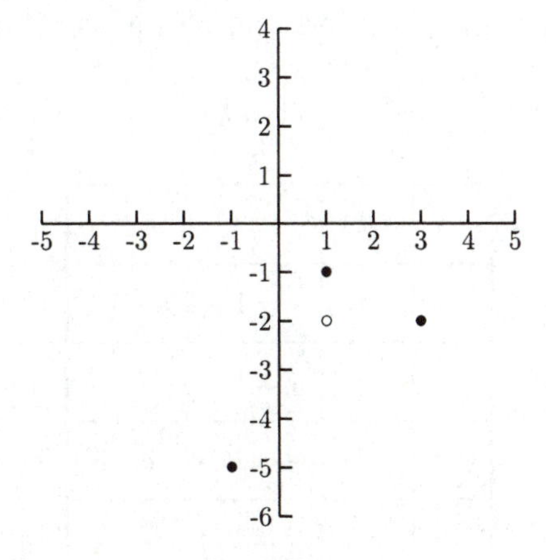

Fig. 6-39.

We will use a solid dot when plotting the point $(1, -1)$ because the inequality for $y = 2x - 3$ is "$x \leq 1$." We will use an open dot for the point $(1, -2)$ because the inequality for $y = -2$ is "$x > 1$." Now we will draw a line starting at $(1, -1)$ through $(-1, -5)$ and another line starting at $(1, -2)$ through $(3, -2)$.

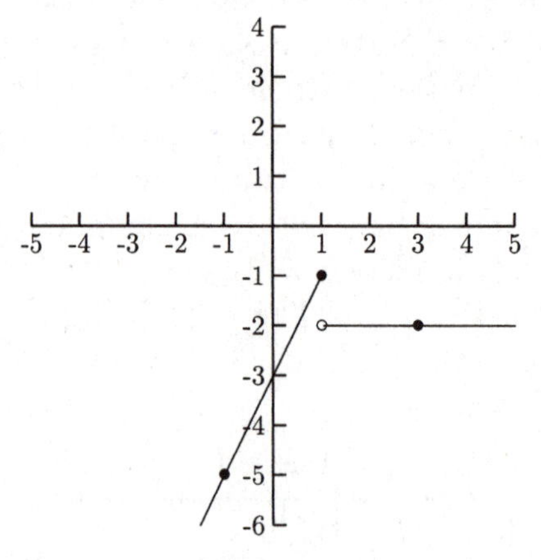

Fig. 6-40.

- $f(x) = \begin{cases} 1 - x & \text{if } x < 2 \\ 2x & \text{if } x \geq 2 \end{cases}$

Each piece of this function is part of a line, so we need to plot two points for each piece. For the piece $y = 1 - x$, we need to plot the point for $x = 2$ (because it is an endpoint) and any point to the left of $x = 2$. We will plot a point for $x = -2$. For the piece $y = 2x$, we need to plot the point for $x = 2$ (because it is an endpoint) and any point to the right of $x = 2$. We will plot a point for $x = 3$ (Table 6-2).

Table 6-2

x	$f(x)$	
2	-1	$y = 1 - 2 = -1$
-2	3	$y = 1 - (-2) = 3$
2	4	$y = 2(2) = 4$
3	6	$y = 2(3) = 6$

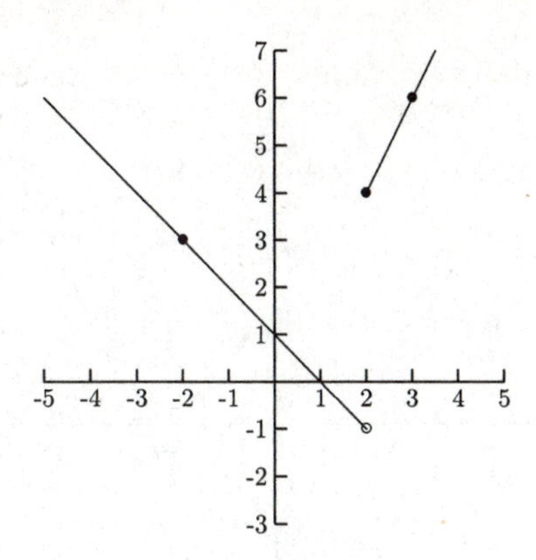

Fig. 6-41.

PRACTICE

Sketch the graphs.

1.

$$f(x) = \begin{cases} \frac{1}{2}x + 1 & \text{if } x \leq 0 \\ x - 2 & \text{if } x > 0 \end{cases}$$

2.

$$f(x) = \begin{cases} -3 & \text{if } x \leq 2 \\ 2x - 5 & \text{if } x > 2 \end{cases}$$

3.

$$f(x) = \begin{cases} -x & \text{if } x < 0 \\ x & \text{if } x \geq 0 \end{cases}$$

This is another way of writing the function $f(x) = |x|$.

SOLUTIONS

1.

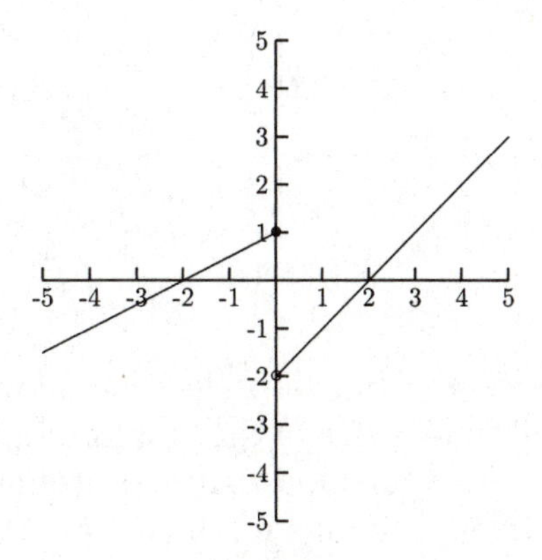

Fig. 6-42.

2.

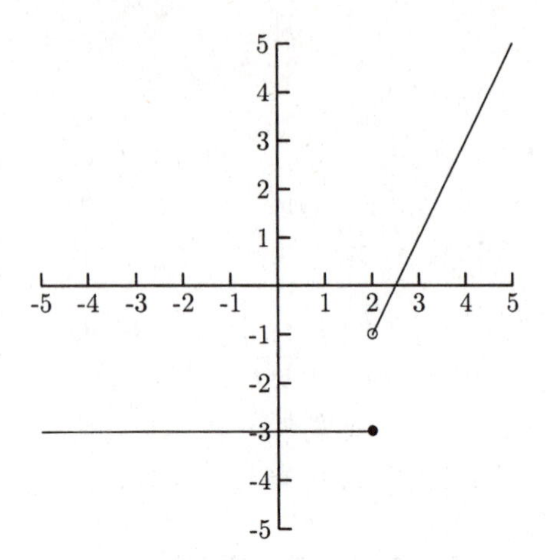

Fig. 6-43.

3.

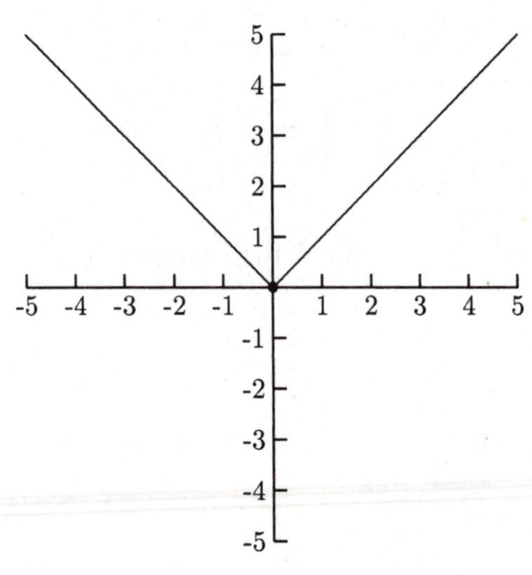

Fig. 6-44.

One or both pieces of the functions in the following will be parts of quadratic functions. Our knowledge of the graphs of quadratic functions will help to graph these piecewise functions. At first, it might be easier to sketch the graph of the entire quadratic function then erase the part that we do not need.

EXAMPLES

- $f(x) = \begin{cases} x^2 - 2x + 1 & \text{if } x \geq 1 \\ 2 - x & \text{if } x < 1 \end{cases}$

  We only need the part of the graph of $y = x^2 - 2x + 1$ to the right of $x = 1$.

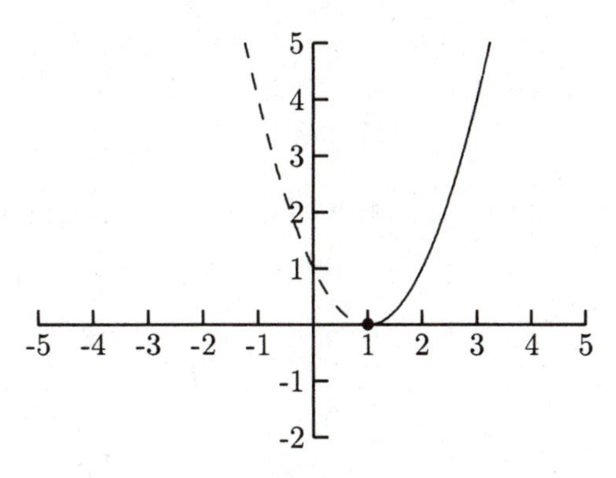

Fig. 6-45.

We need the part of the line $y = 2 - x$ to the left of $x = 1$.

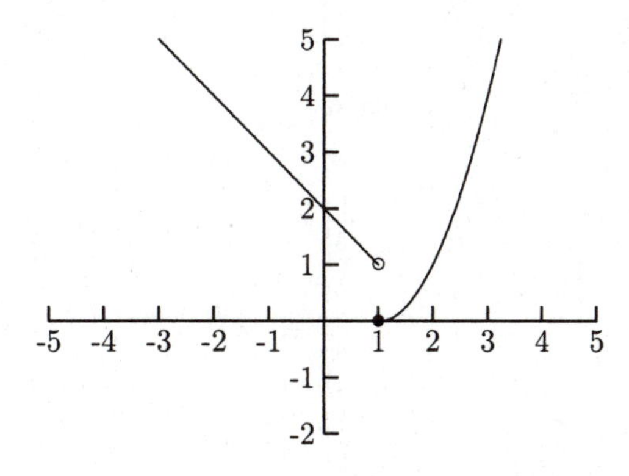

Fig. 6-46.

- $f(x) = \begin{cases} -\frac{1}{2}x^2 + x + 1 & \text{if } x \leq 2 \\ x^2 & \text{if } x > 2 \end{cases}$

  We need the part of the graph of $y = -\frac{1}{2}x^2 + x + 1$ to the left of $x = 2$.

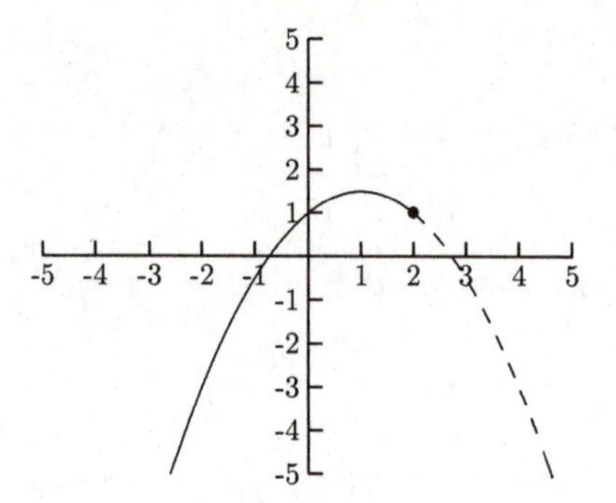

Fig. 6-47.

We need the part of the graph of $y = x^2$ to the right of $x = 2$.

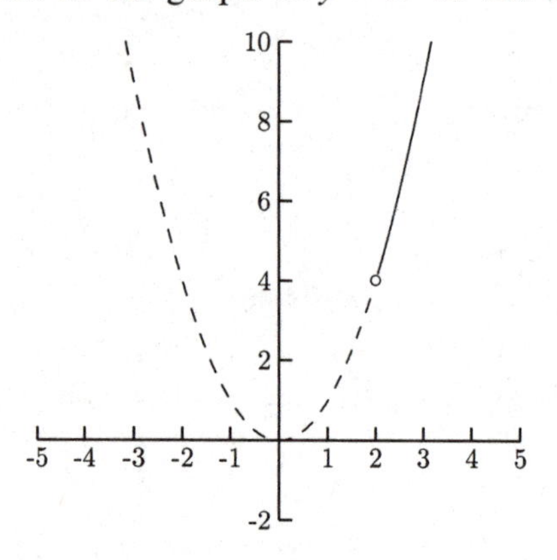

Fig. 6-48.

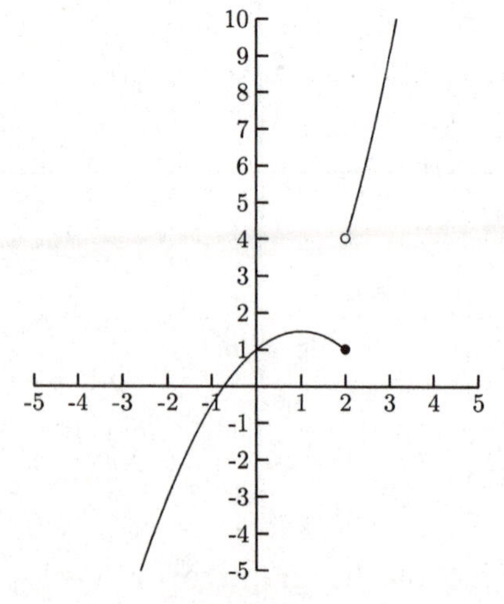

Fig. 6-49.

PRACTICE

Sketch the graphs.

1.

$$f(x) = \begin{cases} -x^2 + 1 & \text{if } x \geq 0 \\ 3 & \text{if } x < 0 \end{cases}$$

2.

$$g(x) = \begin{cases} x^2 + 4x - 2 & \text{if } x < 1 \\ 4x - 5 & \text{if } x \geq 1 \end{cases}$$

3.

$$f(x) = \begin{cases} x + 1 & \text{if } x \leq -2 \\ x^2 + 1 & \text{if } x > 2 \end{cases}$$

SOLUTIONS

1.

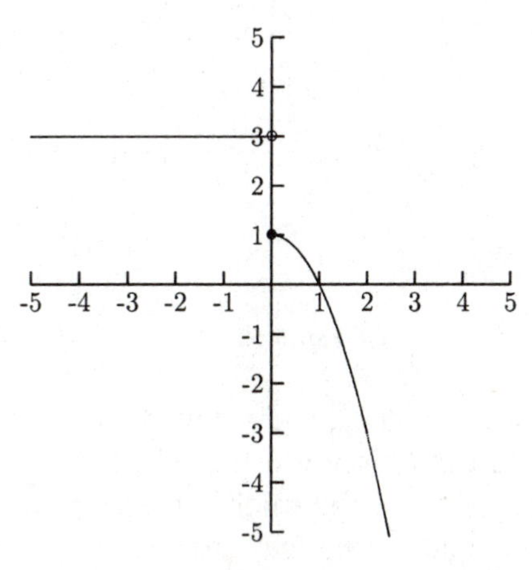

Fig. 6-50.

2.

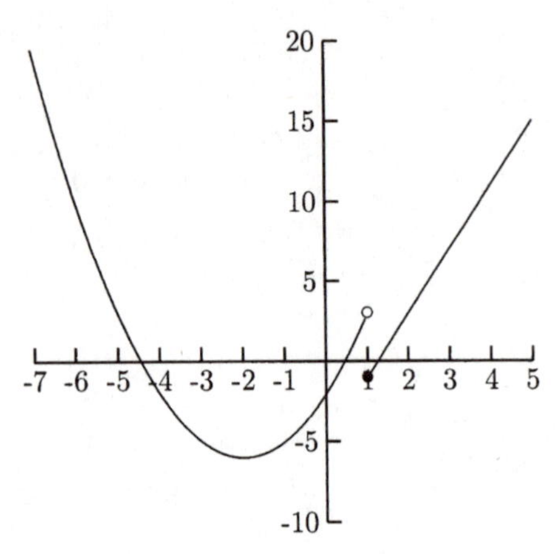

Fig. 6-51.

3.

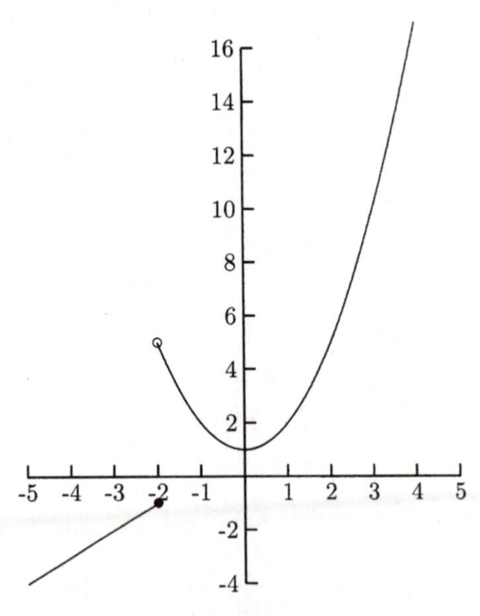

Fig. 6-52.

Piecewise functions can come in any number of pieces. The same rules apply. The endpoints of each piece must be plotted. If the function comes in three pieces, both endpoints of the middle piece must be plotted. The next example comes in three pieces. The two outside pieces are parts of lines, and the middle piece is part of a parabola.

EXAMPLE

- $f(x) = \begin{cases} -x+2 & \text{if } x \leq -2 \\ -x^2+2 & \text{if } -2 < x < 2 \\ x+2 & \text{if } x \geq 2 \end{cases}$

  For the piece $y = -x + 2$, we only need to plot two points, the endpoint $x = -2$ and a point to the left of $x = -2$. The piece $y = -x^2 + 2$ is a parabola between $x = -2$ and $x = 2$.

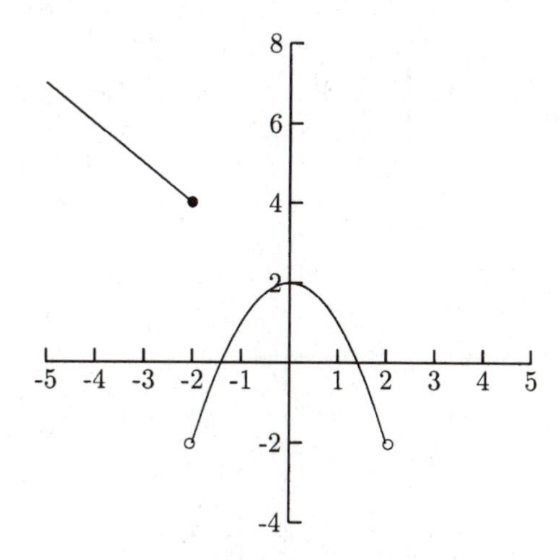

Fig. 6-53.

The last piece, $y = x + 2$, is another line. We need to plot two points, the endpoint $x = 2$ and a point to the right of $x = 2$.

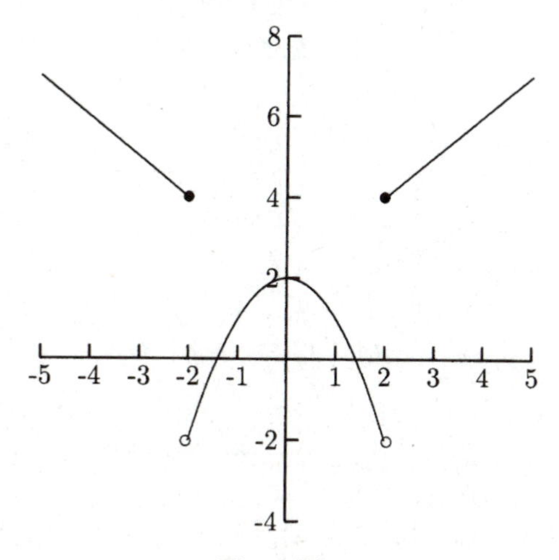

Fig. 6-54.

PRACTICE
Sketch the graphs.

1.

$$f(x) = \begin{cases} x & \text{if } x < -2 \\ 4 & \text{if } -2 \le x < 1 \\ -x & \text{if } x \ge 1 \end{cases}$$

2.

$$f(x) = \begin{cases} -\frac{1}{2}x + 3 & \text{if } x \le 0 \\ x - 1 & \text{if } 0 < x < 2 \\ 2x - 2 & \text{if } x \ge 2 \end{cases}$$

SOLUTIONS

1.

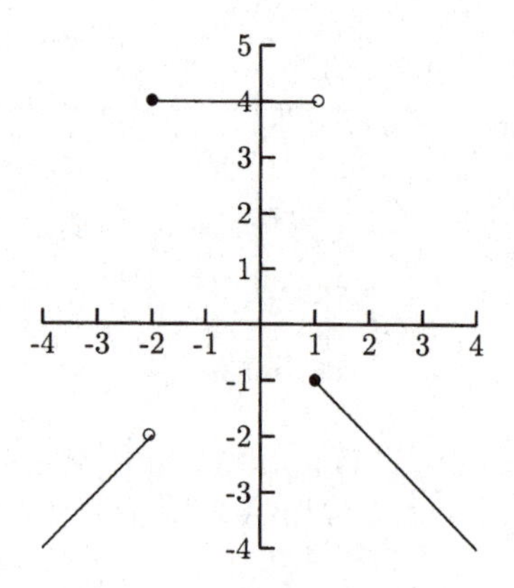

Fig. 6-55.

2.

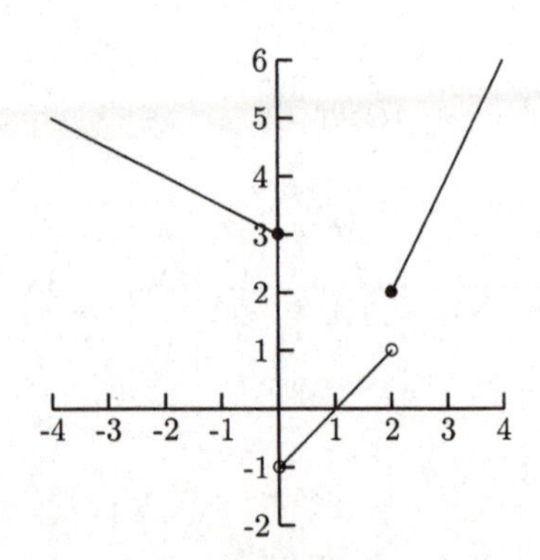

Fig. 6-56.

Chapter 6 Review

1. Evaluate $f(4)$ for $f(x) = \sqrt{x+5}$.
   a) $4\sqrt{x+5}$ b) ± 3 c) 9 d) 3
2. Evaluate $f(-2)$ for
$$f(x) = \begin{cases} 5x+3 & \text{if } x \leq 0 \\ 10 & \text{if } x > 0 \end{cases}$$
   a) -2 b) -7 c) 10 d) -7 and 10
3. What is the domain for $f(x) = (x-5)/(x+6)$?
   a) $(-\infty, -6) \cup (-6, \infty)$ b) $[-6, \infty)$ c) $(-6, -\infty)$
   d) $(-\infty, -6) \cup (-6, 5) \cup (5, \infty)$
4. In the equation $y^2 + (x-8)^2 = 4$, is y a function of x?
   a) Yes b) No c) Cannot be determined
5. What is the domain for $f(x) = \sqrt{x-3}$?
   a) $(-\infty, 3) \cup (3, \infty)$ b) $(3, \infty)$ c) $[3, \infty)$ d) $(-\infty, -3)$
6. Is the graph shown in Fig. 6-57 the graph of a function?
   a) Yes b) No c) Cannot be determined

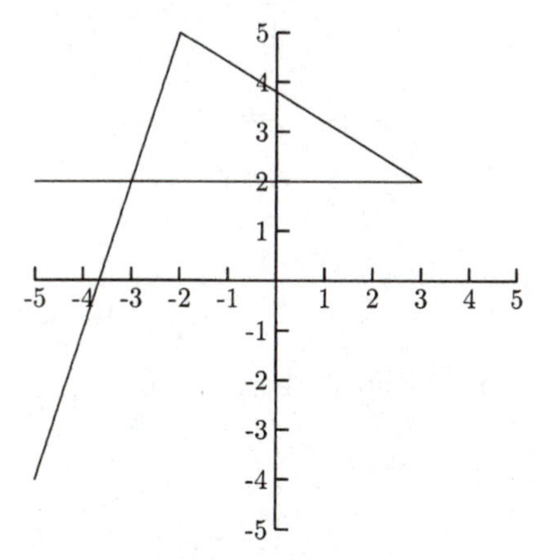

Fig. 6-57.

7. Evaluate $f(v^2)$ for $f(x) = x/(x+1)$.
   a) $v^2 x/(x+1)$ b) $v^2/(x+1)$ c) $v^2/(v^2+1)$ d) 1/2

8. What is $f(a+h) - f(a)$ for $f(x) = x^2 + 2x + 3$?
   a) $h^2 + 2h + 6$ b) $2ah + h^2 + 2a + 6$ c) $h^2 + 2h$
   d) $2ah + h^2 + 2h$

9. For what interval(s) of x is the function $f(x)$ increasing, the graph of which is shown in Fig. 6-58?
   a) $(1, 3)$ b) $(-3, 0)$ c) $(-3, 3)$ d) $(1, 0)$

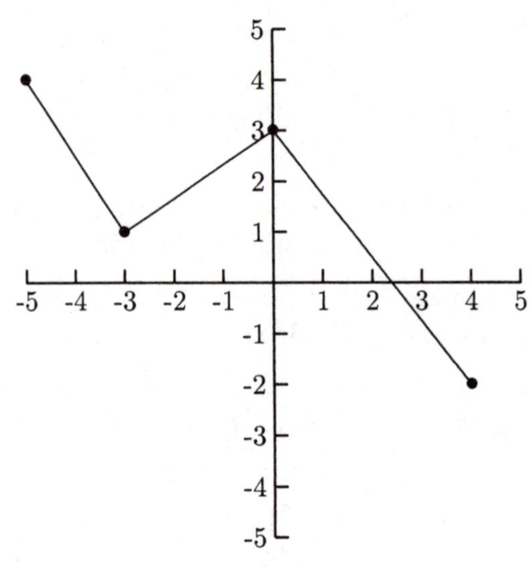

Fig. 6-58.

10. Refer to Fig. 6-58. What is $f(4)$?
    a) -5 b) -2 c) 2 d) Cannot be determined

11. Refer to Fig. 6-58. What is the domain?
    a) $[-5, 4]$ b) $[-5, -2]$ c) $[-2, 4]$
    d) Cannot be determined

12. In the equation $y^3 - 6x^2 + 2x = 5$ is y a function of x?
    a) Yes b) No c) Cannot be determined

13. What is the domain for $f(x) = (x - 10)/(\sqrt{x} - 1)$?
    a) $(-\infty, 1) \cup (1, 10) \cup (10, \infty)$ b) $(-\infty, 1) \cup (1, \infty)$
    c) $(1, \infty)$ d) $[1, \infty)$

14. Find $(f(a+h)-f(a))/h$ for $f(x)=x^2-3x+5$.
 a) h^2+h b) h^2-3h c) $2ah+h^2+h$ d) $2a+h-3$

15. The graph of which function is shown in Fig. 6-59?

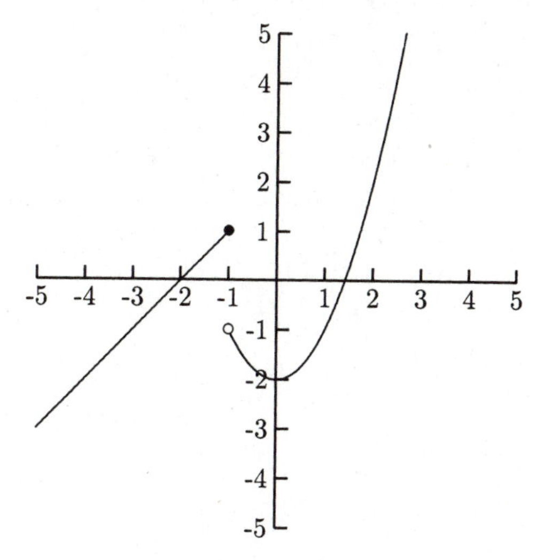

Fig. 6-59.

(a)

$$f(x)=\begin{cases} x^2-2 & \text{if } x<-1 \\ x+2 & \text{if } x\geq -1 \end{cases}$$

(b)

$$f(x)=\begin{cases} x^2-2 & \text{if } x>-1 \\ x+2 & \text{if } x\leq -1 \end{cases}$$

(c)

$$f(x)=\begin{cases} (x-2)^2 & \text{if } x<-1 \\ x+2 & \text{if } x\geq -1 \end{cases}$$

(Continued next page)

(d)

$$f(x) = \begin{cases} (x-2)^2 & \text{if } x > -1 \\ x+2 & \text{if } x \leq -1 \end{cases}$$

SOLUTIONS

1. d) 2. b) 3. a) 4. b) 5. c) 6. b) 7. c) 8. d)
9. b) 10. b) 11. a) 12. a) 13. c) 14. d) 15. b)

CHAPTER 7

Quadratic Functions

The quadratic equations in Chapter 6 are actually quadratic functions. In Chapter 6 the functions are written in the form $y = ax^2 + bx + c$, and here they will be written as $f(x) = ax^2 + bx + c$. Remember that the graph of a quadratic function is a parabola that either opens up or opens down.

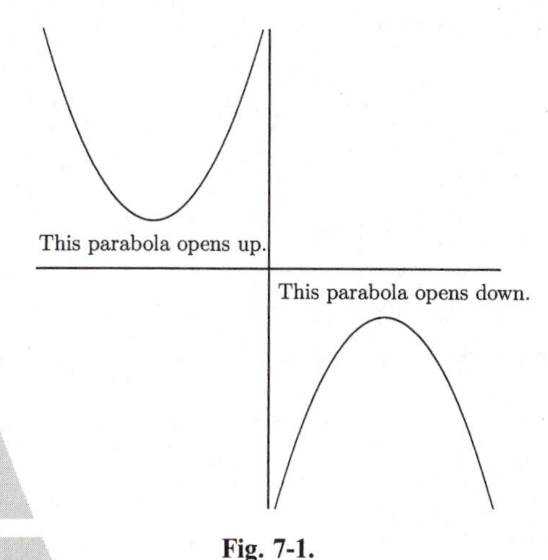

Fig. 7-1.

For a parabola that opens up, the quadratic function is decreasing to the left of the vertex and is increasing to the right of the vertex. For a parabola that opens down, the quadratic function is increasing to the left of the vertex and is decreasing to the right of the vertex. The vertex of a parabola can be found by writing the function in standard form, $f(x) = a(x-h)^2 + k$ (by using completing the square), where (h, k) are the coordinates of the vertex, or by computing $h = -b/2a$, and $k = c - a(b/2a)^2$. This formula for k is not easy to remember (or even to use!). It is easier to find the y-value for $x = -b/2a$.

EXAMPLES

Find the vertex using the fact that $h = -b/2a$.

- $f(x) = 3x^2 - 6x + 1$

  $a = 3$, $b = -6$, and $h = -b/2a = -(-6)/2(3) = 1$. Find k by evaluating $f(1) = 3(1)^2 - 6(1) + 1 = -2$. The vertex is $(1, -2)$.

- $f(x) = -x^2 + 4$

  $a = -1$, $b = 0$, $h = -b/2a = -0/2(-1) = 0$. Find k by evaluating $f(0) = -0^2 + 4 = 4$. The vertex is $(0, 4)$.

- $f(x) = \frac{2}{3}x^2 - 2x - 6$

  $a = \frac{2}{3}$, $b = -2$,

$$h = \frac{-b}{2a} = \frac{-(-2)}{2(2/3)} = \frac{1}{2/3} = \frac{3}{2}$$

  Find k by evaluating $f(\frac{3}{2})$

$$f\left(\frac{3}{2}\right) = \frac{2}{3}\left(\frac{3}{2}\right)^2 - 2\left(\frac{3}{2}\right) - 6 = -\frac{15}{2}$$

  The vertex is $(\frac{3}{2}, -\frac{15}{2})$.

PRACTICE

Find the vertex using the fact that $h = -b/2a$.

1. $f(x) = x^2 + 8x + 3$
2. $f(x) = -5x^2 - 4x - 2$
3. $h(t) = -\frac{1}{2}t^2 + 3t + 5$
4. $r(x) = -0.001x^2 + 2x - 100$

SOLUTIONS

1.

$$h = \frac{-b}{2a} = \frac{-8}{2(1)} = -4$$
$$k = f(-4) = (-4)^2 + 8(-4) + 3 = -13$$

The vertex is $(-4, -13)$.

2.

$$h = \frac{-b}{2a} = \frac{-(-4)}{2(-5)} = \frac{4}{-10} = -\frac{2}{5}$$
$$k = f\left(-\frac{2}{5}\right) = -5\left(-\frac{2}{5}\right)^2 - 4\left(-\frac{2}{5}\right) - 2$$
$$= -5\left(\frac{4}{25}\right) + \frac{8}{5} - 2 = -\frac{6}{5}$$

The vertex is $(-\frac{2}{5}, -\frac{6}{5})$.

3.

$$h = \frac{-b}{2a} = \frac{-3}{2(-1/2)} = \frac{-3}{-1} = 3$$
$$k = h(3) = -\frac{1}{2}(3)^2 + 3(3) + 5 = -\frac{9}{2} + 9 + 5 = \frac{19}{2}$$

The vertex is $(3, \frac{19}{2})$.

4.

$$h = \frac{-b}{2a} = \frac{-2}{2(-0.001)} = \frac{-2}{-0.002} = 1000$$
$$k = r(1000) = -0.001(1000)^2 + 2(1000) - 100 = 900$$

The vertex is $(1000, 900)$.

Finding the range for many functions is not easy—we would probably need to look at their graphs. But finding the range for quadratic functions is not hard. We only need to use the fact that for parabolas that open up, the vertex is the lowest point, and for parabolas that open down, the vertex is the highest point. The range of a quadratic function that opens up is $[k, \infty)$. The range of a quadratic function that opens down is $(-\infty, k]$. We can tell whether a parabola opens up or down by looking at a in $f(x) = ax^2 + bx + c$ or in $f(x) = a(x-h)^2 + k$. If a is positive, the parabola opens up. If a is negative, the parabola opens down.

EXAMPLES

Determine the range for the quadratic functions.

- $f(x) = 3x^2 - 6x + 1$

  Earlier, we found that the vertex is $(1, -2)$. Because the parabola opens up ($a = 3$), the range is $[-2, \infty)$.

- $f(x) = -x^2 + 4$

  We found that the vertex is $(0, 4)$. Because the parabola opens down ($a = -1$), the range is $(-\infty, 4]$.

- $f(x) = \frac{2}{3}x^2 - 2x - 6$

  We found that the vertex is $(\frac{3}{2}, -\frac{15}{2})$. Because the parabola opens up ($a = \frac{2}{3}$), the range is $[-\frac{15}{2}, \infty)$.

PRACTICE

Determine the range for the quadratic functions.

1. $f(x) = x^2 + 8x + 3$
2. $f(x) = -5x^2 - 4x - 2$
3. $h(t) = -\frac{1}{2}t^2 + 3t + 5$
4. $r(x) = -0.001x^2 + 2x - 100$

SOLUTIONS

1. The vertex is $(-4, -13)$. Because a is positive, the parabola opens up, so the range is $[-13, \infty)$.
2. The vertex is $(-\frac{2}{5}, -\frac{6}{5})$. Because a is negative, the parabola opens down, so the range is $(-\infty, -\frac{6}{5}]$.
3. The vertex is $(3, \frac{19}{2})$. Because a is negative, the parabola opens down, so the range is $(-\infty, \frac{19}{2}]$.
4. The vertex is $(1000, 900)$. Because a is negative, the parabola opens down, so the range is $(-\infty, 900]$.

The Maximum/Minimum of a Quadratic Function

An important area of mathematics is concerned with *optimizing* situations. For example, what sales level for a product will give the most profit?

What production level will give the lowest cost per unit? What shape will be the strongest? While calculus is used to solve many of these problems, algebra students can optimize problems involving quadratic functions. Quadratic functions can be maximized (if the parabola opens down) or minimized (if the parabola opens up). The maximum or minimum value of a quadratic function is k, the y-coordinate of the vertex.

EXAMPLES

Find the maximum or minimum value of the quadratic functions.

- $f(x) = -2x^2 - 6x + 7$

  Because $a = -2$ is negative, the parabola opens down, and the function has a maximum value (but no minimum value). We need to find k.

$$h = \frac{-b}{2a} = \frac{-(-6)}{2(-2)} = -\frac{3}{2}$$

$$k = f\left(-\frac{3}{2}\right) = -2\left(-\frac{3}{2}\right)^2 - 6\left(-\frac{3}{2}\right) + 7$$

$$= \frac{23}{2}$$

  The maximum value of the functin is $\frac{23}{2}$. This maximum occurs when $x = -\frac{3}{2}$.

- $C(q) = 0.02q^2 - 5q + 600$

  Because $a = 0.02$ is positive, the parabola opens up, and the function has a minimum value.

$$h = \frac{-b}{2a} = \frac{-(-5)}{2(0.02)} = 125$$

$$k = f(100) = C(125) = 0.02(125)^2 - 5(125) + 600$$

$$= 287.50$$

  The minimum value of the function is 287.50. This minimum occurs when $q = 125$.

PRACTICE

Find the maximum or minimum value of the quadratic functions.

1. $f(x) = x^2 - 8x + 2$
2. $f(t) = -16t^2 + 48t + 25$

3. $f(x) = 100x^2 + 150x + 25$
4. $P(x) = -0.015x^2 + 0.45x + 12$

SOLUTIONS

1. Because $a = 1$ is positive, the parabola opens up, and this function has a minimum value.

$$h = \frac{-(-8)}{2(1)} = 4$$
$$k = f(4) = 4^2 - 8(4) + 2 = -14$$

The minimum value of the function is -14. The minimum occurs at $x = 4$.

2. Because $a = -16$ is negative, the parabola opens down, and this function has a maximum value.

$$h = \frac{-48}{2(-16)} = \frac{3}{2}$$
$$k = f\left(\frac{3}{2}\right) = -16\left(\frac{3}{2}\right)^2 + 48\left(\frac{3}{2}\right) + 25$$
$$= 61$$

The maximum value of this function is 61. The maximum occurs at $t = \frac{3}{2}$.

3. Because $a = 100$ is positive, the parabola opens up, and this function has a minimum value.

$$h = \frac{-150}{2(100)} = -\frac{3}{4}$$
$$k = 100\left(-\frac{3}{4}\right)^2 + 150\left(-\frac{3}{4}\right) + 25 = -\frac{125}{4}$$

The minimum value of the value of the function is $-\frac{125}{4}$. The minimum value occurs at $x = -\frac{3}{4}$.

4. Because $a = -0.015$ is negative, the parabola opens down, and this function has a maximum value.

$$h = \frac{-0.45}{2(-0.015)} = 15$$
$$k = P(15) = -0.015(15)^2 + 0.45(15) + 12 = 15.375$$

The maximum value of the function is 15.375. The maximum value occurs at $x = 15$.

Applied Maximum/Minimum Problems

In the problems for this section, we will be asked to find the maximum or minimum of a situation. The functions that will model these problems will be quadratic functions. Think about what the vertex means in each problem. If the problem is to determine the level of production to maximize profit, h will tell us the production level needed to maximize profit, and k will tell us what the maximum profit is. The function to be optimized will be given in the first problems. Later, we will need to find the function based on information given in the problem.

EXAMPLES

- The profit function for a product is given by $P(x) = -2x^2 + 28x + 150$, where x is the number of units sold, and P is in dollars. What level of production will maximize profit? What is the maximum profit?

  The answer to the first question will be h, and the answer to the second question will be k.

$$h = \frac{-28}{2(-2)} = 7$$
$$k = P(7) = -2(7)^2 + 28(7) + 150 = 248$$

  The level of production which maximizes profit is 7 units, and the maximum profit is \$248.

- The cost per unit of a product is given by the function $C(x) = \frac{1}{4}x^2 - 6x + 40$, where x is the production level (in hundreds of units), and C is the cost in dollars. What level of production will yield the minimum production cost per unit?

  Which will answer the question—h or k? The production level is x, so we need to find h.

$$h = \frac{-(-6)}{2(1/4)} = \frac{6}{1/2} = 6 \div \frac{1}{2} = 6 \cdot 2 = 12$$

  Minimize the cost per unit by producing 12 hundred units.

PRACTICE

1. The daily profit for a vendor of bottled water is given by the function $P(x) = -0.001x^2 + 0.36x - 5$, where x is the number of bottles sold, and P is the profit in dollars. How many bottles should be sold to maximize daily profit? What is the maximum daily profit?

2. The average cost per unit of a product per week is given by the function $C(x) = \frac{1}{3}x^2 - 60x + 5900$, where x is the number of units produced and C is in dollars. What is the minimum cost per unit and how many units should be produced per week to minimize the cost per unit?
3. The weekly revenue of a particular service offered by a company depends on the price—the higher the price, the fewer sales; and the lower the price, the higher the sales. The function describing the revenue is $R(p) = -\frac{1}{10}p^2 + 3p + 125$, where p is the price per hour and R is sales revenue in dollars. What is the revenue-maximizing price? What is the maximum weekly revenue?

SOLUTIONS

1.

$$h = \frac{-0.36}{2(-0.001)} = 180$$

$$k = P(180) = -0.001(180)^2 + 0.36(180) - 5 = 27.4$$

The vendor should sell 180 bottles per day to maximize profit. The maximum daily profit is $27.40.

2.

$$h = \frac{-(-60)}{2(1/3)} = \frac{60}{2/3} = 60 \div \frac{2}{3} = 60 \cdot \frac{3}{2} = 90$$

$$k = C(90) = \frac{1}{3}(90)^2 - 60(90) + 5900 = 3200$$

The minimum average cost is $3200, and 90 units should be produced to minimize the average cost.

3.

$$h = \frac{-3}{2(-1/10)} = \frac{3}{1/5} = 3 \div \frac{1}{5} = 3 \cdot 5 = 15$$

$$k = R(15) = -\frac{1}{10}(15)^2 + 3(15) + 125 = 147.50$$

The revenue-maximizing price is $15 per hour. The maximum weekly revenue is $147.50.

When an object is thrust upward and is free-falling after the initial thrust, its path is in the shape of a parabola. In the following problems, we will be given the height function of these kinds of falling objects. The functions will be in the form $h(x) = ax^2 + bx + c$, where x is the horizontal distance and h is the height. Several types of questions are asked for these problems. We will answer the questions, "What is the object's maximum height?" and "How far has it traveled horizontally to reach its maximum height?"

EXAMPLE

- Suppose the path of a grasshopper's jump is given by the function $h(x) = \frac{-5}{216}x^2 + \frac{5}{3}x$, where both x and h are in inches. What is the maximum height reached by the grasshopper? How far has it traveled horizontally to reach its maximum height?

$$h = \frac{-5/3}{2(-5/216)} = \frac{5/3}{5/108} = \frac{5}{3} \div \frac{5}{108} = \frac{5}{3} \cdot \frac{108}{5} = 36$$

$$k = h(36) = -\frac{5}{216}(36)^2 + \frac{5}{3}(36) = 30$$

The maximum height reached by the grasshopper is 30 inches and it had traveled 36 inches horizontally when it reached it maximum height.

PRACTICE

1. A child throws a ball, its path being given by the function $h(x) = -0.04x^2 + 1.5x + 3$, where x and h are in feet. What is the maximum height of the ball? How far has it traveled horizontally when it reaches its maximum height?
2. A kitten jumped to pounce on a toy mouse. The path of the kitten is given by the function $h(x) = -\frac{5}{72}x^2 + \frac{5}{3}x$, where x and h are in inches. How far had the kitten traveled horizontally when it reached its maximum height? What was the kitten's maximum height?

SOLUTIONS

1.

$$h = \frac{-1.5}{2(-0.04)} = 18.75$$

$$k = h(18.75) = -0.04(18.75)^2 + 1.5(18.75) + 3 = 17.0625$$

The ball's maximum height is 17.0625 feet and it had traveled 18.75 feet horizontally when it reached its maximum height.

2.

$$h = \frac{-5/3}{2(-5/72)} = \frac{5/3}{5/36} = \frac{5}{3} \div \frac{5}{36} = \frac{5}{3} \cdot \frac{36}{5} = 12$$

$$k = h(12) = -\frac{5}{72}(12)^2 + \frac{5}{3}(12) = 10$$

The kitten's maximum height is 10 inches and it had traveled 12 inches horizontally when it reached its maximum height.

Quadratic functions can be used to optimize many types of geometric problems. In the following problems, a number will be fixed (usually the perimeter) and we will be asked to find the maximum enclosed area. The area will be a quadratic function, but getting to this function will require a few steps. The first few problems will involve a fixed amount of fencing to be used to enclose a rectangular area.

EXAMPLES

- A farmer has 600 feet of fencing available to enclose a rectangular pasture and then subdivide the pasture into two equal rectangular yards. What dimensions will yield the maximum area? What is the maximum area?

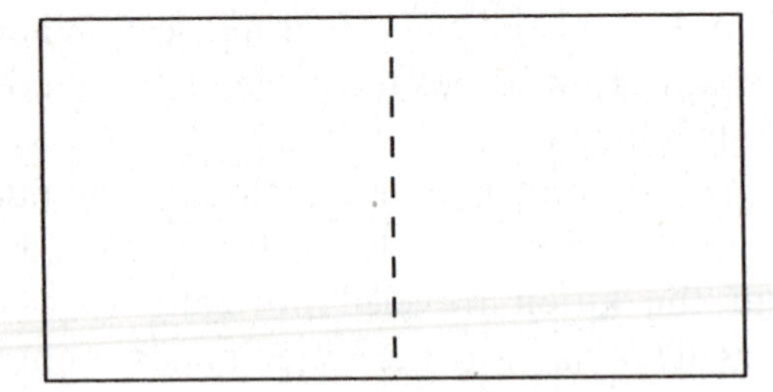

Fig. 7-2.

The formula for the area of a rectangle is $A = LW$. This formula has two variables (other than A), and we must reduce it to one. There are 600 feet of fencing available, so $L + W + W + W + L = 2L + 3W$ must equal 600. This gives us the equation $2L + 3W = 600$. Solve for either L or W.

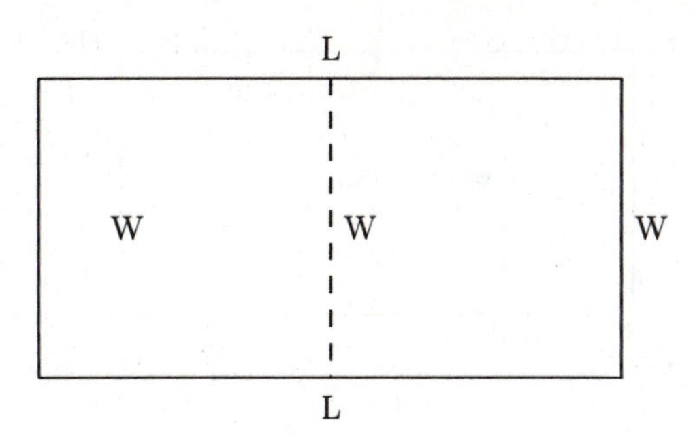

Fig. 7-3.

$$2L + 3W = 600$$
$$L = \frac{600 - 3W}{2} = 300 - 1.5W$$

Now substitute $L = 300 - 1.5W$ in the formula $A = LW$.

$$A = LW = (300 - 1.5W)W = 300W - 1.5W^2$$
$$= -1.5W^2 + 300W$$

This quadratic function has a maximum value, the maximum area.

$$h = \frac{-300}{2(-1.5)} = 100$$
$$k = -1.5(100)^2 + 300(100) = 15,000$$

The maximum area is 15,000 square feet. This occurs when the width is 100 feet, and the length is $300 - 1.5(100) = 150$ feet.

- A zoo has 1100 meters of fencing available to create four rectangular pens. What dimensions will enclose the maximum area? What is the maximum area?

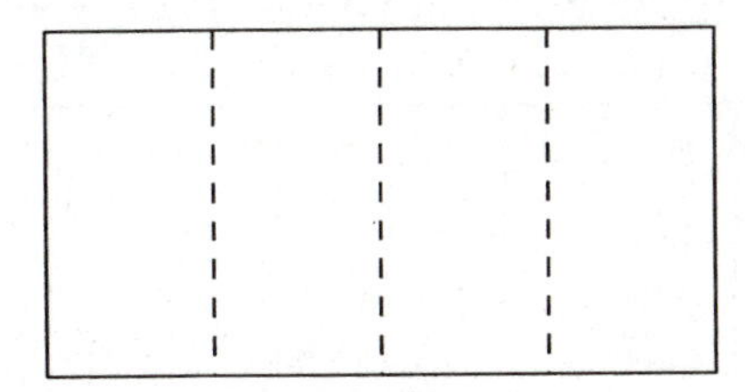

Fig. 7-4.

We will label the figure with L and W.

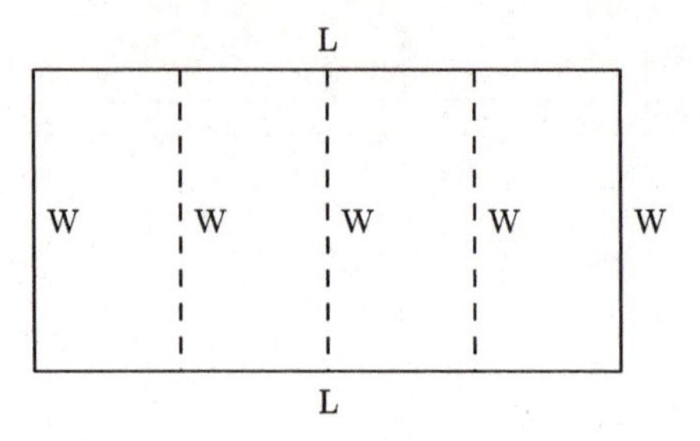

Fig. 7-5.

We want to maximize the area, $A = LW$. The 1100 meters of fencing must be divided among 2 Ls and 5 Ws, so we have the equation $2L + 5W = 1100$. Solve for either L or W and use the new equation to reduce the number of variables in $A = LW$.

$$2L + 5W = 1100$$

$$L = \frac{1100 - 5W}{2} = 550 - 2.5W$$

$A = LW$ becomes $A = (550 - 2.5W)W = 550W - 2.5W^2 = -2.5W^2 + 550W$.

$$h = \frac{-550}{2(-2.5)} = 110$$

$$k = -2.5(110)^2 + 550(110) = 30,250$$

To maximize the area, let $W = 110$ meters, $L = 550 - 2.5(110) = 275$ meters. The maximum area is 30,250 square meters.

PRACTICE

1. A parks department wants to enclose three adjacent rectangular playing fields. It has 1600 feet of fencing available. What dimensions will yield the maximum area? What is the maximum area?

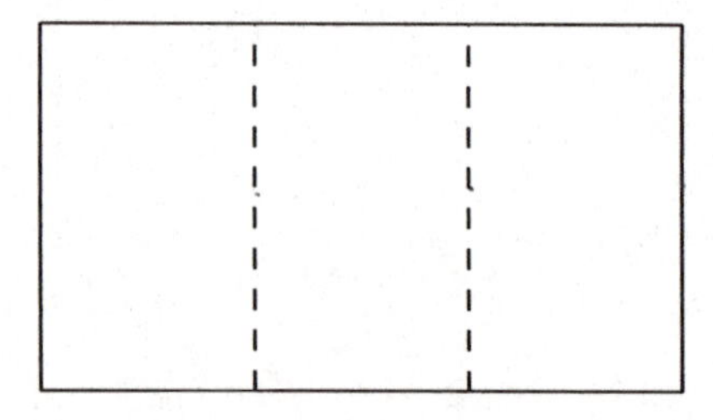

Fig. 7-6.

2. A farmer has 450 meters of fencing available. The farmer decides to fence two rectangular pastures. What dimensions will maximize the area? What is the maximum area?

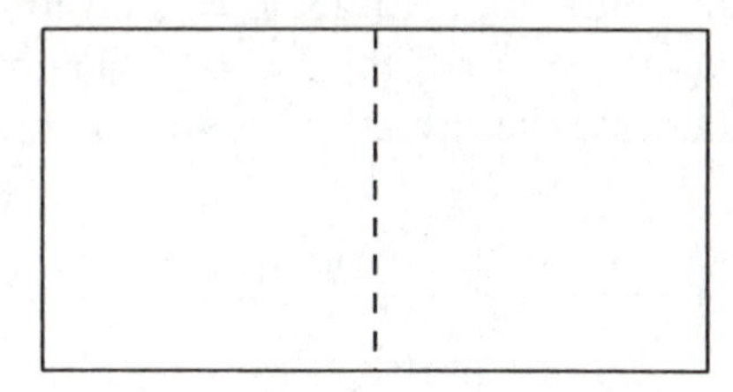

Fig. 7-7.

SOLUTIONS

1. We can see from Fig. 7-6 that $2L + 4W = 1600$.

$$2L + 4W = 1600$$
$$L = \frac{1600 - 4W}{2} = 800 - 2W$$
$$A = LW = (800 - 2W)W = 800W - 2W^2$$
$$= -2W^2 + 800W$$
$$h = \frac{-800}{2(-2)} = 200$$
$$k = -2(200)^2 + 800(200) = 80,000$$

Maximize the area by letting $W = 200$ feet and $L = 800 - 2(200) = 400$ feet. The maximum area is 80,000 square feet.

2. We can see from Fig. 7-7 that $2L + 3W = 450$.

$$2L + 3W = 450$$
$$L = \frac{450 - 3W}{2} = 225 - 1.5W$$
$$A = LW = (225 - 1.5W)W = 225W - 1.5W^2$$
$$= -1.5W^2 + 225W$$
$$h = \frac{-225}{2(-1.5)} = 75$$
$$k = -1.5(75)^2 + 225(75) = 8437.5$$

Maximize the area by letting $W = 75$ meters and $L = 225 - 1.5(75) = 112.5$ meters. The maximum area is 8437.5 square meters.

There is another common problem where we are asked to find the dimensions which will maximize a rectangular area. The area needs to be fenced but only on three sides. We can see from Fig. 7-8 that $2W + L =$ amount of fencing. Solve this for L to get $L =$ amount of fencing $-2W$. As before, substitute this quantity for L in $A = LW$. Then find h, which will be the width that will maximize the area, and k will be the maximum area.

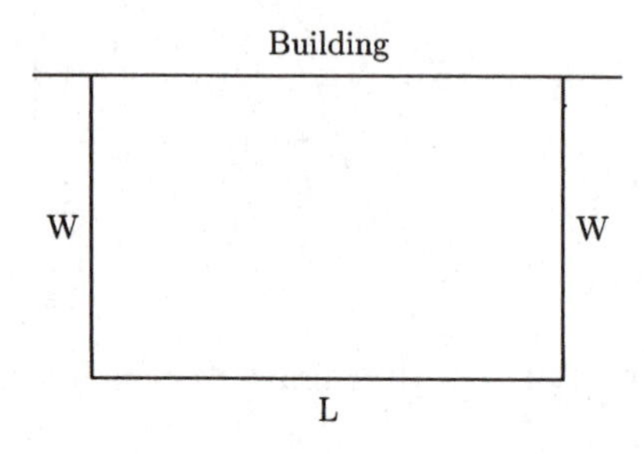

Fig. 7-8.

EXAMPLE

- A business needs to enclose an area behind its offices for storage. It has 240 feet of fencing available. If the side of the building is not fenced, what dimensions will maximize the enclosed area?

  We can see from Fig. 7-8 that $2W + L = 240$.

$$L + 2W = 240$$

$$L = 240 - 2W$$

$$A = LW = (240 - 2W)W = 240W - 2W^2$$

$$h = \frac{-240}{2(-2)} = 60$$

$$k = 240(60) - 2(60)^2 = 7200$$

Maximize the area by letting $W = 60$ feet and $L = 240 - 2(60) = 120$ feet. The maximum area is 7200 square feet.

PRACTICE

1. A rancher wants to enclose a rectangular pasture that borders a stream. The rancher has 500 feet of fencing available and will not fence the side along the stream. What dimensions will maximize the area? What is the maximum area?

2. The manager of an office complex wants to provide extra parking behind the office building. The contractor has 150 meters of fencing available. If the side along the building will not be fenced, what dimensions will maximize the enclosed area? What is the maximum enclosed area?

SOLUTIONS

1.

$$2W + L = 500$$
$$L = 500 - 2W$$
$$A = LW = (500 - 2W)W = 500W - 2W^2$$
$$h = \frac{-500}{2(-2)} = 125$$
$$k = 500(125) - 2(125)^2 = 31,250$$

Maximize the area by letting $W = 125$ feet and $L = 500 - 2(125) = 250$ feet. The maximum area is 31,250 square feet.

2.

$$2W + L = 150$$
$$L = 150 - 2W$$
$$A = LW = (150 - 2W)W = 150W - 2W^2$$
$$h = \frac{-150}{2(-2)} = 37.5$$
$$k = 150(37.5) - 2(37.5)^2 = 2812.5$$

Maximize the area by letting $W = 37.5$ meters and $L = 150 - 2(37.50) = 75$ meters. The maximum area is 2812.5 square meters.

Often in business, revenue depends on the price in two ways. Obviously, if the price is raised, more money will be collected for each unit sold, but the number of units sold might drop. In general, the lower the price, the higher the demand (number of units sold); and the higher the price, the lower the demand. In the following problems, the demand for a certain price will be given. Then we will be told how many sales are lost from a price increase or how many sales are gained from a price decrease. With this information, we can find the price to charge to maximize revenue. Let n represent the number of increases or decreases in the price, so if the price is raised in \$10 increments, then the increase in price would be $10n$ and the price would be "old price $+10n$." If 5 sales are lost for each \$10 increase in price, then the number sold would be "old sales level $-5n$." The total revenue would be "$R =$ (old price $+ 10n$)(old sales level $- 5n$)." The revenue function is a quadratic function. Revenue is maximized when $n = h$, and the maximum revenue is k.

EXAMPLES

- A store sells an average of 345 pounds of apples per day when the price is \$0.85 per pound. The manager thinks that for every increase of \$0.10 in the price, 30 fewer pounds of apples will be sold each day. What price will maximize revenue from the sale of apples? What is the maximum revenue?

  Let n represent the number of \$0.10 increases in the price per pound. The price is then represented by $0.85 + 0.10n$. The number of pounds of apples sold per day would be $345 - 30n$. This makes the revenue $R = (0.85 + 0.10n)(345 - 30n)$. The revenue equation is a quadratic function. The vertex of this function tells us two things. First, h will tell us how many times we need to raise the price by \$0.10 in order to maximize revenue, and k will tell us what the maximum revenue is.

  $$R = (0.85 + 0.10n)(345 - 30n) = 293.25 + 9n - 3n^2$$
  $$h = \frac{-9}{2(-3)} = 1.5$$
  $$k = 293.25 + 9(1.5) - 3(1.5)^2 = 300$$

  Maximize revenue by charging $0.85 + 0.10(1.5) = \$1$ per pound. The maximum revenue is \$300 per day.

- An apartment manager is leasing 60 of 75 apartments in her apartment complex with the monthly rent at \$1950. For each \$25 decrease in the monthly rent, she believes that one more apartment can be rented. What monthly rent will maximize revenue? What is the maximum monthly revenue?

  Let n represent the number of \$25 decreases in the rent. Then monthly rent is represented by $1950 - 25n$, and the number of apartments rented is $60 + n$. The revenue function is $R = (1950 - 25n)(60 + n)$. The coordinates of the vertex will help us to answer the questions.

  $$R = (1950 - 25n)(60 + n) = 117,000 + 450n - 25n^2$$
  $$h = \frac{-450}{2(-25)} = 9$$
  $$k = 117,000 + 450(9) - 25(9)^2 = 119,025$$

Maximize revenue by charging $1950 - 25(9) = \$1725$ for the monthly rent. The maximum revenue is \$119,025.

PRACTICE

1. At a small college, 1200 tickets can be sold during a football game when the ticket price is \$9. The athletic director learns that for each \$0.75 decrease in the ticket price, 200 more people will attend the game. What should the ticket price be in order to maximize ticket revenue? What is the maximum ticket revenue?
2. The owner of a concession stand sells 10,000 soft drinks for \$3.70 per drink during baseball games. A survey reveals that for each \$0.20 decrease in the price of the drinks, 800 more will be sold. What should the price be in order to maximize revenue? What is the maximum revenue?
3. The manager of an apartment complex can rent all 60 apartments in a building if the monthly rent is \$2800, and that for each \$50 increase in the monthly rent one tenant will be lost and will not likely be replaced. What should the monthly rent be to maximize revenue? What is the maximum revenue?

SOLUTIONS

1. Let n represent the number of \$0.75 decreases in the ticket price. This makes the new ticket price $9 - 0.75n$ and the number of tickets sold $1200 + 200n$. Ticket revenue is $R = (9 - 0.75n)(1200 + 200n)$.

$$R = (9 - 0.75n)(1200 + 200n) = 10{,}800 + 900n - 150n^2$$

$$h = \frac{-900}{2(-150)} = 3$$

$$k = 10{,}800 + 900(3) - 150(3)^2 = 12{,}150$$

Maximize ticket revenue by charging $\$9.00 - 0.75(3) = \$6.75$ per ticket. The maximum ticket revenue is \$12,150.

2. Let n represent the number of \$0.20 decreases in the drink price. This makes the new drink price $3.70 - 0.20n$ and the number of drinks sold $10{,}000 + 800n$. Revenue is $R = (3.70 - 0.20n)(10{,}000 + 800n)$.

$$R = (3.70 - 0.20n)(10{,}000 + 800n) = 37{,}000 + 960n - 160n^2$$

$$h = \frac{-960}{2(-160)} = 3$$

$$k = 37{,}000 + 960(3) - 160(3)^2 = 38{,}440$$

Maximize revenue by charging $\$3.70 - 0.20(3) = \$3.10$ per drink. The maximum revenue is $38,440.

3. Let n represent the number of $50 increases in the monthly rent. This makes the monthly rent $2800 + 50n$ and the number of tenants $60 - 1n = 60 - n$. Monthly revenue is $R = (2800 + 50n)(60 - n)$.

$$R = (2800 - 50n)(60 - n) = 168,000 + 200n - 50n^2$$

$$h = \frac{-200}{2(-50)} = 2$$

$$k = 168,000 + 200(2) - 50(2)^2 = 168,200$$

Maximize revenue by charging $\$2800 + 50(2) = \$2900$ monthly rent. The maximum revenue is $168,200.

Maximizing/Minimizing Other Functions

Algebra students can use graphing calculators to approximate the maximum and/or minimum values of other kinds of functions. For example, the volume of a certain box is given by the function $V = 4x^3 - 40x^2 + 100x$, where x is the height (in inches) of the box and V is the volume (in cubic inches) of the box, and conditions make it necessary for $0 < x < 5$. The graph of this function is shown in Fig. 7-9.

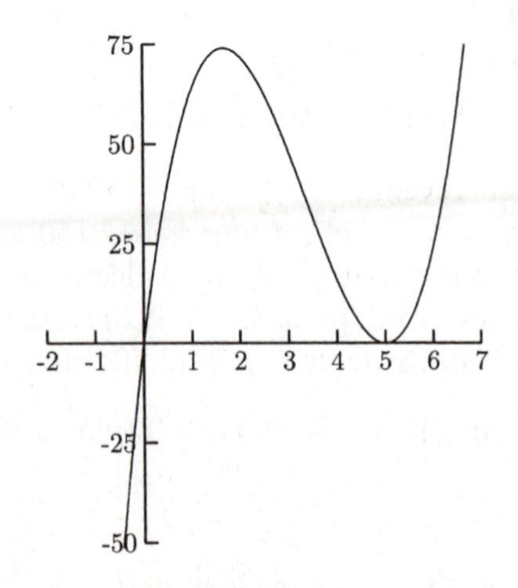

Fig. 7-9.

Because the domain of this applied function is $(0, 5)$, we need consider only this part of the graph.

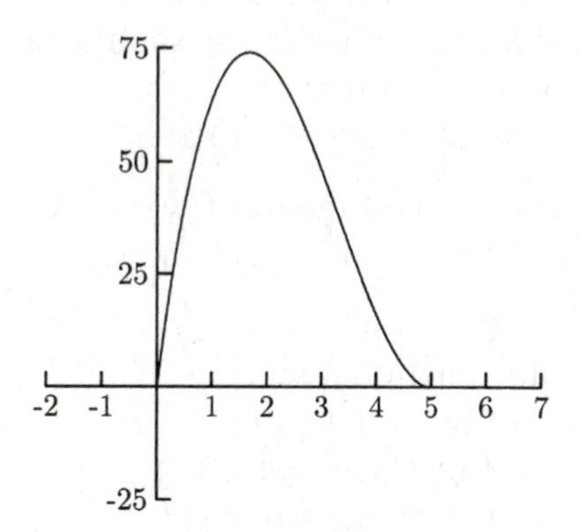

Fig. 7-10.

We can use a graphing calculator to approximate the highest point, $(1.67, 74.07)$. The maximum volume is approximately 74.07 cubic inches and the height at which the box's volume is maximum is about 1.67 inches. Calculus is necessary to find the exact values.

Chapter 7 Review

1. What is the range for the function $f(x) = 2x^2 - 6x + 1$?
   a) $[\frac{3}{2}, \infty)$ b) $(-\infty, \frac{3}{2})$ c) $[-\frac{7}{2}, \infty)$ d) $(-\infty, -\frac{7}{2}]$
2. What is the maximum value for the function $f(x) = -x^2 + 3x + 10$?
   a) 5, -2 b) $\frac{3}{2}$ c) $\frac{49}{4}$ d) There is no maximum value
3. The profit function for a product is $P(x) = -0.01x^2 + 3x + 500$, where x is the number produced and P is in dollars. How many units must be produced to maximize profit?
   a) 150 b) 725 c) 419 d) 119
4. A parks department wants to enclose a rectangular playing field and subdivide it into two fields (see Fig. 7-2). There are 1800 feet of fencing available. What is the maximum area?
   a) 405,000 square feet b) 150,000 square feet
   c) 450 square feet d) 135,000 square feet

5. The revenue of a certain product depends on the amount spent on advertising. The function is $R(x) = -0.001x^2 + 240x - 13,500,000$, where R is the revenue (in dollars) and x is the amount spent on advertising (in dollars). How much should be spent on advertising in order to maximize revenue?
   a) \$120,000 b) \$900,000 c) \$90,000 d) \$150,000

6. What is the range for the function $f(x) = -x^2 - 10x + 8$?
   a) $[33, \infty)$ b) $(-\infty, 33]$ c) $[-5, \infty)$ d) $(-\infty, -5]$

7. For the quadratic function $g(t) = -\frac{1}{16}t^2 + t + 10$
   a) the maximum functional value is 14.
   b) the minimum functional value is 14.
   c) the maximum functional value is 8.
   d) the minimum functional value is 8.

8. A cotton candy vendor at a small fair sells 270 cones on average when the price is \$1 each. The vendor believes that for each \$0.10 increase in the price, sales will drop by 15. What is the maximum revenue?
   a) \$294 b) \$356 c) \$400 d) Cannot be determined

SOLUTIONS

1. c) 2. c) 3. a) 4. d) 5. a) 6. b) 7. a) 8. a)

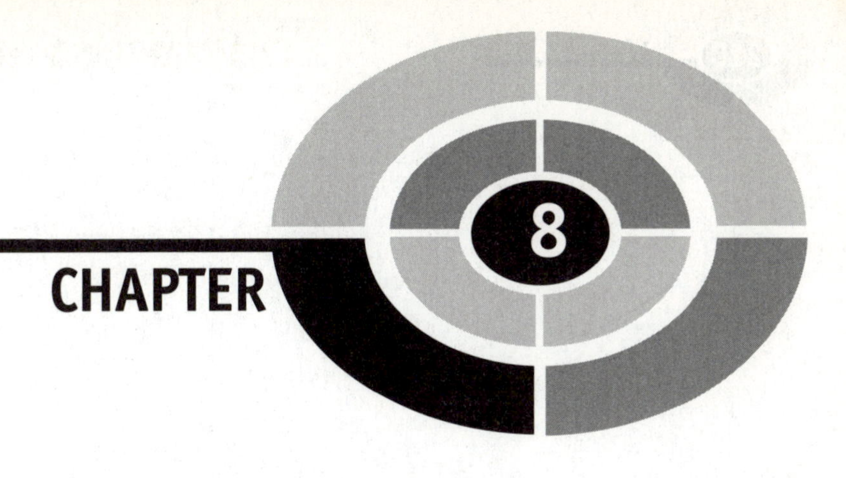

Transformations and Combinations

Many important graphs come in families. We have already studied three families: graphs of circles, lines, and parabolas. There is a lot we can tell about the graph of an equation by the equation itself. From the equation of a circle in the form $(x-h)^2+(x-k)^2=r^2$, we know that its center is at (h,k) and its radius is r. From the equation of a line in the form $y=mx+b$, we know its slope is m and its y-intercept is b. From the equation of a quadratic function in the form $y=a(x-h)^2+k$, we know that its vertex is (h,k) and that it opens up if a is positive or opens down if a is negative.

Let us begin with a closer look at quadratic functions. The graph of every quadratic function is more or less the graph of $y=x^2$. For example, the vertex for the function $y=(x-2)^2$ is $(2,0)$. Another way of looking at this is to say that the vertex moved from $(0,0)$ to $(2,0)$. That is, the vertex moved to the right two units.

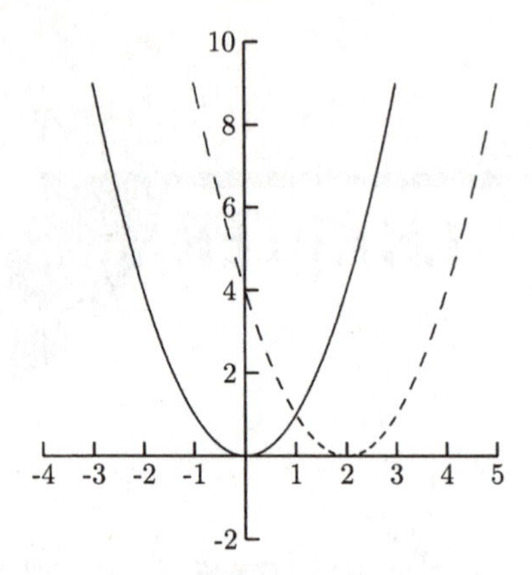

Fig. 8-1.

The function $y = (x - 2)^2$ is really the function $f(x) = x^2$ evaluated at $x - 2 : f(x - 2) = (x - 2)^2$. Evaluating *any* function at $x - 2$ shifts the entire function two units to the right. For any function $f(x)$, the graph of $f(x - 2)$ is the graph of $f(x)$ shifted to the right two units. For any positive number k, the graph $y = f(x - k)$ is the graph of $f(x)$ shifted to the right k units, no matter what function $f(x)$ is.

EXAMPLES

- The graph of $f(x - 5)$ is the graph of $f(x)$ shifted to the right 5 units.
- The graph of $f(x - 20)$ is the graph of $f(x)$ shifted to the right 20 units.
- The graph of $f(x - \frac{1}{2})$ is the graph of $f(x)$ shifted to the right $\frac{1}{2}$ units.

What would be the effect on the graph of $f(x)$ by adding positive k to x? The vertex for $y = (x + 3)^2$ is $(-3, 0)$. This is the graph of $f(x) = x^2$ shifted to the *left* 3 units. The graph of any function $f(x + k)$ is the graph of $f(x)$ shifted to the left k units.

EXAMPLES

- The graph of $f(x + 12)$ is the graph of $f(x)$ shifted to the left 12 units.
- The graph of $g(x + 3)$ is the graph of $g(x)$ shifted to the left 3 units.

PRACTICE

Compare the graph of the functions with the graph of $f(x)$.

1. $f(x + 5)$
2. $f(x - 0.10)$
3. $f(x - 35)$

SOLUTIONS

1. The graph of $f(x+5)$ is the graph of $f(x)$ shifted to the left 5 units.
2. The graph of $f(x-0.10)$ is the graph of $f(x)$ shifted to the right 0.10 units.
3. The graph of $f(x-35)$ is the graph of $f(x)$ shifted to the right 35 units.

The vertex for the quadratic function $y = x^2 + 2$ is $(0, 2)$, which is shifted two units up from the vertex of $f(x) = x^2$. Adding a positive number to a function has the effect of shifting its graph upward. Subracting a positive number from a function has the effect of shifting its graph downward. If k is a positive number, the graph of $f(x) + k$ is the graph of $f(x)$ shifted up k units, and the graph of $f(x) - k$ is the graph of $f(x)$ shifted down k units.

EXAMPLES

- The graph of $f(x) - 4$ is the graph if $f(x)$ shifted down 4 units.
- The graph of $h(x) + 9$ is the graph of $h(x)$ shifted up 9 units.

PRACTICE

Compare the graph of the functions with the graph of $f(x)$.

1. $f(x) + 1$
2. $f(x) + 15$
3. $f(x) - 8$
4. $f(x) - 1$
5. $f(x+1)$
6. $f(x-6)$

SOLUTIONS

1. The graph of $f(x) + 1$ is the graph of $f(x)$ shifted up 1 unit.
2. The graph of $f(x) + 15$ is the graph of $f(x)$ shifted up 15 units.
3. The graph of $f(x) - 8$ is the graph of $f(x)$ shifted down 8 units.
4. The graph of $f(x) - 1$ is the graph of $f(x)$ shifted down 1 unit.
5. The graph of $f(x+1)$ is the graph of $f(x)$ shifted left 1 unit.
6. The graph of $f(x-6)$ is the graph of $f(x)$ shifted right 6 units.

Functions can have a combination of vertical and horizontal shifts. If h and k are positive numbers, the graph of $f(x-h) + k$ is the graph of $f(x)$ shifted to the right h units and up k units. The graph of $f(x+k) - h$ is the graph of $f(x)$ shifted to the left k units and down h units.

EXAMPLES

- The graph of $y = (x-2)^2 + 1$ is the graph of $f(x) = x^2$ shifted to the right 2 units and up 1 unit.

- The graph of $f(x+2)+3$ is the graph of $f(x)$ shifted to the left 2 units and up 3 units.

PRACTICE

Compare the graph of the functions with the graph of $f(x)$.

1. $f(x+\frac{1}{2})+3$
2. $f(x-4)-5$
3. $f(x+6)-8$
4. $f(x+10)+15$
5. $f(x-1)+9$

SOLUTIONS

1. The graph of $f(x+\frac{1}{2})+3$ is the graph of $f(x)$ shifted to the left $\frac{1}{2}$ units and up 3 units.
2. The graph of $f(x-4)-5$ is the graph of $f(x)$ shifted to the right 4 units and down 5 units.
3. The graph of $f(x+6)-8$ is the graph of $f(x)$ shifted to the left 6 units and down 8 units.
4. The graph of $f(x+10)+15$ is the graph of $f(x)$ shifted to the left 10 units and up 15 units.
5. The graph of $f(x-1)+9$ is the graph of $f(x)$ shifted to the right 1 unit and up 9 units.

In the following we compare the graph of a function with its transformation. The solid graphs are the graphs of $f(x)$, and the dashed graphs are the transformations of $f(x)$.

EXAMPLES

Compare the graph of the functions with the graph of $f(x)$. Then write the transformed function.

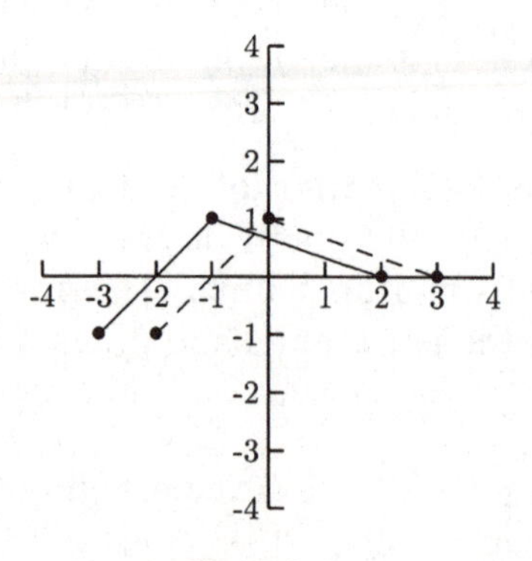

Fig. 8-2.

The dashed graph is the graph of $f(x)$ shifted to the right 1 unit, so this is the graph of $f(x - 1)$.

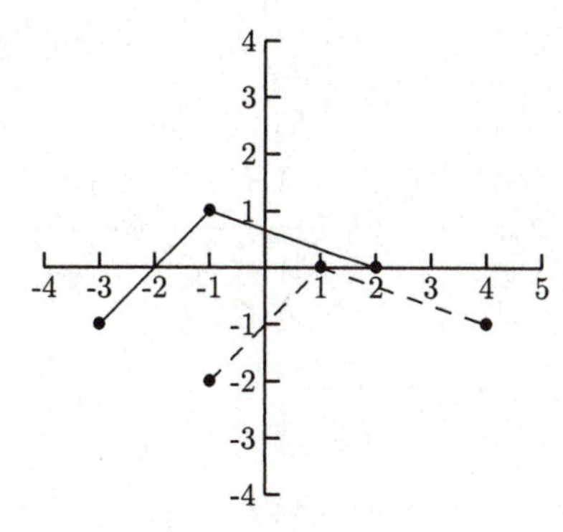

Fig. 8-3.

The dashed graph is the graph of $f(x)$ shifted to the right 2 units and down 1 unit, so this is the graph of $f(x - 2) - 1$.

PRACTICE

Compare the graph of the transformations with the graph of $f(x)$. Then write the transformed function.

1.

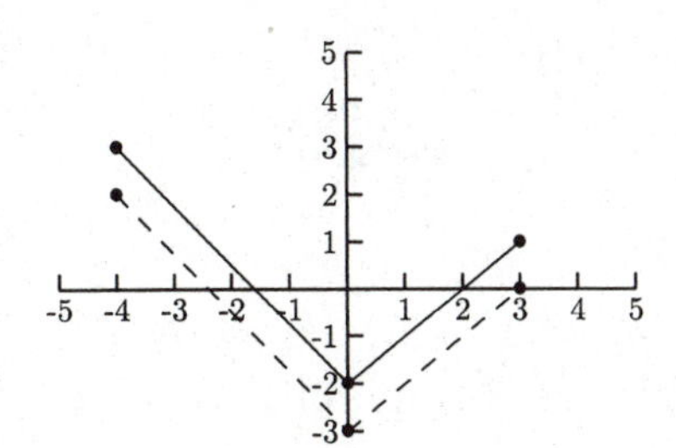

Fig. 8-4.

2.

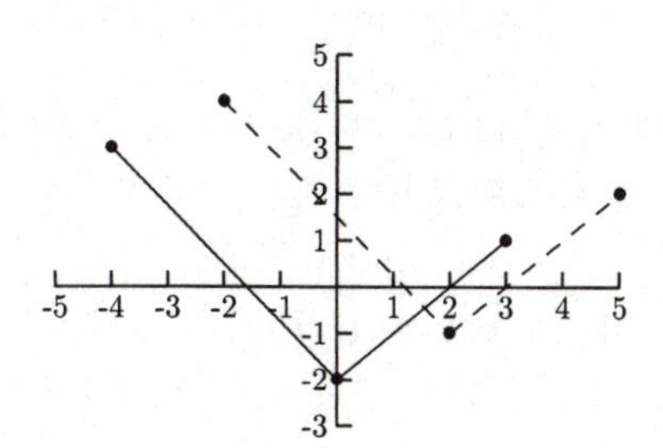

Fig. 8-5.

3.

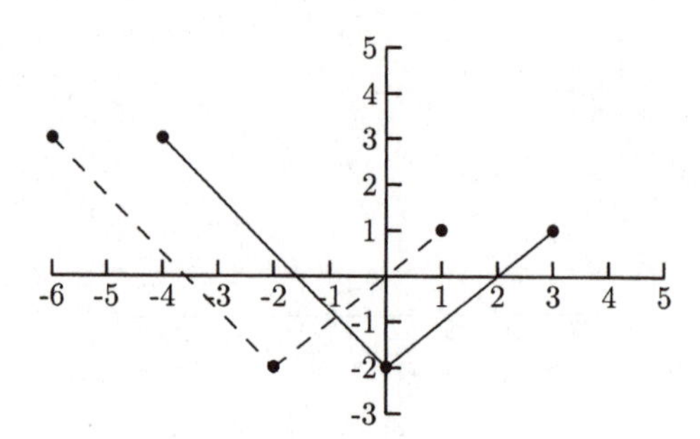

Fig. 8-6.

4.

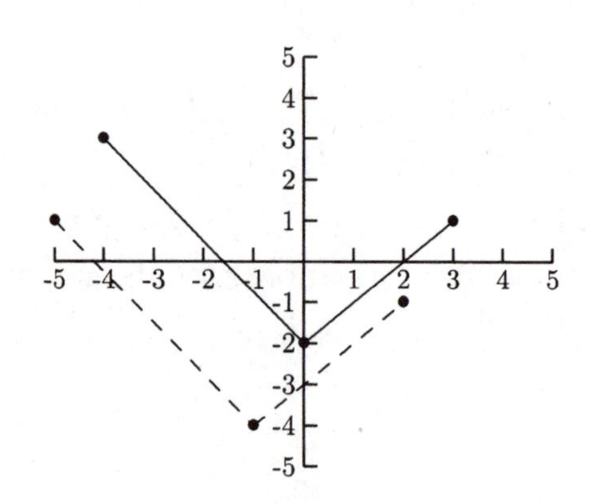

Fig. 8-7.

SOLUTIONS

1. The dashed graph is the graph of $f(x)$ shifted down 1 unit. It is the graph of $f(x) - 1$.
2. The dashed graph is the graph of $f(x)$ shifted to the right 2 units and up 1 unit. It is the graph of $f(x-2)+1$.
3. The dashed graph is the graph of $f(x)$ shifted to the left 2 units. It is the graph of $f(x+2)$.
4. The dashed graph is the graph of $f(x)$ shifted to the left 1 unit and down 2 units. It is the graph of $f(x+1) - 2$.

For any function $f(x)$, the graph of $-f(x)$ is the graph of $f(x)$ flipped upside down, or in more technical terms, "reflected about the x-axis." For example, the graph of $y = -x^2$ is a reflection of the graph of $f(x) = x^2$.

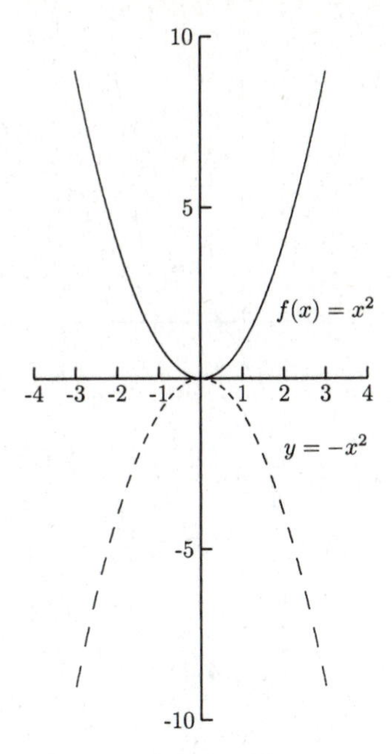

Fig. 8-8.

EXAMPLES

The dashed graphs are the reflections about the x-axis of the solid graphs.

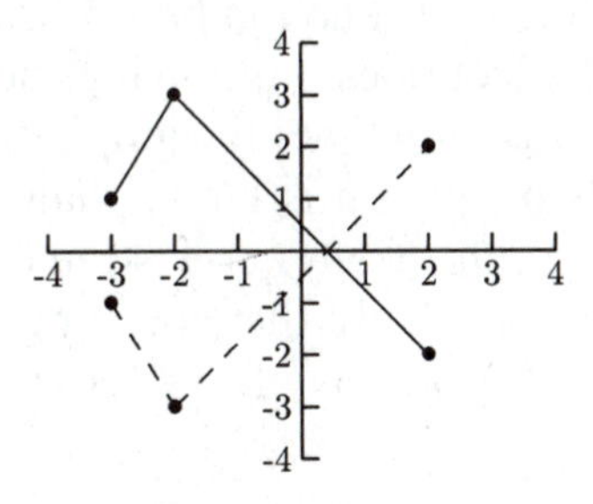

Fig. 8-9.

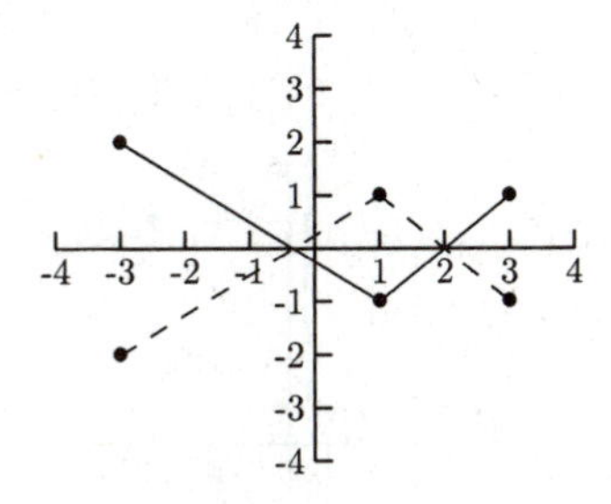

Fig. 8-10.

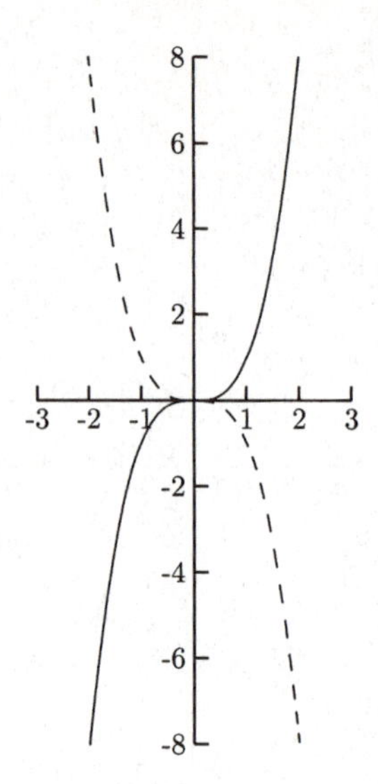

Fig. 8-11.

With these reflections, the x-values do not change but the y-values of $-f(x)$ are the opposite of the y-values for $f(x)$. The x-intercepts (where $y = 0$) do not change.

If multiplying the y-values of a function by -1 has the effect of turning the graph upside down, what effect does multiplying the y-values by some other number have? In other words, how does the graph of $af(x)$ compare with the graph of $f(x)$? It depends on a. If a is larger than 1, the graph of $af(x)$ is vertically stretched. The graph of $50f(x)$ is stretched more than the graph of $3f(x)$. If a is between 0 and 1, then the graph of $af(x)$ is vertically compressed, or flattened. The graph of $\frac{1}{10}f(x)$ is flattened more than the graph of $\frac{2}{3}f(x)$.

EXAMPLES

•

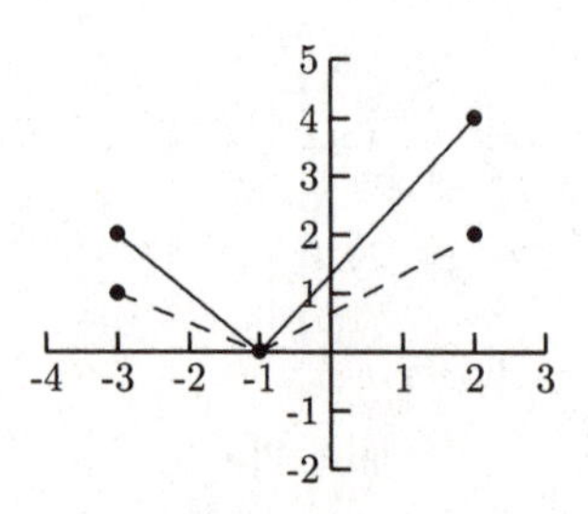

Fig. 8-12.

The dashed graph in Fig. 8-12 is the graph of $\frac{1}{2}f(x)$. The y-value for each point on the dashed graph is half of the corresponding y-value in the solid graph. For example, the point $(-3, 2)$ on the graph of $f(x)$ is moved to $(-3, 1)$ on the graph of $\frac{1}{2}f(x)$. The point $(2, 4)$ on the solid graph is moved to $(2, 2)$ on the dashed graph.

•

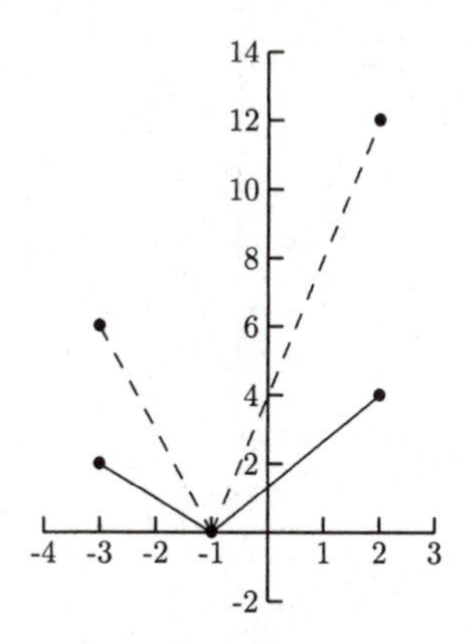

Fig. 8-13.

The dashed graph in Fig. 8-13 is the graph of $3f(x)$. The y-values for each point on the dashed graph are three times the y-values on the solid graph. For example, the point $(2, 4)$ on the solid graph is moved to $(2, 12)$ on the dashed graph. The point $(-1, 0)$ on the solid graph did not move because $3 \cdot 0 = 0$.

When a is a negative number, other than -1, the effect of $af(x)$ is a combination of the changes above. First, the graph will be turned upside down (reflected about the x-axis). Then it will either be vertically compressed or stretched. In the following examples, the solid graphs are the graphs of $f(x)$ and the dashed graphs are the graphs of $af(x)$.

EXAMPLES

•

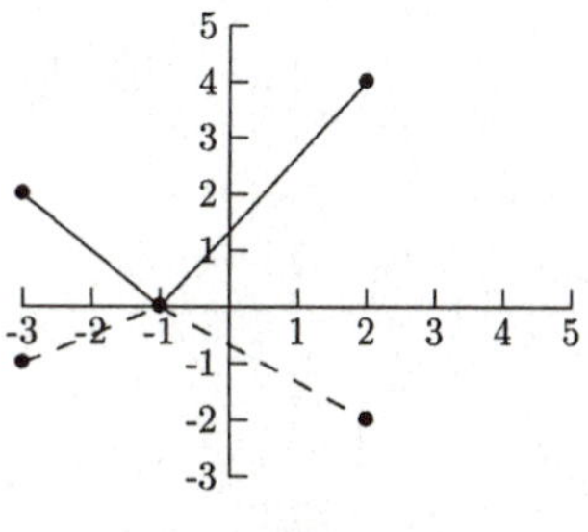

Fig. 8-14.

The dashed graph is the graph of $-\frac{1}{2}f(x)$. The point (2, 4) on the solid graph moved to (2, − 2) because $4 \cdot (-\frac{1}{2}) = -2$. The point (−3, 2) on the solid graph moved to (−3, − 1) because $2 \cdot (-\frac{1}{2}) = -1$. The point (−1, 0) on the solid graph did not move because $0 \cdot (-\frac{1}{2}) = 0$.

•

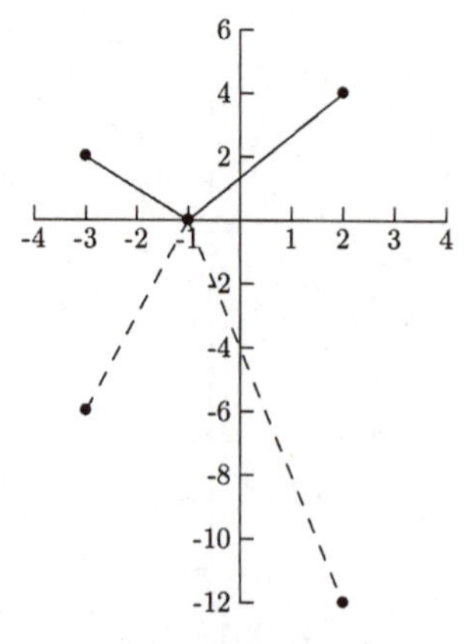

Fig. 8-15.

The dashed graph is the graph of $-3f(x)$.

To summarize, if $a > 1$, the graph of $af(x)$ is vertically stretched. If $0 < a < 1$, the graph of $af(x)$ is vertically flattened. If $a < -1$, the graph is reflected about the x-axis and vertically stretched. If $-1 < a < 0$, the graph is reflected about the x-axis and vertically flattened.

PRACTICE

For problems 1–4, determine whether the dashed graph is a vertically stretched or flattened version of the solid graph and whether or not it is reflected about the x-axis.

1.

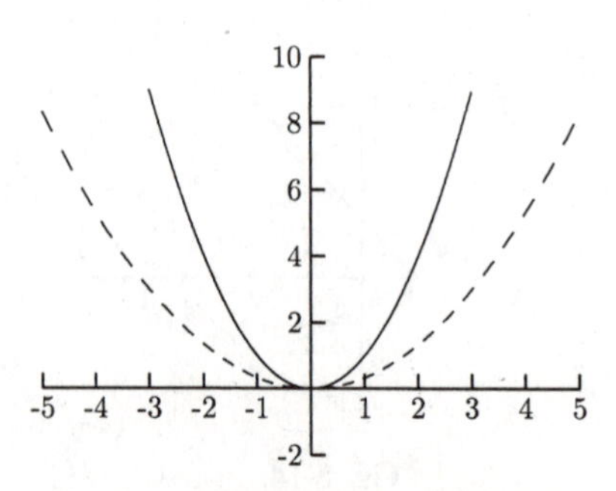

Fig. 8-16.

2.

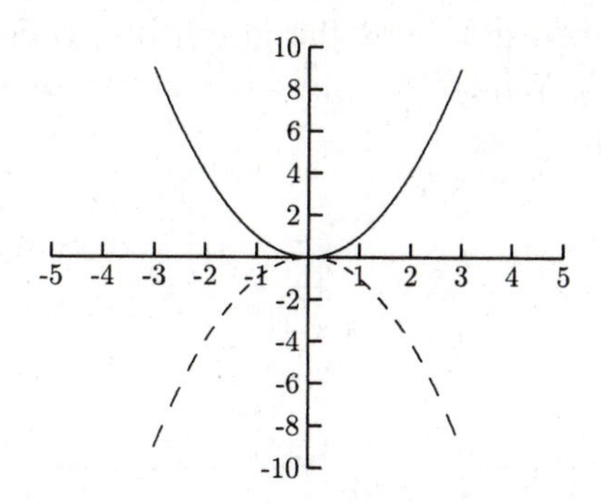

Fig. 8-17.

3.

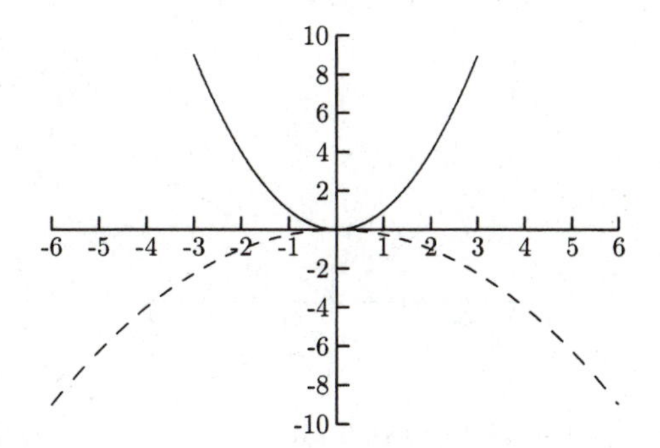

Fig. 8-18.

4.

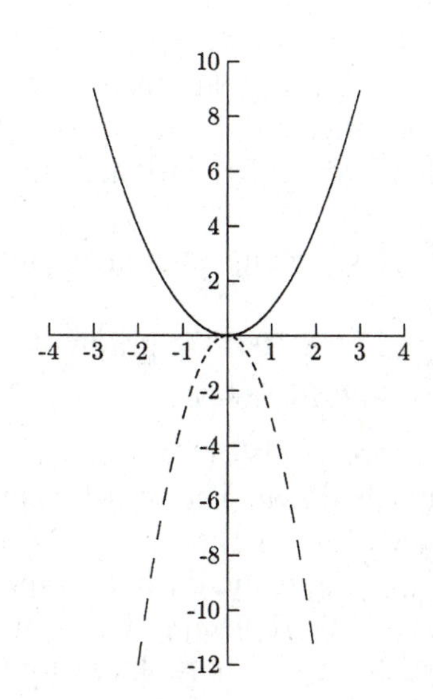

Fig. 8-19.

5. Below are two pairs of graphs. The solid graphs are the graph of $f(x)$. One of the dashed graphs is the graph of $\frac{3}{2}f(x)$ and the other is the graph of $4f(x)$. Which graph is the graph of $\frac{3}{2}f(x)$? Which is the graph of $4f(x)$?

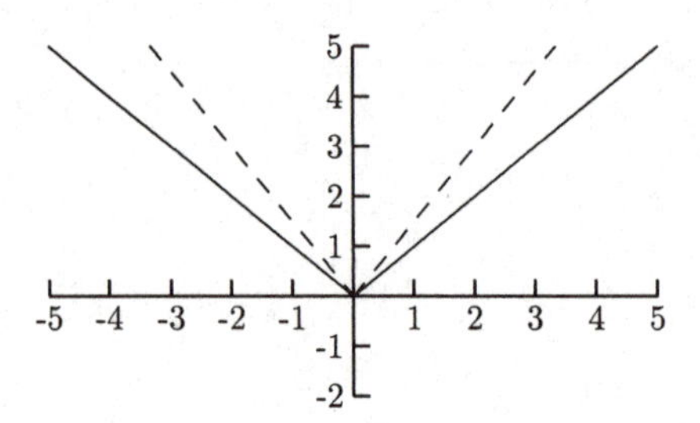

Fig. 8-20.

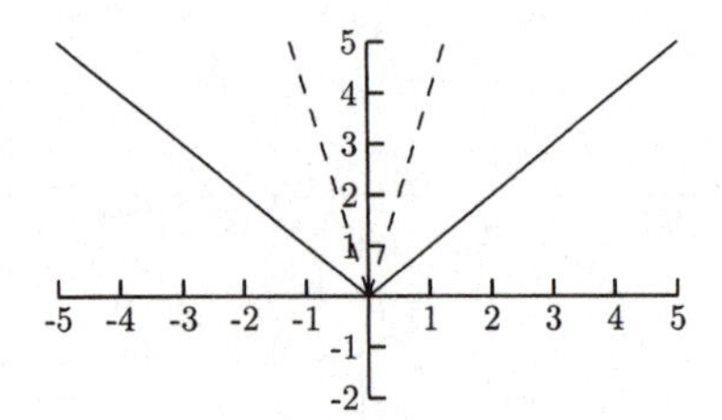

Fig. 8-21.

SOLUTIONS

1. The dashed graph is vertically flattened.
2. The dashed graph is reflected about the x-axis.
3. The dashed graph is vertically flattened and is reflected about the x-axis.
4. The dashed graph is vertically stretched and is reflected about the x-axis.
5. The dashed graph in Fig. 8-20 is the graph of $\frac{3}{2}f(x)$. The dashed graph in Fig. 8-21 is the graph of $4f(x)$.

We will look at one more transformation, $f(-x)$. The transformation $-f(x)$ turned the graph upside down. The transformation $f(-x)$ will turn the graph sideways, or "reflected about the y-axis." The solid graph shown in Fig. 8-22 is the graph of $f(x)$, and the dashed graph is the graph of $f(-x)$. We can get the graph of $f(-x)$ by replacing the x-values with their opposites. For example, the point $(4, 2)$ on $f(x)$ is replaced by $(-4, 2)$.

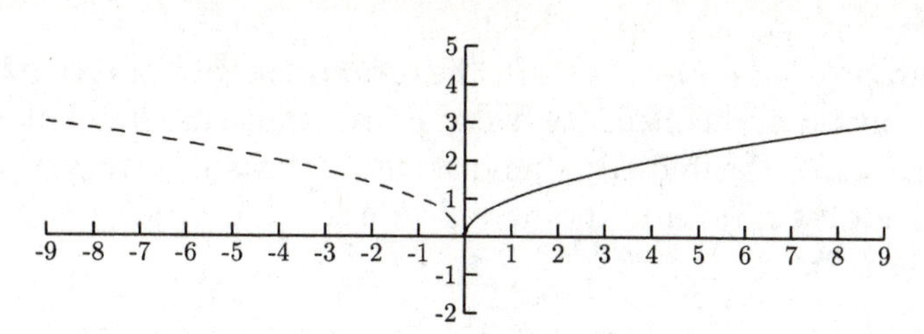

Fig. 8-22.

We are ready to sketch transformations of a given graph. The graph of $f(x)$ will be given and we will be asked to sketch a given transformation. Some of the transformations can be done with no extra work, but we will need to be careful with others. To help with the more complicated transformations, we will use tables of values.

EXAMPLES

Sketch the transformations $f(x+1)-3$, $-f(x)+1$, $-f(x+2)-3$, $2f(x)$, $f(-x)$, $f(-x)-1$, and $f(2-x)$.

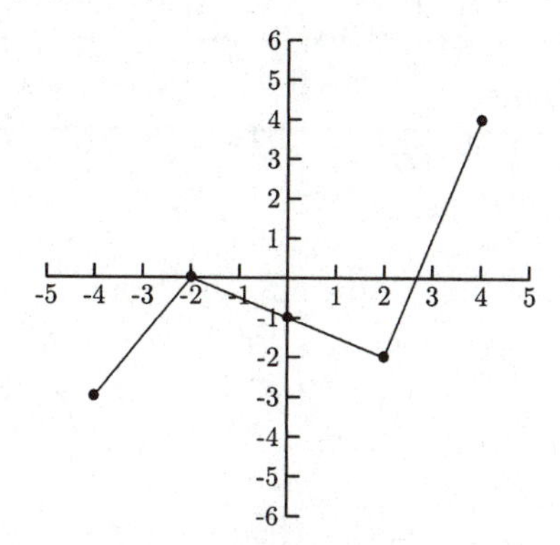

Fig. 8-23.

Table 8-1 gives the values for this function.

Table 8-1

x	$f(x)$
-4	-3
-2	0
0	-1
2	-2
4	4

- To graph $f(x+1)-3$, shift the graph of $f(x)$ to the left 1 unit and down 3 units. Because every point is moving left 1 unit, the new x-values are the old x-values minus 1. Because every point is also moving down 3 units, the new y-values are the old y-values minus 3 (Table 8-2).

Table 8-2

$x-1$	$y-3$	**Plot this point**
$-4-1=-5$	$-3-3=-6$	$(-5,-6)$
$-2-1=-3$	$0-3=-3$	$(-3,-3)$
$0-1=-1$	$-1-3=-4$	$(-1,-4)$
$2-1=1$	$-2-3=-5$	$(1,-5)$
$4-1=3$	$4-3=1$	$(3,1)$

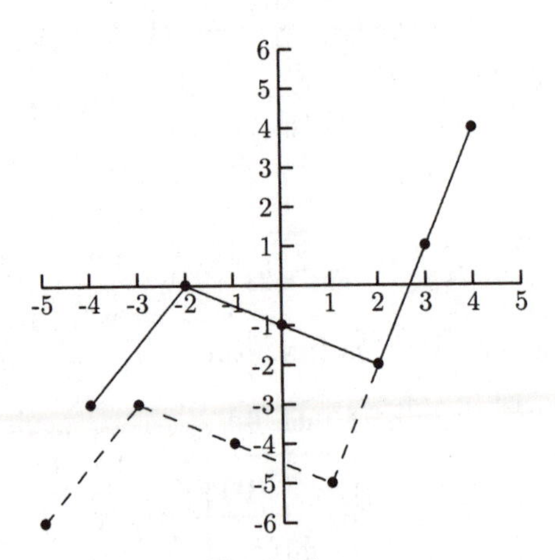

Fig. 8-24.

- For the graph of $-f(x)+1$, we will not change the x-values. The y-values have two changes. First, we will take the negative of the old y-values; second, we will add 1 to them (Table 8-3).

Table 8-3

x	$-y+1$	Plot this point
−4	$-(-3)+1=4$	$(-4, 4)$
−2	$-0+1=1$	$(-2, 1)$
0	$-(-1)+1=2$	$(0, 2)$
2	$-(-2)+1=3$	$(2, 3)$
4	$-4+1=-3$	$(4, -3)$

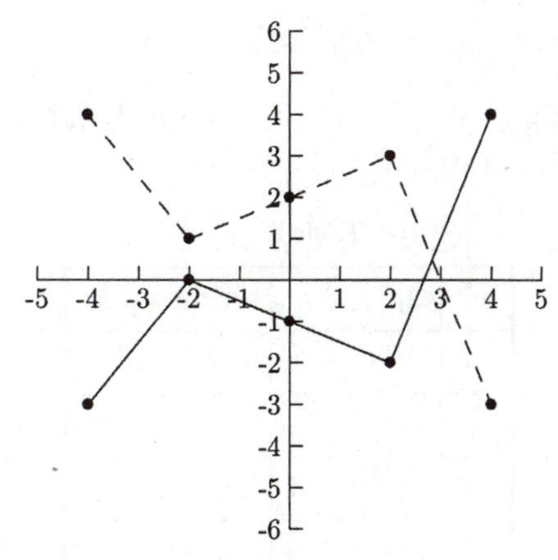

Fig. 8-25.

- To find the points for $-f(x+2)-3$, subtract 2 from each x-value, and subtract 3 from the opposite of the y-values (Table 8-4).

Table 8-4

$x-2$	$-y-3$	Plot this point
$-4-2=-6$	$-(-3)-3=0$	$(-6, 0)$
$-2-2=-4$	$-0-3=-3$	$(-4, -3)$
$0-2=-2$	$-(-1)-3=-2$	$(-2, -2)$
$2-2=0$	$-(-2)-3=-1$	$(0, -1)$
$4-2=2$	$-4-3=-7$	$(2, -7)$

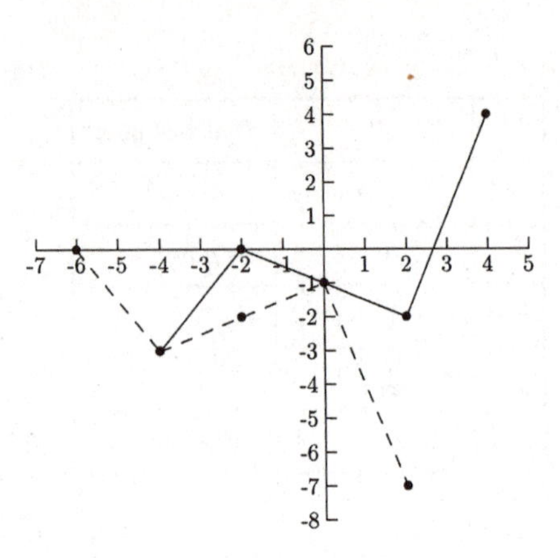

Fig. 8-26.

- Sketch the graph of $2f(x)$ by multiplying each y-value by 2. The x-values do not change (Table 8-5).

Table 8-5

x	$2y$	Plot this point
-4	$2(-3)=-6$	$(-4,-6)$
-2	$2(0)=0$	$(-2,0)$
0	$2(-1)=-2$	$(0,-2)$
2	$2(-2)=-4$	$(2,-4)$
4	$2(4)=8$	$(4,8)$

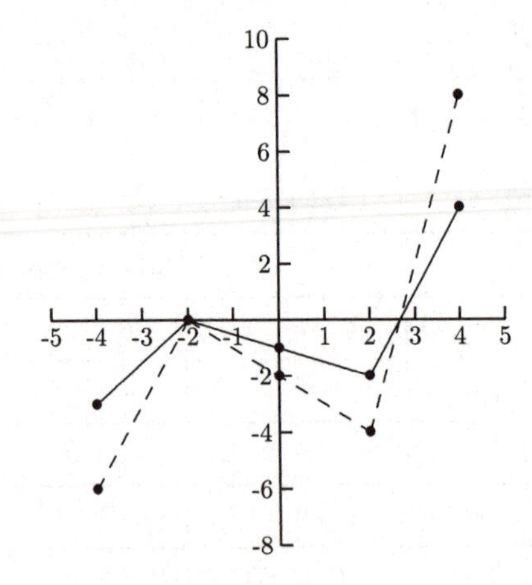

Fig. 8-27.

- Sketch the graph of $f(-x)$ by replacing each x-value with its opposite. The y-values do not change (Table 8-6).

Table 8-6

$-x$	y	Plot this point
$-(-4)=4$	-3	$(4,-3)$
$-(-2)=2$	0	$(2,0)$
$-0=0$	-1	$(0,-1)$
-2	-2	$(-2,-2)$
-4	4	$(-4,4)$

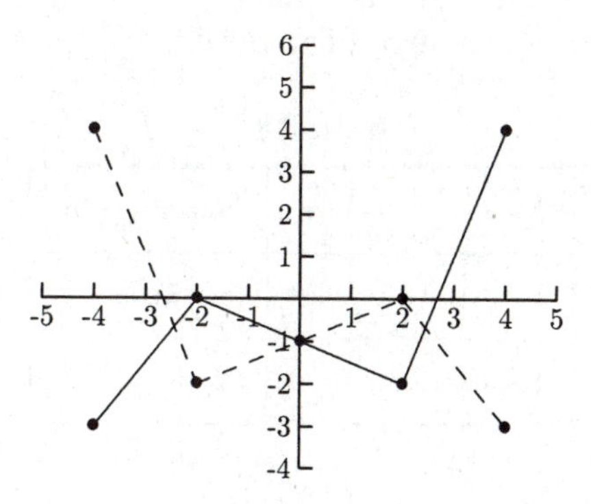

Fig. 8-28.

- Sketch the graph of $f(-x)-1$ by taking the the opposite of each x-value and by subtracting 1 from each y-value (Table 8-7).

Table 8-7

$-x$	$y-1$	Plot this point
$-(-4)=4$	$-3-1=-4$	$(4,-4)$
$-(-2)=2$	$0-1=-1$	$(2,-1)$
$-0=0$	$-1-1=-2$	$(0,-2)$
-2	$-2-1=-3$	$(-2,-3)$
-4	$4-1=3$	$(-4,3)$

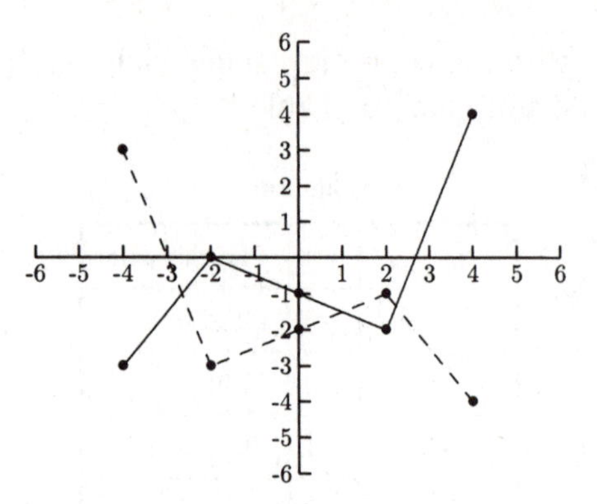

Fig. 8-29.

- The graph of $f(2-x)$ is a little more difficult. This transformation is first a reflection about the y-axis, then a shift to the right 2 units. The y-values do not change (Table 8-8).

Table 8-8

$2-x$	y	Plot this point
$2-(-4)=6$	-3	$(6,-3)$
$2-(-2)=4$	0	$(4,0)$
$2-0=2$	-1	$(2,-1)$
$2-2=0$	-2	$(0,-2)$
$2-4=-2$	4	$(-2,4)$

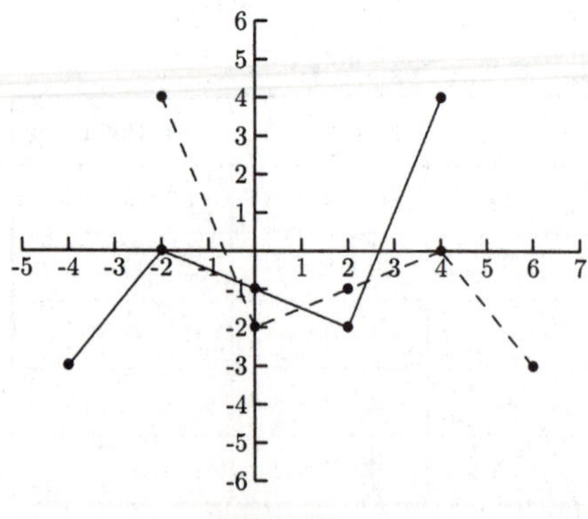

Fig. 8-30.

PRACTICE

Sketch the transformations of the function $f(x)$ whose graph is given in Fig. 8-31.

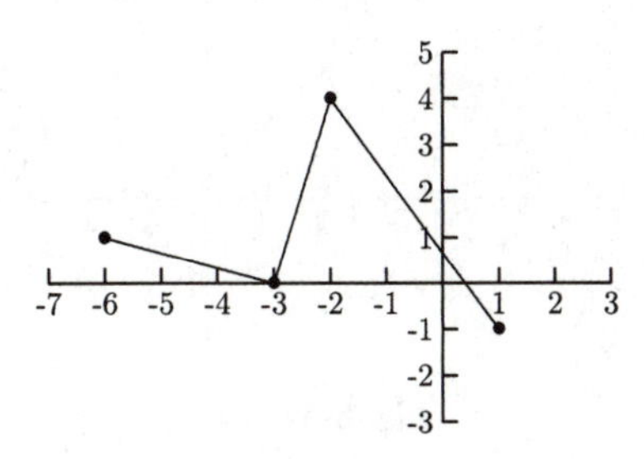

Fig. 8-31.

1. $f(x+2)$
2. $f(x-1)$
3. $\frac{2}{3}f(x)$
4. $f(-x)$
5. $f(x-1)+2$
6. $-f(x)+1$
7. $2f(x-2)$

SOLUTIONS

1.

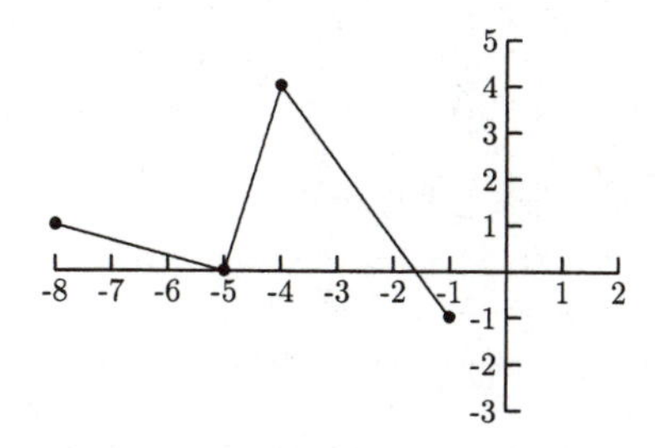

Fig. 8-32.

2.

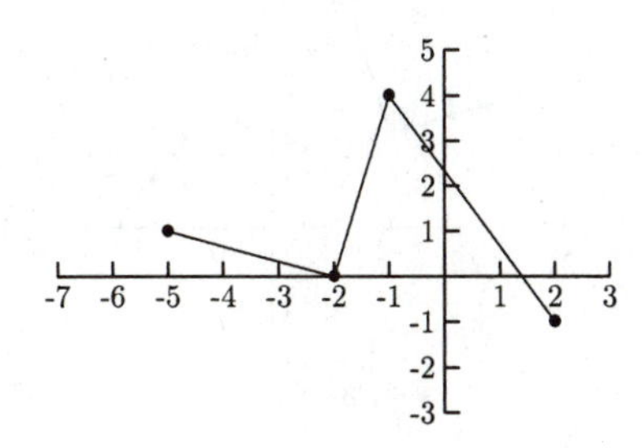

Fig. 8-33.

3.

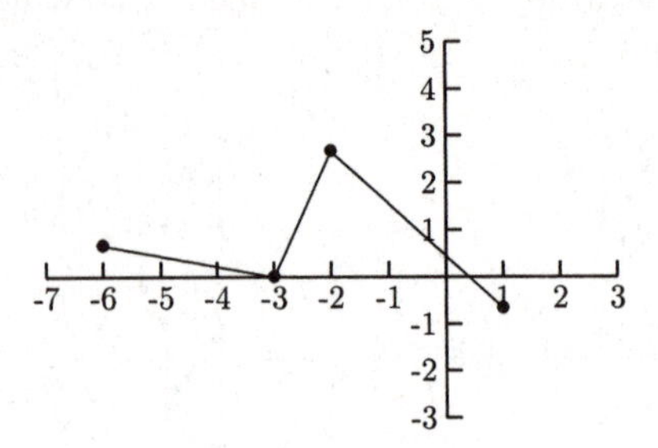

Fig. 8-34.

4.

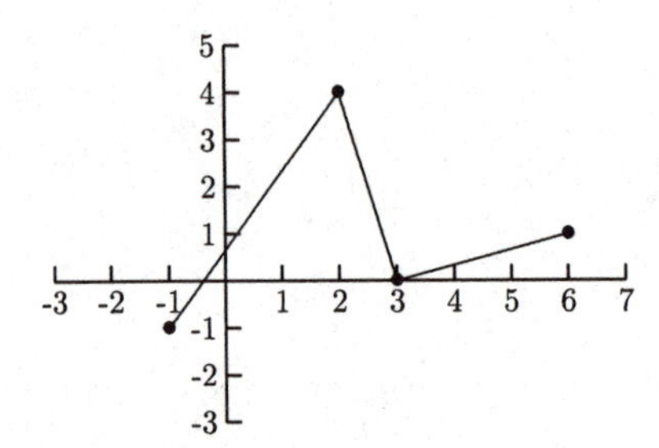

Fig. 8-35.

5.

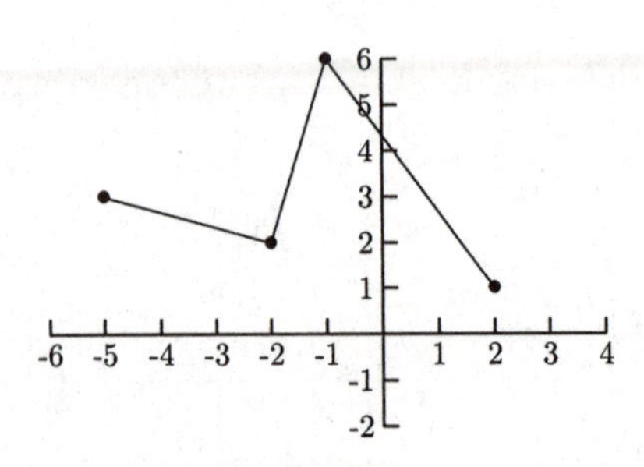

Fig. 8-36.

6.

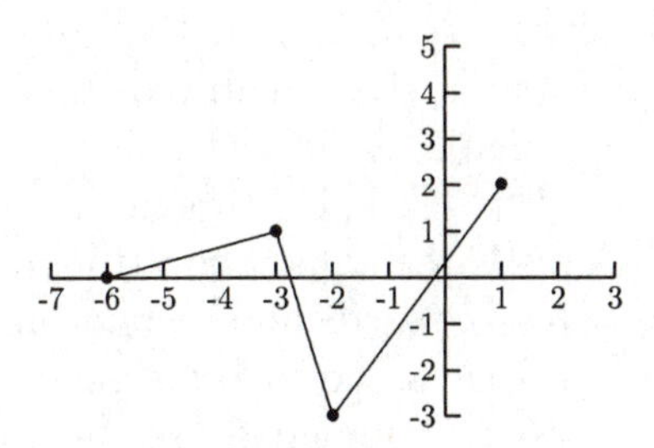

Fig. 8-37.

7.

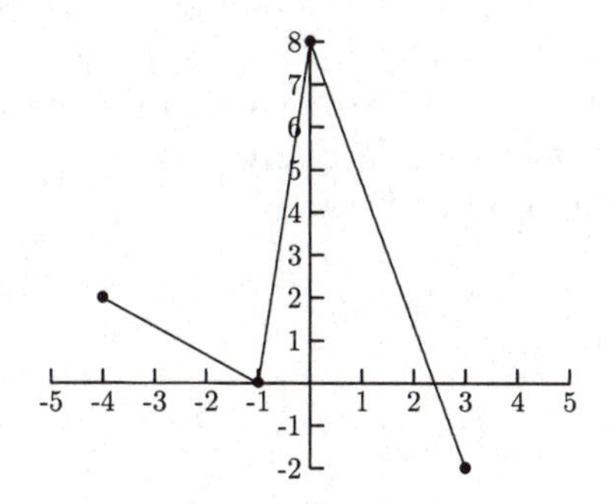

Fig. 8-38.

It might seem odd that in the transformation $f(x+1)+2$, "$x+1$" has the effect of moving x in the negative direction while "+2" has the effect of moving y in the positive direction. The reason we subtract 1 from each x is so that when we evaluate $f(x+1)$ at $x-1$, we end up with $f(x)$: $f((x-1)+1) = f(x)$. It is for this reason that to sketch the graph of the transformation $f(dx)$, we would compute the x-values by dividing them by d. For example, if we need to sketch the graph of $f(2x)$, we would need to divide each x-value by 2. This is so that $f(2(\frac{1}{2}x)) = f(x)$.

Special Functions

There are several families of functions whose graphs college algebra students should know.

1. $f(x)=c$ This is the constant function. Its graph is a horizontal line.

2. $f(x)=mx+b$ This is the linear function. Its graph is a nonvertical line.
3. $f(x)=x^2$ This is the quadratic function. Its graph is a parabola.
4. $f(x)=x^3$ This is the cubic function.
5. $f(x)=\sqrt{x}$ This is the square root function.
6. $f(x)=|x|$ This is the absolute value function.
7. $f(x)=a^x$ This is the exponential function.
8. $f(x)=\log_a x$ This is the logarithmic function.

The graphs of the first three are covered in Chapter 4, the second three are covered in this chapter, and the exponential function is covered in Chapter 11. Once you know the basic shape of the graphs of these functions, you can use what you learned earlier in this chapter to sketch the graphs of many functions with only a little work. This information can also help you use a graphing calculator more effectively.

Figure 8-39 shows the graph of $y=x^3$.

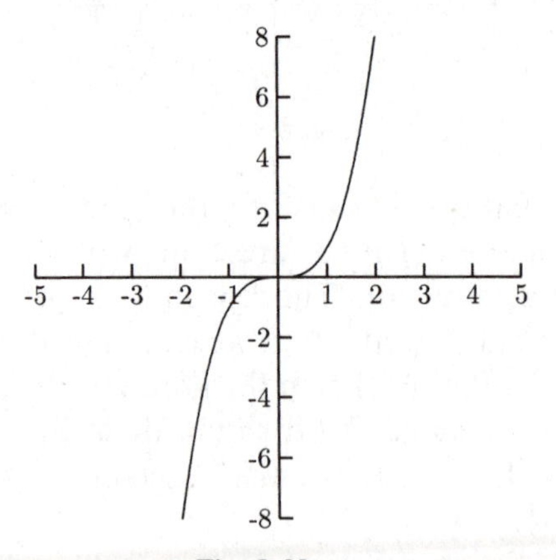

Fig. 8-39.

The solid graphs in Figs. 8-39–8-46 are the graphs of $y=x^3$ and the dashed graphs are transformations of $y=x^3$. The transformation in Fig. 8-40 is $f(x)=-x^3$, a reflection of $y=x^3$ about the x-axis. Coincidentally, both reflections about the x-axis and y-axis are the same. In other words, $y=-x^3$ and $y=(-x)^3$ are the same function.

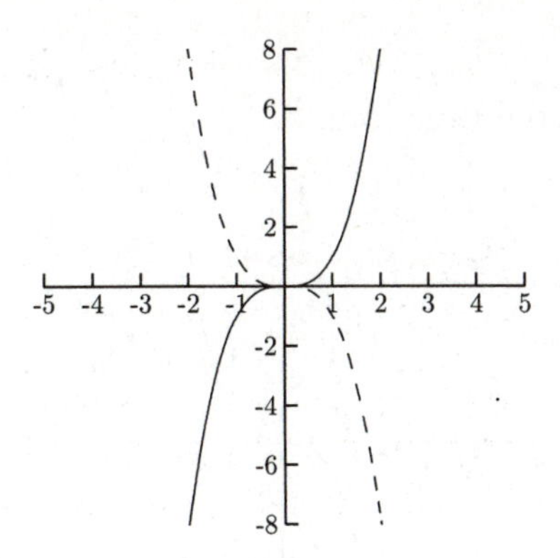

Fig. 8-40.

The graph $f(x) = 2x^3$ is the graph $y = x^3$ stretched vertically.

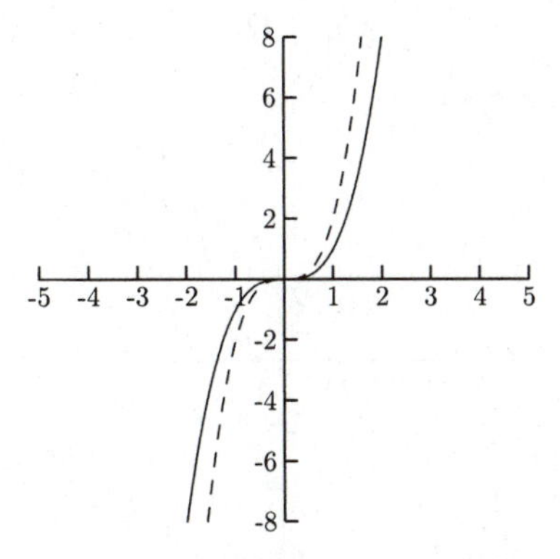

Fig. 8-41.

The graph of $f(x) = (x - 2)^3 - 1$ is the graph of $y = x^3$ shifted to the right 2 units and down 1 unit.

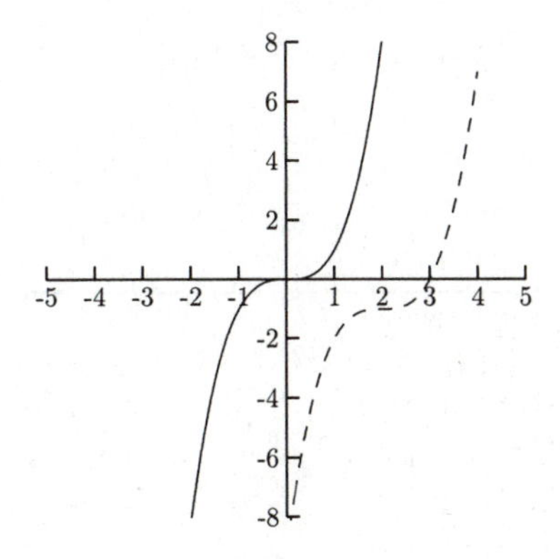

Fig. 8-42.

PRACTICE

Match the graph with the function.

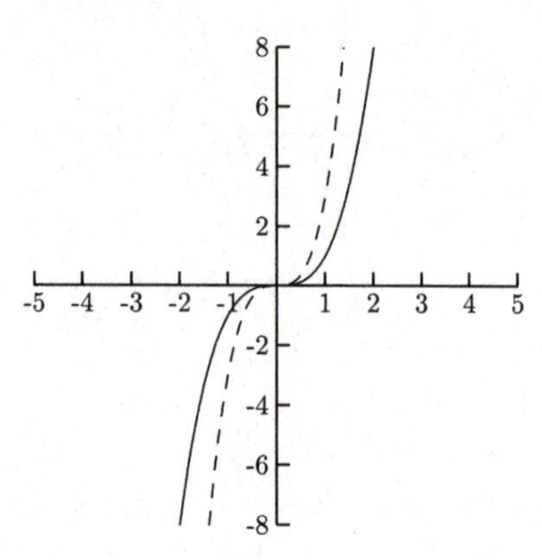

Fig. 8-43.

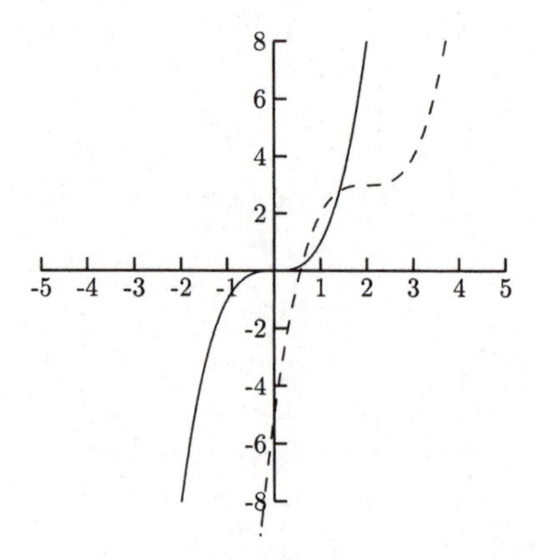

Fig. 8-44.

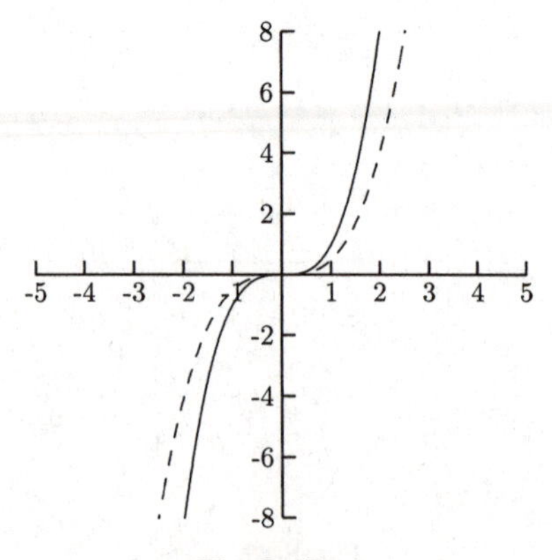

Fig. 8-45.

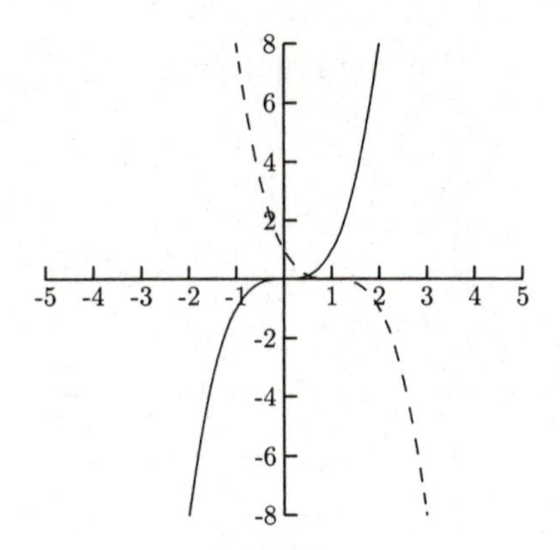

Fig. 8-46.

1. $f(x) = \frac{1}{2}x^3$
2. $f(x) = -(x-1)^3$
3. $f(x) = 3x^3$
4. $f(x) = (x-2)^3 + 3$

SOLUTIONS

1. Figure 8-45
2. Figure 8-46
3. Figure 8-43
4. Figure 8-44

The graph of $y = \sqrt{x}$ is part of a parabola. Imagine turning a parabola on its side and cutting off the bottom half. What would be left is the graph of $y = \sqrt{x}$. The graph of $y = \sqrt{x}$ is shown in Fig. 8-47. The solid graphs in Figs. 8-48–8-56 are the graph of $y = \sqrt{x}$. The dashed graphs are transformations.

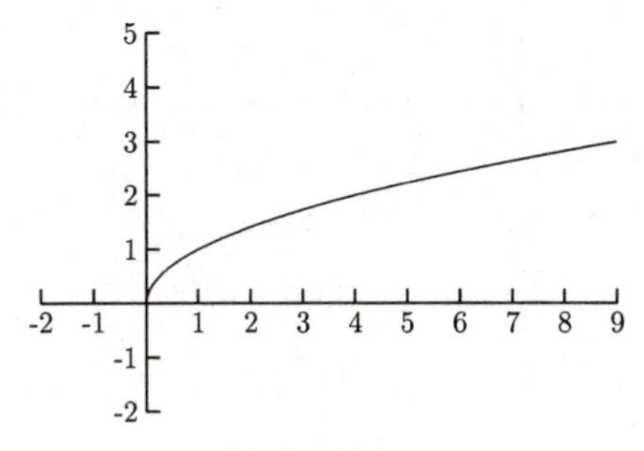

Fig. 8-47.

The graph of $f(x) = \sqrt{x} + 2$ is the graph of $y = \sqrt{x}$ shifted up 2 units.

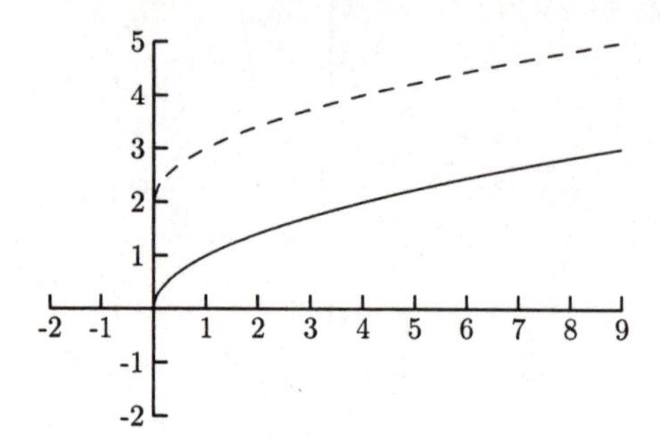

Fig. 8-48.

The graph of $f(x) = \sqrt{-x}$ is the graph of $y = \sqrt{x}$ reflected about the y-axis.

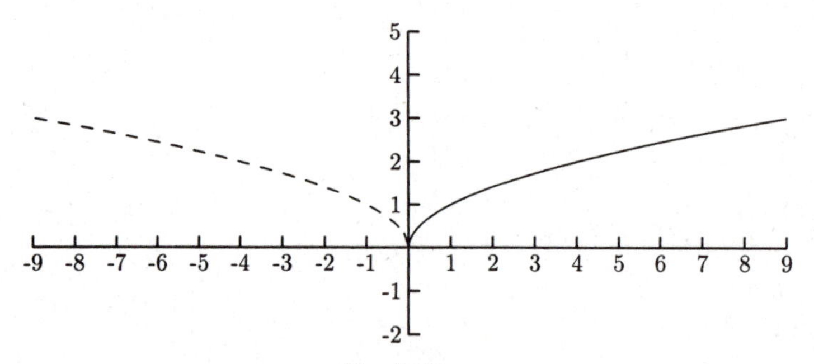

Fig. 8-49.

The graph of $f(x) = -\sqrt{x}$ is the graph of $y = \sqrt{x}$ reflected about the x-axis.

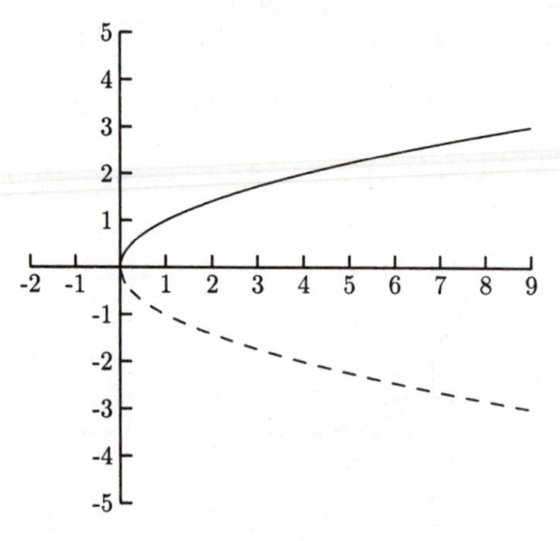

Fig. 8-50.

The graph of $f(x) = \sqrt{2-x}$ is the graph of $y = \sqrt{x}$ first reflected across the y-axis then shifted to the right 2 units.

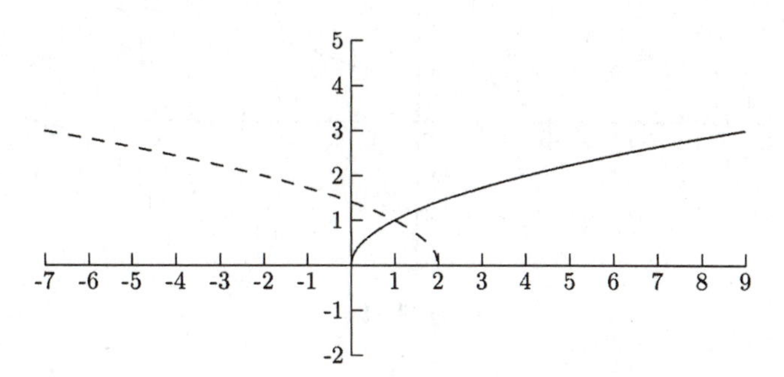

Fig. 8-51.

PRACTICE
Match the graph with the function.

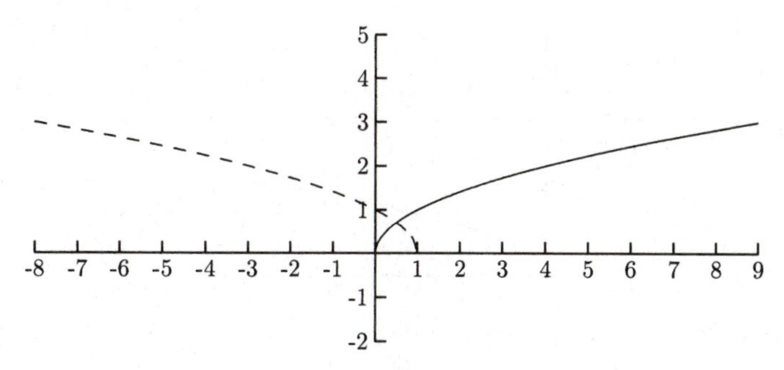

Fig. 8-52.

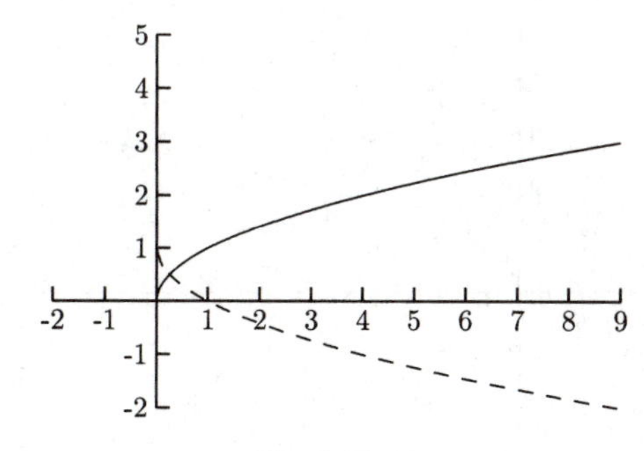

Fig. 8-53.

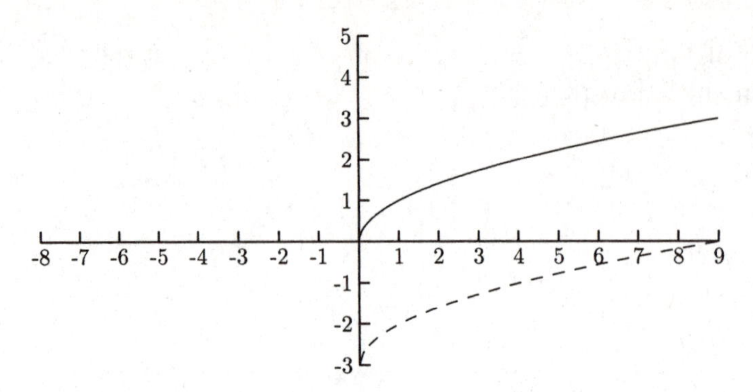

Fig. 8-54.

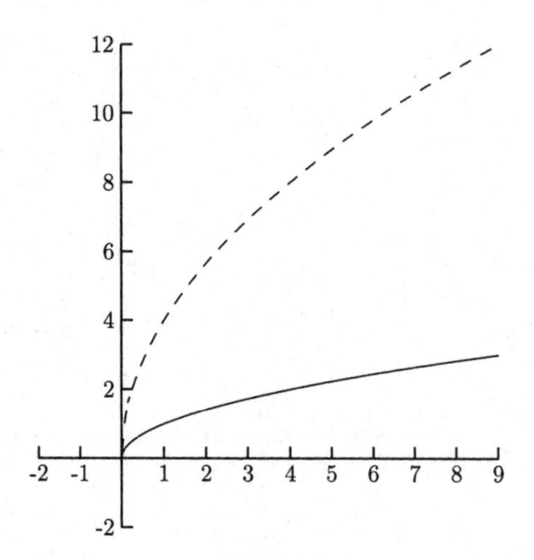

Fig. 8-55.

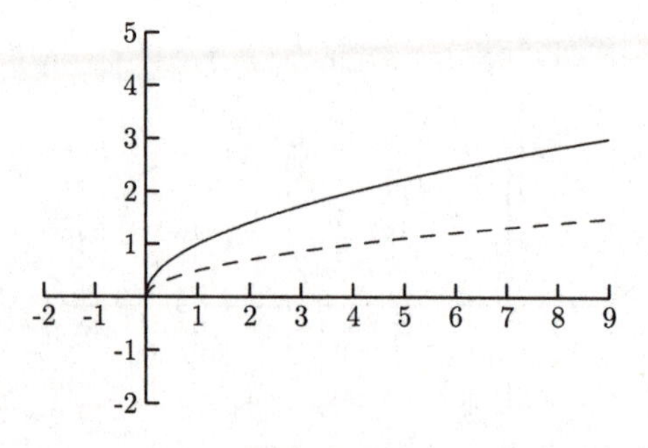

Fig. 8-56.

1. $f(x) = \sqrt{1-x}$
2. $f(x) = -3 + \sqrt{x}$
3. $f(x) = 4\sqrt{x}$
4. $f(x) = 1 - \sqrt{x}$
5. $f(x) = \frac{1}{2}\sqrt{x}$

SOLUTIONS

1. Figure 8-52
2. Figure 8-54
3. Figure 8-55
4. Figure 8-53
5. Figure 8-56

The last new function in this section is the absolute value function, $y = |x|$. Its graph is in the shape of a "V." The graph of $y = |x|$ is shown in Fig. 8-57. As before, the solid graphs in Figs. 8-58–8-65 are the graphs of $y = |x|$ and the dashed graphs are transformations.

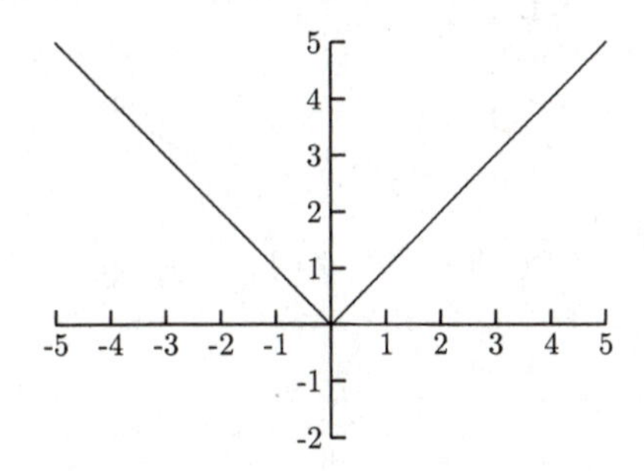

Fig. 8-57.

The graph of $f(x) = 3|x|$ is the graph of $y = |x|$ stretched vertically.

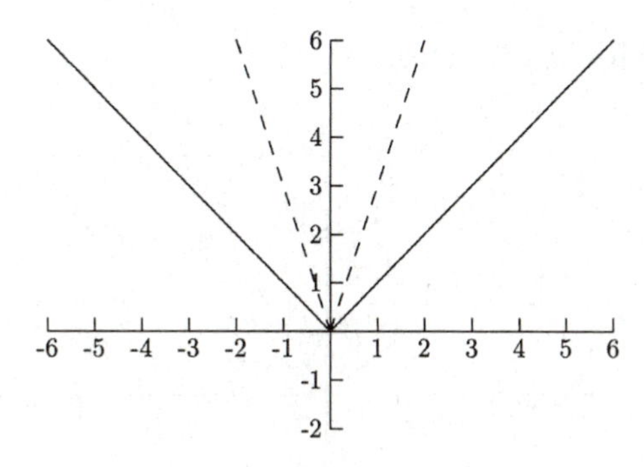

Fig. 8-58.

The graph of $f(x) - \frac{1}{4}|x|$ is the graph of $y = |x|$ reflected across the x-axis and vertically flattened.

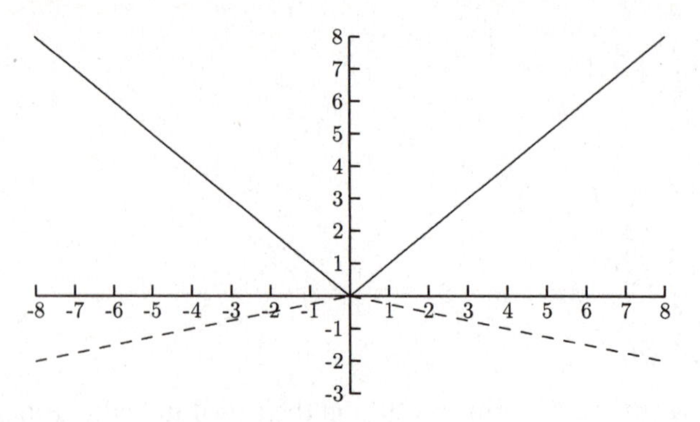

Fig. 8-59.

The graph of $f(x) = 4|x| - 6$ is the graph of $y = |x|$ first stretched vertically then shifted down 6 units.

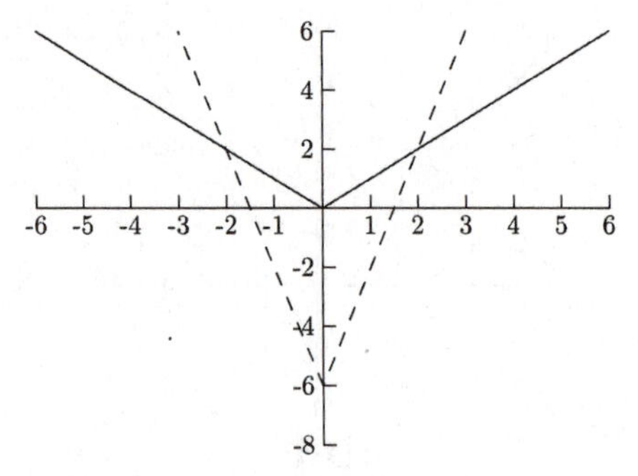

Fig. 8-60.

PRACTICE

Match the graph with the function.

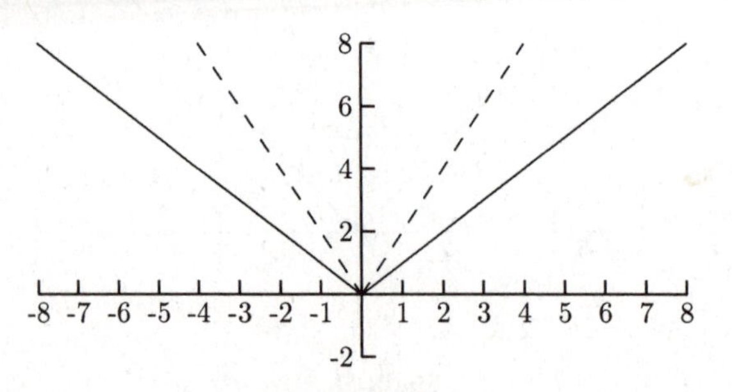

Fig. 8-61.

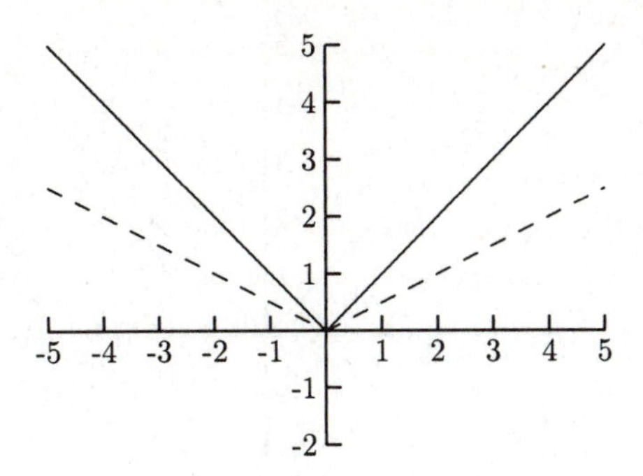

Fig. 8-62.

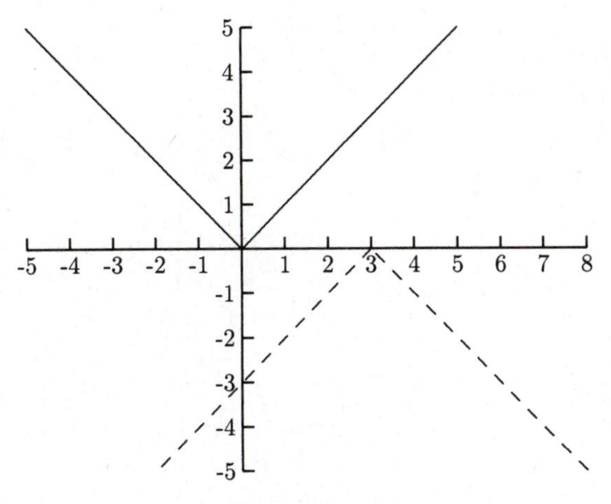

Fig. 8-63.

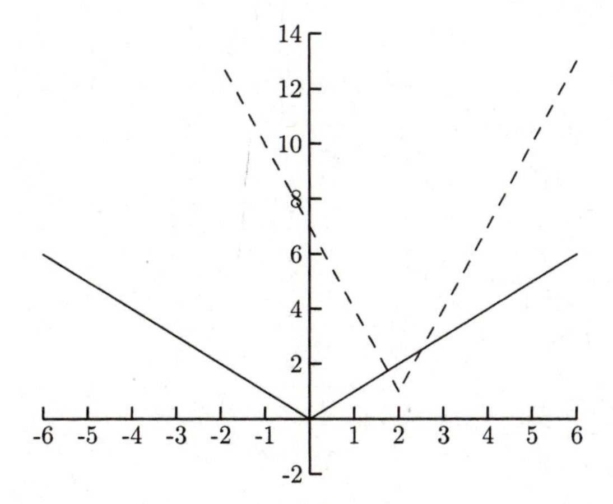

Fig. 8-64.

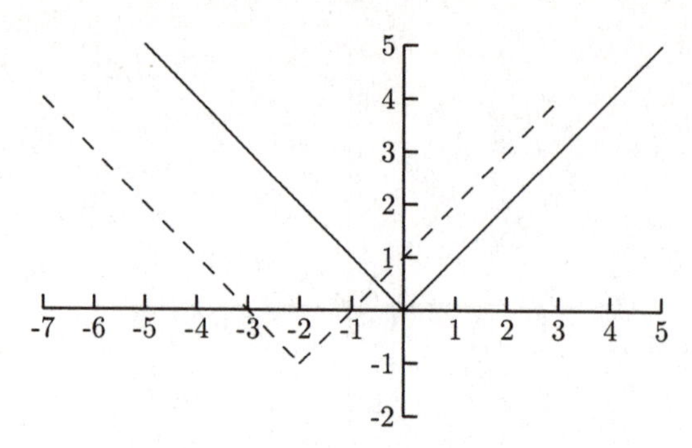

Fig. 8-65.

1. $f(x) = -|x-3|$
2. $f(x) = |x+2| - 1$
3. $f(x) = \frac{1}{2}|x|$
4. $f(x) = 2|x|$
5. $f(x) = 3|x-2| + 1$

SOLUTIONS

1. Figure 8-63
2. Figure 8-65
3. Figure 8-62
4. Figure 8-61
5. Figure 8-64

The next set of practice problems is another set of matching problems but the reference graphs will not be given. These will be transformations of $y = x^2$, $y = x^3$, $y = \sqrt{x}$, $y = |x|$.

PRACTICE

Match the graph to its function.

1. $f(x) = (x-1)^3$
2. $f(x) = 4\sqrt{x}$
3. $f(x) = -x^2$
4. $f(x) = |x+2| + 2$
5. $f(x) = (x+1)^3 - 4$
6. $f(x) = 2 + \sqrt{x}$
7. $f(x) = \sqrt{x-3}$
8. $f(x) = \frac{1}{2}x^2 - 4$
9. $f(x) = -|x| - 4$

SOLUTIONS

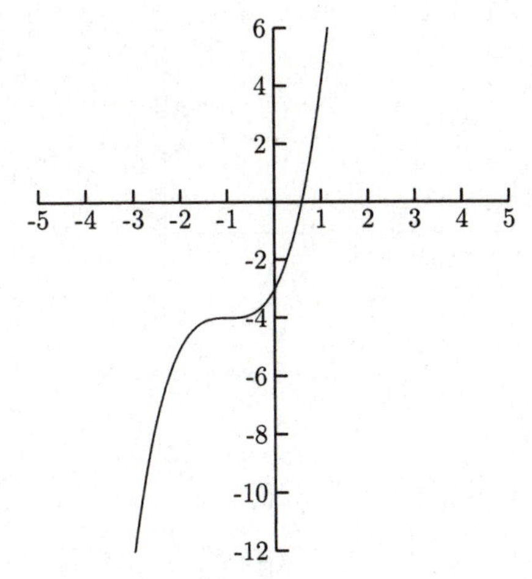

Fig. 8-66.

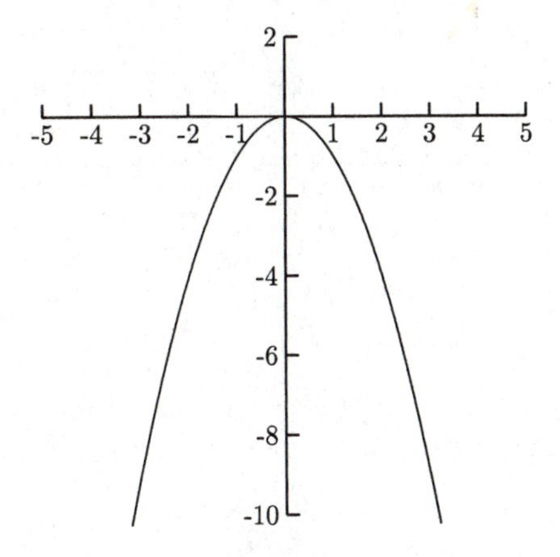

Fig. 8-67.

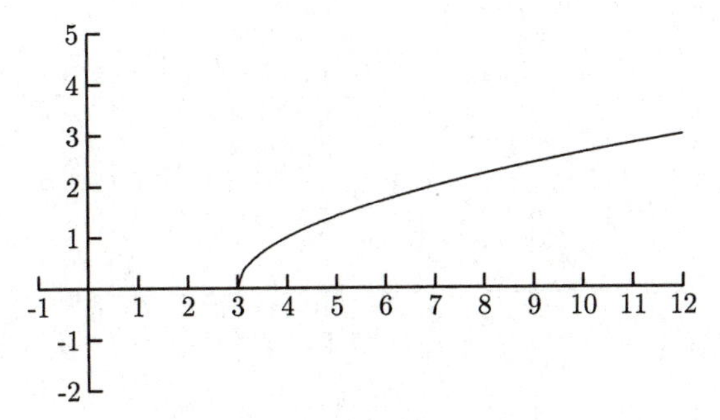

Fig. 8-68.

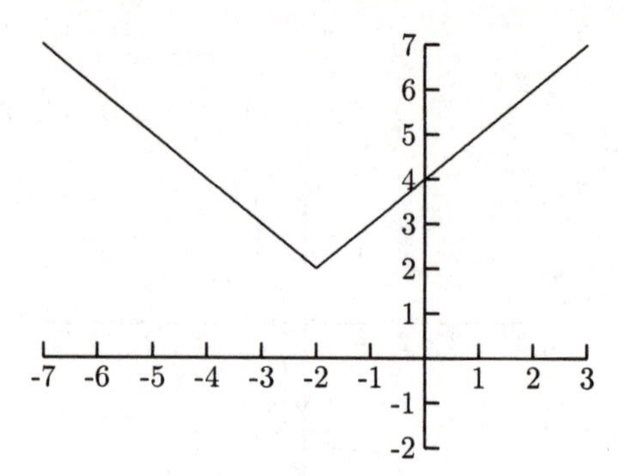

Fig. 8-69.

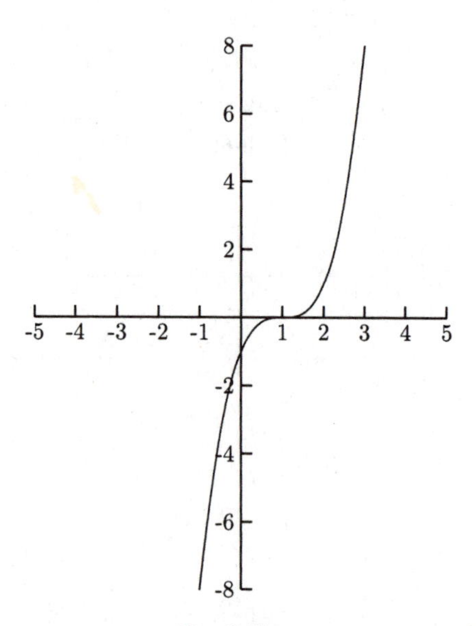

Fig. 8-70.

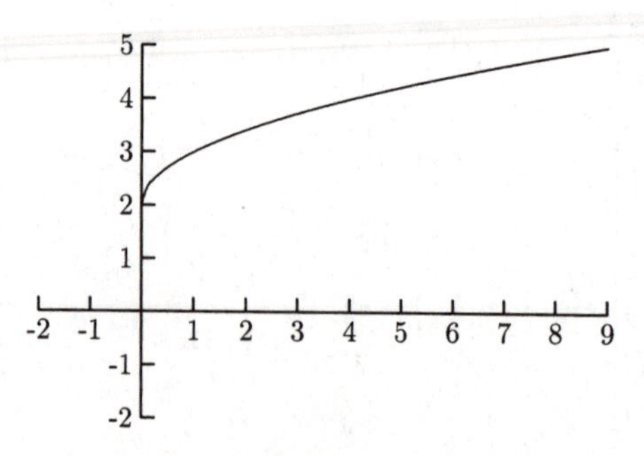

Fig. 8-71.

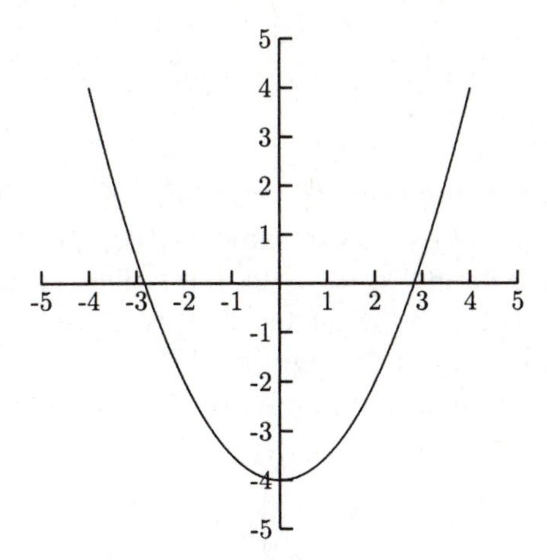

Fig. 8-72.

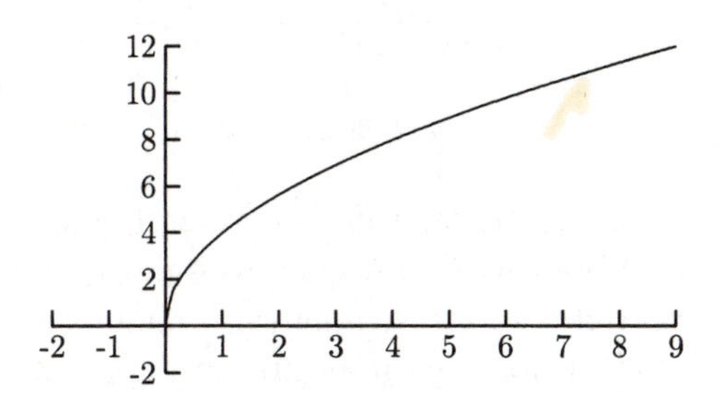

Fig. 8-73.

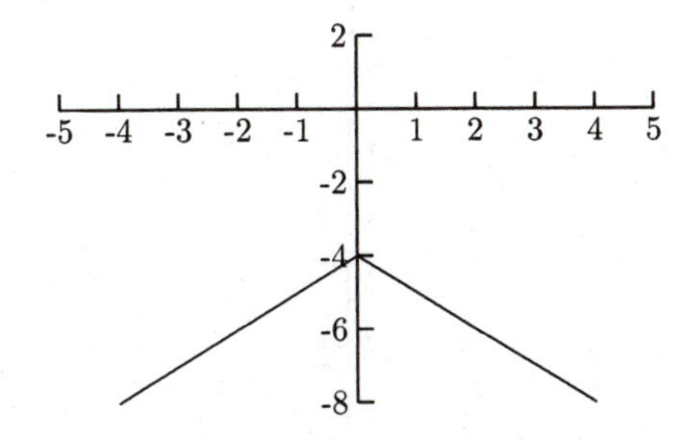

Fig. 8-74.

1. Figure 8-70
2. Figure 8-73
3. Figure 8-67
4. Figure 8-69
5. Figure 8-66
6. Figure 8-71

7. Figure 8-68
8. Figure 8-72
9. Figure 8-74

Sometimes the transformations $f(-x)$ do not change the graph at all. For example, for $f(x) = x^2$, $f(-x)$ is the same as $f(x)$: $f(-x) = (-x)^2 = x^2$, and $g(x) = |x|$ is the same as $g(-x) = |-x| = |x|$. This is because these graphs are *symmetric* with respect to the y-axis. That is, the left half of the graph is a reflection (or mirror image) of the right half. The dashed part of the graph in Fig. 8-75 is a reflection of the solid part of the graph.

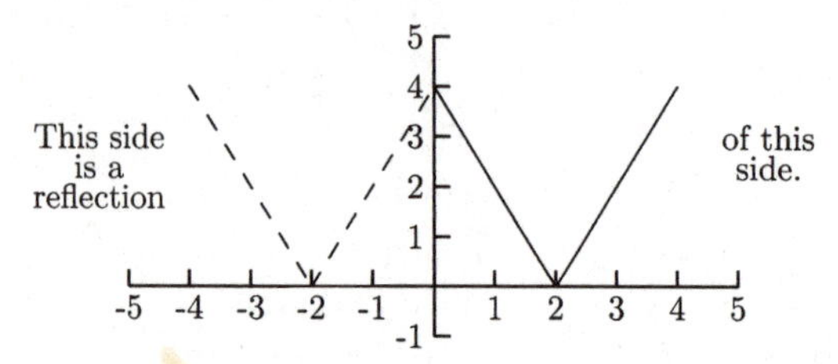

Fig. 8-75.

A function whose vertical reflection $(-f(x))$ is the same as its horizontal reflection $(f(-x))$ is symmetric with respect to the origin. Origin symmetry is a little harder to see than y-axis symmetry. Imagine folding the graph in Fig. 8-76 along the x-axis then again along the y-axis; the upper right-hand part of the graph will be the same as the lower left-hand part of the graph.

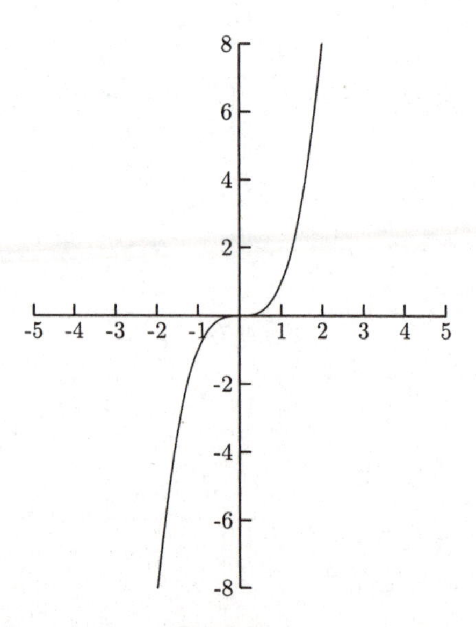

Fig. 8-76.

The graph of $y^2 = x$ in Fig. 8-77 has x-axis symmetry. This symmetry is not as important as y-axis symmetry and origin symmetry because only one function has this kind of symmetry ($y = 0$).

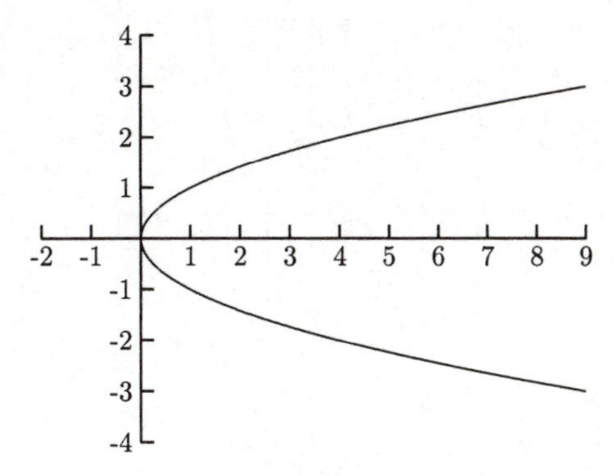

Fig. 8-77.

PRACTICE

Determine whether the graphs are symmetric with respect to the y-axis, x-axis, or origin.

1.

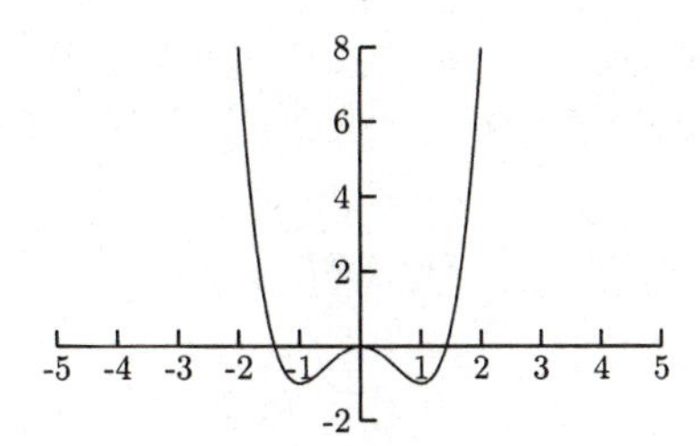

Fig. 8-78.

2.

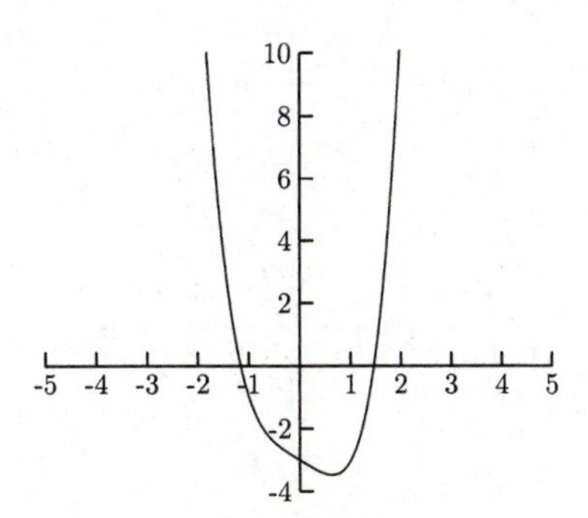

Fig. 8-79.

3.

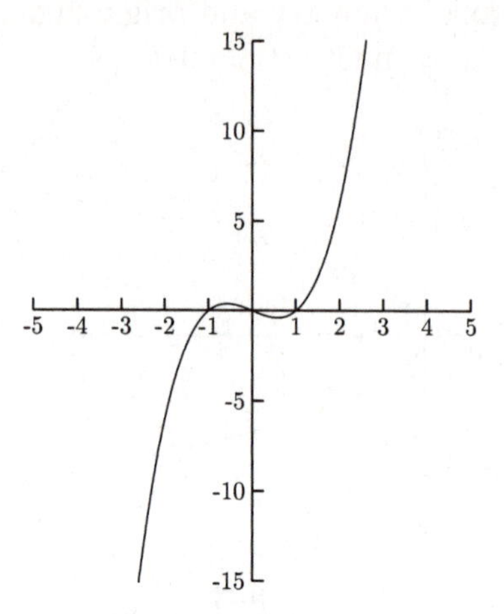

Fig. 8-80.

4.

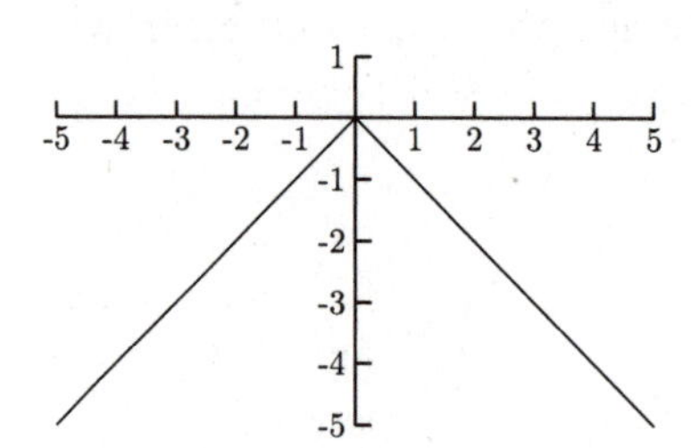

Fig. 8-81.

5.

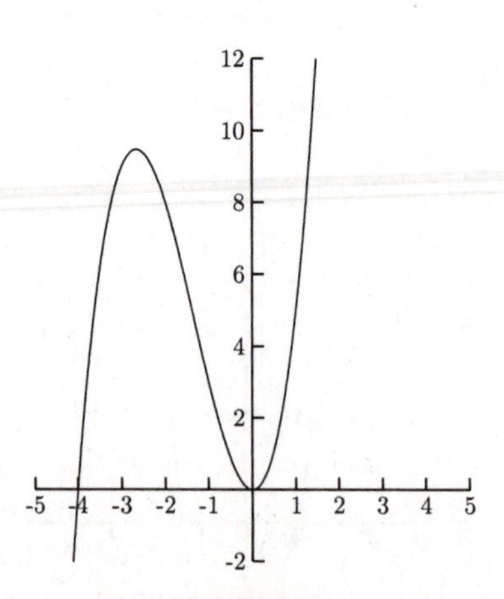

Fig. 8-82.

6.

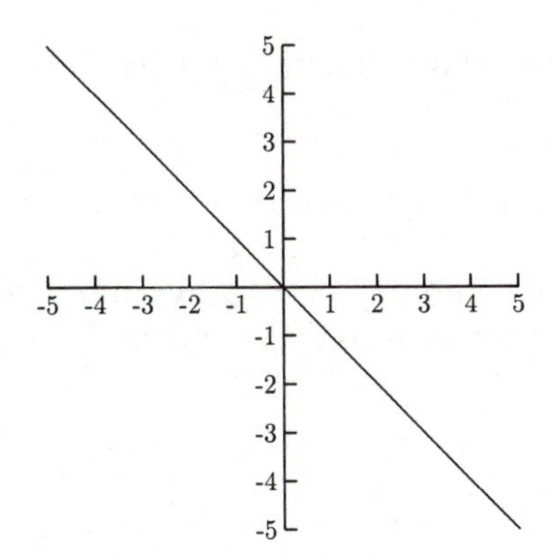

Fig. 8-83.

SOLUTIONS

1. y-axis symmetry
2. No symmetry
3. Origin symmetry
4. y-axis symmetry
5. No symmetry
6. Origin symmetry

We can tell if the graph of a function has y-axis symmetry or origin symmetry by looking at its equation. If we evaluate the function at $-x$ (replace x with $-x$) and the y-values do not change, then the function has y-axis symmetry. Knowing that the graph of a function has y-axis symmetry is very useful when sketching the graph by hand. This is because if (x, y) is on the graph, then $(-x, y)$ is also on the graph. For example, if a function is symmetric with respect to the y-axis and the point $(2, 3)$ is on the graph, then we automatically know that $(-2, 3)$ is also on the graph. The graph of a function has origin symmetry when evaluating the function at $-x$ also changes the sign on the y-values. If a function is symmetric with respect to the origin and the point $(2, 3)$ is on the graph, then the point $(-2, -3)$ is also on the graph.

Functions whose graphs have y-axis symmetry are called *even* functions. Functions whose graphs have origin symmetry are called *odd* functions. A function is even if $f(-x) = f(x)$. This is the mathematical notation for the idea that replacing an x-value with its opposite will not change the y-value. A function is odd if $f(-x) = -f(x)$. This is the mathematical notation for the idea that replacing an x-value with its opposite replaces the y-value with its opposite.

Because so many important functions involve x to powers, the following facts will be useful.

$$(-x)^{\text{even power}} = x^{\text{even power}} \quad \text{and} \quad (-x)^{\text{odd power}} = -x^{\text{odd power}}$$

The product of an even number of negative numbers is positive, and the product of an odd number of negative numbers is negative.

EXAMPLES

- $2(-x)^3 = 2(-1)x^3 = -2x^3$
- $-3(-x)^7 = -3(-1)x^7 = 3x^7$
- $1-(-x) = 1+x$
- $1-(-x)^3 = 1+x^3$
- $5(-x)^6 = 5x^6$
- $-4(-x)^8 = -4x^8$
- $1-(-x)^2 = 1-x^2$
- $(-x)^2-(-x)^3 = x^2+x^3$

PRACTICE

Simplify.

1. $(-x)^3$
2. $(-x)^2$
3. $4(-x)^{100}$
4. $7(-x)^{15}$
5. $-2(-x)^5$
6. $-8(-x)^4$
7. $(-x)^3+2(-x)^2+4(-x)+1$
8. $5(-x)^4-3(-x)^3-6(-x)^2+9(-x)+16$

SOLUTIONS

1. $(-x)^3 = -x^3$
2. $(-x)^2 = x^2$
3. $4(-x)^{100} = 4x^{100}$
4. $7(-x)^{15} = -7x^{15}$
5. $-2(-x)^5 = 2x^5$
6. $-8(-x)^4 = -8x^4$
7. $(-x)^3+2(-x)^2+4(-x)+1 = -x^3+2x^2-4x+1$
8. $5(-x)^4-3(-x)^3-6(-x)^2+9(-x)+16 = 5x^4+3x^3-6x^2-9x+16$

Evaluating a function at $-x$ is often the most difficult part of determining if a function is even, odd, or neither. Once we have evaluated a function at $-x$ and simplified it, we will compare this both to $f(x)$ and to $-f(x)$. If the simplified equation is the same as $f(x)$, then the function is even. If it is the same as $-f(x)$, then the function is odd.

EXAMPLES

Determine whether the functions are even, odd, or neither.

- $f(x) = x^3 + x$

  First we will multiply both sides of $f(x)$ by -1 so that we can compare $f(-x)$ with $f(x)$ and with $-f(x)$.

  $$-f(x) = -(x^3 + x) = -x^3 - x$$

  Now we will find and simplify $f(-x)$ and compare it to $f(x) = x^3 + x$ and to $-f(x) = -x^3 - x$.

  $$\begin{aligned} f(-x) &= (-x)^3 + (-x) \\ &= -x^3 - x \end{aligned}$$

  Because $-f(x)$ and $f(-x)$ are the same, $f(x)$ is an odd function.

- $f(x) = 5x^3 - 4$

  We will multiply both sides of the equation by -1 to find $-f(x)$.

  $$-f(x) = -(5x^3 - 4) = -5x^3 + 4$$

  Now we will find and simplify $f(-x)$.

  $$f(-x) = 5(-x)^3 - 4 = -5x^3 - 4$$

  $f(-x)$ is not the same as $f(x)$ and not the same as $-f(x)$, so $f(x)$ is neither even nor odd.

- $f(x) = -6x^2 + 1$

  $$-f(x) = -(-6x^2 + 1) = 6x^2 - 1$$

  $$f(-x) = -6(-x)^2 + 1 = -6x^2 + 1$$

  $f(x)$ and $f(-x)$ are the same, so $f(x)$ is an even function.

- $f(x) = -9$

  Because the y-value is -9 no matter what x is, $f(-x) = -9$, so $f(x)$ is an even function.

- $f(x) = |2x| + 3$

  $$-f(x) = -(|2x| + 3) = -|2x| - 3$$

  $$f(-x) = |2(-x)| + 3 = |-2x| + 3 = |2x| + 3 \qquad (|-x| = |x|)$$

  $f(x)$ and $f(-x)$ are the same, so $f(x)$ is an even function.

- $g(x) = \dfrac{3}{x^4 - x^2 + 2}$

$$-g(x) = -\frac{3}{x^4 - x^2 + 2}$$

$$g(-x) = \frac{3}{(-x)^4 - (-x)^2 + 2} = \frac{3}{x^4 - x^2 + 2}$$

$g(x)$ and $g(-x)$ are the same, so $g(x)$ is an even function.

PRACTICE

Determine whether the functions are even, odd, or neither.

1. $f(x) = 2x^5 - 6x^3 + x$
2. $f(x) = 8 - x^2$
3. $g(x) = x^3 - x^2 + x - 1$
4. $f(x) = 2x + 1$
5. $h(x) = \dfrac{-8}{(x+1)}$
6. $f(x) = \dfrac{-2x}{(5x^3 + x)}$
7. $g(x) = \sqrt{x^2 + 4}$
8. $h(x) = |3x| - 8$
9. $f(x) = 25$

SOLUTIONS

1. $f(x) = 2x^5 - 6x^3 + x$

$$-f(x) = -(2x^5 - 6x^3 + x) = -2x^5 + 6x^3 - x$$

$$f(-x) = 2(-x)^5 - 6(-x)^3 + (-x) = -2x^5 + 6x^3 - x$$

$-f(x)$ and $f(-x)$ are the same, so $f(x)$ is an odd function.

2. $f(x) = 8 - x^2$

$$-f(x) = -(8 - x^2) = -8 + x^2$$

$$f(-x) = 8 - (-x)^2 = 8 - x^2$$

$f(x)$ and $f(-x)$ are the same, so $f(x)$ is an even function.

3. $g(x) = x^3 - x^2 + x - 1$

$$-g(x) = -(x^3 - x^2 + x - 1) = -x^3 + x^2 - x + 1$$

$$g(-x) = (-x)^3 - (-x)^2 + (-x) - 1 = -x^3 - x^2 - x - 1$$

$g(-x)$ is not the same as $g(x)$ or $-g(x)$, so $g(x)$ is neither even nor odd.

4. $f(x) = 2x + 1$

$$-f(x) = -(2x + 1) = -2x - 1$$
$$f(-x) = 2(-x) + 1 = -2x + 1$$

$f(-x)$ is not the same as $f(x)$ or $-f(x)$, so $f(x)$ is neither even nor odd.

5. $h(x) = \dfrac{-8}{(x+1)}$

$$-h(x) = -\left(\frac{-8}{x+1}\right) = \frac{8}{x+1}$$
$$h(-x) = \frac{-8}{-x+1}$$

$h(-x)$ is not the same as $h(x)$ or $-h(x)$, so $h(x)$ is neither even nor odd.

6. $f(x) = \dfrac{-2x}{(5x^3 + x)}$

$$-f(x) = -\frac{-2x}{5x^3 + x} = \frac{2x}{5x^3 + x}$$
$$f(-x) = \frac{-2(-x)}{5(-x)^3 + (-x)} = \frac{2x}{-5x^3 - x}$$
$$= \frac{2x}{-(5x^3 + x)} = \frac{-2x}{5x^3 + x}$$

$f(-x)$ is the same as $f(x)$, so $f(x)$ is an even function.

7. $g(x) = \sqrt{x^2 + 4}$

$$-g(x) = -\sqrt{x^2 + 4}$$
$$g(-x) = \sqrt{(-x)^2 + 4} = \sqrt{x^2 + 4}$$

$g(-x)$ and $g(x)$ are the same, so $g(x)$ is an even function.

8. $h(x) = |3x| - 8$

$$-h(x) = -(|3x| - 8) = -|3x| + 8$$
$$h(-x) = |3(-x)| - 8 = |-3x| - 8 = |3x| - 8$$

$h(-x)$ and $h(x)$ are the same, so $h(x)$ is an even function.

9. $f(x) = 25$. Because $f(x) = 25$ for *every* x, $f(-x) = 25$ also, so $f(x)$ is an even function.

Combining Functions

The vast majority of functions studied in algebra and calculus are some combination of only a handful of basic functions, most of them introduced in this book. The most obvious combination of two or more functions are arithmetic combinations: adding, subtracting, multiplying, and dividing. Suppose two functions $f(x)$ and $g(x)$ are given.

- $(f+g)(x)$ means $f(x)+g(x)$.
- $(f-g)(x)$ means $f(x)-g(x)$.
- $(fg)(x)$ means $f(x)g(x)$.
- $(f/g)(x)$ means $f(x)/g(x)$.

EXAMPLE

- $f(x) = 3x - 4$ and $g(x) = x^2 + x$

$$\begin{aligned}(f+g)(x) &= (3x-4)+(x^2+x)\\ &= x^2+4x-4\end{aligned}$$

$$\begin{aligned}(f-g)(x) &= (3x-4)-(x^2+x) = 3x-4-x^2-x\\ &= -x^2+2x-4\end{aligned}$$

$$\begin{aligned}(fg)(x) &= (3x-4)(x^2+x) = 3x^3+3x^2-4x^2-4x\\ &= 3x^3-x^2-4x\end{aligned}$$

$$\frac{f}{g}(x) = \frac{3x-4}{x^2+x}$$

PRACTICE

Find $(f+g)(x)$, $(f-g)(x)$, $(fg)(x)$, and $(f/g)(x)$.

1. $f(x) = x+6$ and $g(x) = -2x+4$
2. $f(x) = \sqrt{x+6}$ and $g(x) = x-2$
3. $f(x) = \dfrac{x+2}{3x-1}$ and $g(x) = \dfrac{3x+1}{x-2}$

SOLUTIONS

1.

$$(f+g)(x) = (x+6)+(-2x+4) = -x+10$$

$$(f-g)(x) = (x+6)-(-2x+4) = x+6+2x-4 = 3x+2$$

$$(fg)(x) = (x+6)(-2x+4) = -2x^2 + 4x - 12x + 24$$
$$= -2x^2 - 8x + 24$$
$$\frac{f}{g}(x) = \frac{x+6}{-2x+4}$$

2.

$$(f+g)(x) = \sqrt{x+6} + x - 2$$
$$(f-g)(x) = \sqrt{x+6} - (x-2) = \sqrt{x+6} - x + 2$$
$$(fg)(x) = \sqrt{x+6}(x-2) = (x-2)\sqrt{x+6}$$
$$\frac{f}{g}(x) = \frac{\sqrt{x+6}}{x-2}$$

3.

$$(f+g)(x) = \frac{x+2}{3x-1} + \frac{3x+1}{x-2} = \frac{x+2}{3x-1} \cdot \frac{x-2}{x-2} + \frac{3x+1}{x-2} \cdot \frac{3x-1}{3x-1}$$
$$= \frac{x^2-4}{(3x-1)(x-2)} + \frac{9x^2-1}{(3x-1)(x-2)} = \frac{10x^2-5}{3x^2-7x+2}$$
$$(f-g)(x) = \frac{x+2}{3x-1} - \frac{3x+1}{x-2}$$
$$= \frac{x^2-4}{(3x-1)(x-2)} - \frac{9x^2-1}{(3x-1)(x-2)}$$
$$= \frac{x^2-4-(9x^2-1)}{3x^2-7x+2} = \frac{-8x^2-3}{3x^2-7x+2}$$
$$(fg)(x) = \frac{x+2}{3x-1} \cdot \frac{3x+1}{x-2} = \frac{(x+2)(3x+1)}{(3x-1)(x-2)}$$
$$= \frac{3x^2+7x+2}{3x^2-7x+2}$$
$$\frac{f}{g}(x) = \frac{(x+2)/(3x-1)}{(3x+1)/(x-2)}$$
$$= \frac{x+2}{3x-1} \div \frac{3x+1}{x-2} = \frac{x+2}{3x-1} \cdot \frac{x-2}{3x+1}$$
$$= \frac{x^2-4}{9x^2-1}$$

Function Composition

Two (or more) functions can be combined by composing one function with another. We have performed function composition without calling it by its name. The basic idea behind function composition is that one function is evaluated at another. For example, if $f(x) = 4x + 7$ and $g(x) = 2x - 3$, then "to evaluate f at g" means to compute $f(2x - 3)$. Remember that to "evaluate a function" means to substitute the quantity in the parentheses for x.

$$f(2x - 3) = 4(2x - 3) + 7 = 8x - 12 + 7 = 8x - 5$$

The notation for this operation is $f \circ g(x)$. By definition, $f \circ g(x)$ means $f(g(x))$. This operation is not commutative. That is, $f \circ g(x)$ is usually not the same as $g \circ f(x)$.

EXAMPLES

Find $f \circ g(x)$ and $g \circ f(x)$.

- $f(x) = x^2 + 4x - 3$ and $g(x) = 2x - 5$

$f \circ g(x)$

$= f(g(x))$	This is the definition of $f \circ g(x)$
$= f(2x - 5)$	Replace $g(x)$ with $2x - 5$
$= (2x - 5)^2 + 4(2x - 5) - 3$	Replace x with $2x - 5$ in $f(x)$
$= (2x - 5)(2x - 5) + 4(2x - 5) - 3$	
$= 4x^2 - 20x + 25 + 8x - 20 - 3$	
$= 4x^2 - 12x + 2$	

$g \circ f(x) = g(f(x))$	This is the definition of $g \circ f(x)$
$= g(x^2 + 4x - 3)$	Replace $f(x)$ with $x^2 + 4x - 3$
$= 2(x^2 + 4x - 3) - 5$	Replace x with $x^2 + 4x - 3$ in $g(x)$
$= 2x^2 + 8x - 6 - 5$	
$= 2x^2 + 8x - 11$	

- $f(x) = 8 - 5x$ and $g(x) = x + 4$

$$
\begin{aligned}
f \circ g(x) &= f(g(x)) \\
&= f(x+4) \\
&= 8 - 5(x+4) \\
&= 8 - 5x - 20 = -5x - 12 \\
g \circ f(x) &= g(f(x)) = g(8 - 5x) \\
&= 8 - 5x + 4 = -5x + 12
\end{aligned}
$$

- $f(x) = \sqrt{x}$ and $g(x) = x^2 + 2x + 2$

$$
\begin{aligned}
f \circ g(x) &= f(g(x)) = f(x^2 + 2x + 2) \\
&= \sqrt{x^2 + 2x + 2} \\
g \circ f(x) &= g(f(x)) = g(\sqrt{x}) \\
&= (\sqrt{x})^2 + 2\sqrt{x} + 2 = x + 2\sqrt{x} + 2
\end{aligned}
$$

- $f(x) = 16x - 1$ and $g(x) = 1/(x+2)$

$$
\begin{aligned}
f \circ g(x) &= f(g(x)) = f\left(\frac{1}{x+2}\right) \\
&= 16\left(\frac{1}{x+2}\right) - 1 \\
&= \frac{16}{x+2} - 1 \text{ or } \frac{14 - x}{x+2} \\
g \circ f(x) &= g(f(x)) = g(16x - 1) \\
&= \frac{1}{(16x-1)+2} = \frac{1}{16x+1}
\end{aligned}
$$

PRACTICE

Find $f \circ g(x)$ and $g \circ f(x)$.

1. $f(x) = x + 2$ and $g(x) = x^2 - 4$
2. $f(x) = x^2$ and $g(x) = \sqrt{2x - 4}$
3. $f(x) = 3x^2 + x$ and $g(x) = 1/x$
4. $f(x) = 1/x$ and $g(x) = 2/(x - 1)$

SOLUTIONS

1.

$$f \circ g(x) = f(g(x)) = f(x^2 - 4)$$
$$= (x^2 - 4) + 2 = x^2 - 2$$
$$g \circ f(x) = g(f(x)) = g(x + 2)$$
$$= (x + 2)^2 - 4 = (x + 2)(x + 2) - 4$$
$$= x^2 + 4x + 4 - 4 = x^2 + 4x$$

2.

$$f \circ g(x) = f(g(x)) = f(\sqrt{2x - 4})$$
$$= (\sqrt{2x - 4})^2 = 2x - 4$$
$$g \circ f(x) = g(f(x)) = g(x^2)$$
$$= \sqrt{2x^2 - 4}$$

3.

$$f \circ g(x) = f(g(x)) = f\left(\frac{1}{x}\right)$$
$$= 3\left(\frac{1}{x}\right)^2 + \frac{1}{x} = \frac{3}{x^2} + \frac{1}{x}$$
$$g \circ f(x) = g(f(x)) = g(3x^2 + x)$$
$$= \frac{1}{3x^2 + x}$$

4.

$$f \circ g(x) = f(g(x)) = f\left(\frac{2}{x - 1}\right)$$
$$= \frac{1}{2/(x - 1)} = 1 \div \frac{2}{x - 1}$$
$$= 1 \cdot \frac{x - 1}{2} = \frac{x - 1}{2}$$

$$g \circ f(x) = g(f(x)) = g\left(\frac{1}{x}\right)$$
$$= \frac{2}{1/x - 1} = \frac{2}{1/x - x/x}$$
$$= \frac{2}{(1-x)/x} = 2 \div \frac{1-x}{x}$$
$$= 2 \cdot \frac{x}{1-x} = \frac{2x}{1-x}$$

There is no reason a function cannot be evaluated at itself. In other words, we can compute $f \circ f(x)$ for some function $f(x)$. For example, suppose $f(x) = 3x^2 + 5$.

$$f \circ f(x) = f(f(x)) = f(3x^2 + 5)$$
$$= 3(3x^2 + 5)^2 + 5 = 3(3x^2 + 5)(3x^2 + 5) + 5$$
$$= 3(9x^2 + 30x + 25) + 5 = 27x^2 + 90x + 80$$

Sometimes we need only to compose functions at a single x-value. For example, if $f(x) = 8 - 5x$ and $g(x) = x + 4$, we might only need to find $f \circ g(x)$ for $x = 2$. To do this, let $x = 2$ in $g(x)$: $g(2) = 2 + 4 = 6$. Now let $x = 6$ in $f(x)$: $f(6) = 8 - 5(6) = -22$. We have just found that $f \circ g(2) = -22$. Of course, if we know that $f \circ g(x) = -5x - 10$ (as we computed earlier), we could evaluate $f \circ g(2)$ by letting $x = 2$ in $-5x - 12$. For the following examples and practice problems, both $f \circ g(x)$ and $g \circ f(x)$ were computed as above.

EXAMPLES

- Find $f \circ g(1)$ and $g \circ f(-2)$ for $f(x) = x^2 + 4x - 3$ and $g(x) = 2x - 5$.

  First find $g(1)$: $g(1) = 2(1) - 5 = -3$. Now let $x = -3$ in $f(x)$: $f(-3) = (-3)^2 + 4(-3) - 3 = -6$, so $f \circ g(1) = f(g(1)) = f(-3) = -6$.

$$g \circ f(-2) = g(f(-2))$$
$$= g(-7) \quad (f(-2) = (-2)^2 + 4(-2) - 3 = -7)$$
$$= -19 \quad (g(-7) = 2(-7) - 5 = -19)$$

- Find $f \circ g(0)$ and $g \circ f(9)$ for $f(x) = \sqrt{x}$ and $g(x) = x^2 + 2x + 2$.

$$f \circ g(0) = f(g(0))$$
$$= f(2) \quad (g(0) = 0^2 + 2(0) + 2 = 2)$$
$$= \sqrt{2}$$

$$g \circ f(9) = g(f(9)) = g(3) \quad (f(9) = \sqrt{9} = 3)$$
$$= 3^2 + 2(3) + 2 = 17$$

PRACTICE

1. Find $f \circ g(-3)$ and $g \circ f(5)$ for $f(x) = x + 2$ and $g(x) = x^2 - 4$.
2. Find $f \circ g(5)$ and $g \circ f(2)$ for $f(x) = x^2$ and $g(x) = \sqrt{2x - 4}$.
3. Find $f \circ g(-2)$ and $g \circ f(2)$ for $f(x) = 1/x$ and $g(x) = 2/(x - 1)$.

SOLUTIONS

1.

$$\begin{aligned} f \circ g(-3) &= f(g(-3)) \\ &= f(5) \qquad (g(-3) = (-3)^2 - 4 = 5) \\ &= 5 + 2 = 7 \\ g \circ f(5) &= g(f(5)) \\ &= g(7) \qquad (f(5) = 5 + 2 = 7) \\ &= 7^2 - 4 = 45 \end{aligned}$$

2.

$$\begin{aligned} f \circ g(5) &= f(g(5)) \\ &= f(\sqrt{6}) \qquad (g(5) = \sqrt{2(5) - 4} = \sqrt{6}) \\ &= (\sqrt{6})^2 = 6 \\ g \circ f(2) &= g(4) \qquad (f(2) = 2^2 = 4) \\ &= \sqrt{2(4) - 4} = \sqrt{4} = 2 \end{aligned}$$

3.

$$\begin{aligned} f \circ g(-2) &= f(g(-2)) \\ &= f\left(-\frac{2}{3}\right) \qquad \left(g(-2) = \frac{2}{-2 - 1} = -\frac{2}{3}\right) \\ &= \frac{1}{-2/3} = 1 \div -\frac{2}{3} = 1 \cdot -\frac{3}{2} = -\frac{3}{2} \\ g \circ f(2) &= g(f(2)) \\ &= g\left(\frac{1}{2}\right) \qquad \left(f(2) = \frac{1}{2}\right) \\ &= \frac{2}{1/2 - 1} = \frac{2}{-1/2} \\ &= 2 \div \left(-\frac{1}{2}\right) = 2 \cdot (-2) = -4 \end{aligned}$$

Graphs can be used to evaluate the composition of functions at a particular x-value. For $f \circ g(x)$, the y-value of $g(x)$ becomes the x-value for $f(x)$. In other words, $f \circ g(x)$ is the y-value for $f(x)$ whose x-value is $g(x)$. For example, if we are asked to find $f \circ g(1)$, we need to look on the graph of $g(x)$ for the point whose x-coordinate is 1, and put this y-coordinate in for x in $f(x)$.

EXAMPLE

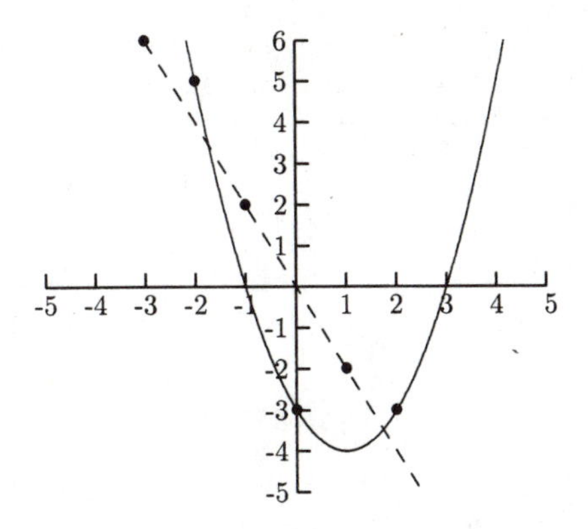

Fig. 8-84.

The solid graph is the graph of $f(x)$ and the dashed graph is the graph of $g(x)$.

- Find $f \circ g(1)$, $f \circ g(-1)$, $g \circ f(0)$

  For $f \circ g(1) = f(g(1))$, we need to look on the graph of $g(x)$ for the point whose x-coordinate is 1. This point is $(1, -2)$. Now we will use $y = -2$ as $x = -2$ in $f(x)$. We need to look on the graph of $f(x)$ for the point whose x-coordinate is -2. That point is $(-2, 5)$. This means that $f \circ g(1) = 5$.

  For $f \circ g(-1) = f(g(-1))$, we will look on the graph of $g(x)$ for the point whose x-coordinate is -1. This point is $(-1, 2)$. Now we will use $y = 2$ as $x = 2$ in $f(x)$. We need to look on the graph of $f(x)$ for the point whose x-coordinate is 2. That point is $(2, -3)$. This means that $f \circ g(-1) = -3$.

  For $g \circ f(0) = g(f(0))$, we need to look on the graph of $f(x)$ for the point whose x-coordinate is 0. This point is $(0, -3)$. Now we will look on the graph of $g(x)$ for the point whose x-coordinate is -3. That point is $(-3, 6)$. This means that $g \circ f(0) = 6$.

PRACTICE

The graph of $f(x)$ is the solid graph, and the graph of $g(x)$ is the dashed graph.

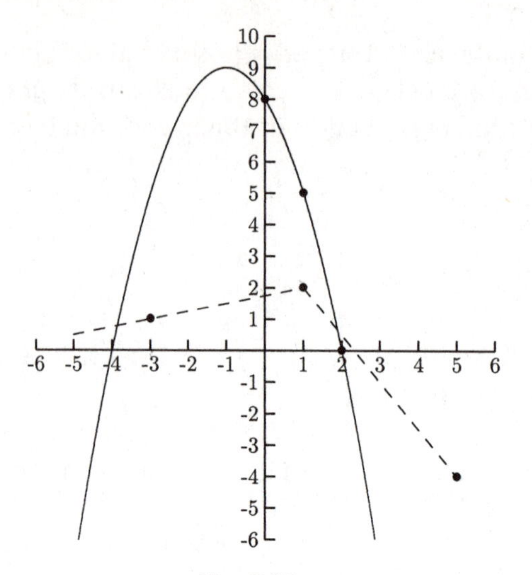

Fig. 8-85.

1. Find $f \circ g(-3)$
2. Find $g \circ f(1)$
3. Find $f \circ g(1)$
4. Find $f \circ f(2)$

SOLUTIONS

1. The point $(-3, 1)$ is on the graph of $g(x)$. The point on the graph of $f(x)$ whose x-coordinate is 1 is $(1, 5)$. This makes $f \circ g(-3) = 5$.
2. The point $(1, 5)$ is on the graph of $f(x)$. The point on the graph of $g(x)$ whose x-coordinate is 5 is $(5, -4)$. This makes $g \circ f(1) = -4$.
3. The point $(1, 2)$ is on the graph of $g(x)$. The point on the graph of $f(x)$ whose x-coordinate is 2 is $(2, 0)$. This makes $f \circ g(1) = 0$.
4. The point $(2, 0)$ is on the graph of $f(x)$. The point on the graph of $f(x)$ whose x-coordinate is 0 is $(0, 8)$. This makes $f \circ f(2) = 8$.

THE DOMAIN OF THE COMPOSITION OF FUNCTIONS

Finding the domain of the composition of functions is a little more complicated than finding the domain for the other combinations. For example, $x = -1$ is in the domain of both $f(x) = 1/x$ and $g(x) = x + 1$ but not in the domain of $f \circ g(x)$. Why not? We cannot let $f(x)$ be evaluated at $x = 0$, and $g(-1) = -1 + 1 = 0$, so $f \circ g(-1)) = f(g(-1)) = f(0) = \frac{1}{0}$ is not defined.

For any functions $f(x)$ and $g(x)$, the domain of $f \circ g(x)$ is the domain of $g(x)$ after deleting any x-values whose y-values are not allowed for $f(x)$.

In the above example, $x = -1$ is in the domain of $g(x)$ but $x = g(-1)$ is not in the domain of $f(x)$. To find the domain of $f \circ g(x)$, we first need to find the domain of $g(x)$. Next, we will evaluate $f(x)$ at $g(x)$. Before simplifying, see which, if any, x-values need to be removed from the domain of $g(x)$.

EXAMPLES

Find the domain of $f \circ g(x)$.

- $f(x) = 1/(x-1)$ and $g(x) = x^2$
  The domain for $g(x)$ is all x. Are there any x-values that need to be removed? This means, are there any y-values for $g(x)$ that cause a zero in the denominator for $f(x)$? Evaluate $f(g(x))$.

$$f(g(x)) = f(x^2) = \frac{1}{x^2 - 1}$$

  Because $x = 1$ and $x = -1$ cause the denominator to be 0, we must remove them from the domain of $g(x)$. The domain of $f \circ g(x)$ is all x except 1 and -1. In interval notation, this set is $(-\infty, -1) \cup (-1, 1) \cup (1, \infty)$.
- $f(x) = x^2$ and $g(x) = \sqrt{x+1}$
  The domain for $g(x)$ is all $x \geq -1$. Are there any x-values that need to be removed from $[-1, \infty)$? We need for $f \circ g(x) = f(g(x)) = f(\sqrt{x+1}) = (\sqrt{x+1})^2$ to be defined. Since $\sqrt{x+1}$ is defined for all $x \geq -1$, we do not need to remove any numbers from $[-1, \infty)$, so the domain for $f \circ g(x)$ is $[-1, \infty)$.
- $f(x) = 1/(x+1)$ and $g(x) = 1/(x-1)$
  The domain for $g(x)$ is all x except 1.

$$f \circ g(x) = f(g(x)) = \frac{1}{1/(x-1) + 1}$$

  We cannot allow the denominator, $1/(x-1) + 1$, to be 0, so we need to remove any x-value from the domain of $g(x)$ that makes $1/(x-1) + 1 = 0$.

$$\frac{1}{x-1} + 1 = 0$$

$$\frac{1}{x-1} + 1 \cdot \frac{x-1}{x-1} = 0$$

$$\frac{1 + (x-1)}{x} = \frac{x}{x-1} = 0$$

The fraction $x/(x-1)$ is 0 when the numerator, x, is 0. This makes the domain for $f \circ g(x)$ all x except 1 and 0. In interval notation, this is $(-\infty, 0) \cup (0, 1) \cup (1, \infty)$.

This is a good example of why we need to look at $f \circ g(x)$ *before* it is simplified because $f \circ g(x)$ simplifies to $(x-1)/x$. This simplification hides the fact that we cannot allow x to equal 1.

PRACTICE

Find the domain for $f \circ g(x)$. Give solutions in interval notation.

1. $f(x) = 4x - 5$ and $g(x) = \sqrt{x+3}$
2. $f(x) = x^2$ and $g(x) = \sqrt{3x+5}$
3. $f(x) = 1/(x+1)$ and $g(x) = x - 1$
4. $f(x) = 1/x$ and $g(x) = \sqrt{x}$

SOLUTIONS

1. The domain for $g(x)$ is $x \geq -3$. Evaluate $f \circ g(x)$: $f(\sqrt{x+3}) = 4\sqrt{x+3} - 5$. We do not need to remove any x-values, so the domain is $[-3, \infty)$.
2. The domain for $g(x)$ is $x \geq -\frac{5}{3}$. Evaluate $f \circ g(x)$: $f(\sqrt{3x+5}) = (\sqrt{3x+5})^2 = 3x + 5$. We do not need to remove any x-values. The domain is $[-\frac{5}{3}, \infty)$.
3. The domain for $g(x)$ is all x. Evaluate $f \circ g(x)$

$$f \circ g(x) = f(g(x)) = f(x-1) = \frac{1}{(x-1)+1} = \frac{1}{x}$$

We cannot allow x to be 0, so we need to remove $x = 0$ from the domain of $g(x)$. The domain is $(-\infty, 0) \cup (0, \infty)$.

4. The domain for $g(x)$ is $x \geq 0$. Evaluate $f \circ g(x)$.

$$f \circ g(x) = f(\sqrt{x}) = \frac{1}{\sqrt{x}}$$

We cannot allow $\sqrt{x} = 0$, so we need to remove $x = 0$ from the domain of $g(x)$. The domain is $(0, \infty)$.

A chain of three or more functions can be composed together. The steps in evaluating these chains are the same as in evaluating the composition of two functions. By definition, $f \circ g \circ h(x) = f \circ g(h(x)) = f(g(h(x)))$. In this expression, we call f the *outside function*, g the *middle function*, and

h the *inside function*. In $f \circ g \circ h(x)$, we begin by finding $g(h(x))$, then evaluating $f(x)$ at $g(h(x))$.

EXAMPLES

Find $f \circ g \circ h(x)$.

- $f(x) = \sqrt{x}$, $g(x) = 5x + 6$, and $h(x) = 8x^2 + x + 4$

  First we will find $g \circ h(x) = g(h(x))$.

$$g(h(x)) = g(8x^2 + x + 4) = 5(8x^2 + x + 4) + 6$$
$$= 40x^2 + 5x + 26$$

  Now we can evaluate $f(x)$ at $40x^2 + 5x + 26$.

$$f \circ g \circ h(x) = f(40x^2 + 5x + 26) = \sqrt{40x^2 + 5x + 26}$$

- $f(x) = x^{10}$, $g(x) = 2x^2 - 9$, and $h(x) = 3x + 8$

$$g \circ h(x) = g(h(x)) = g(3x + 8)$$
$$= 2(3x + 8)^2 - 9 = 2(3x + 8)(3x + 8) - 9$$
$$= 18x^2 + 96x + 119$$

  Evaluate $f(x)$ at $18x^2 + 96x + 119$

$$f \circ g \circ h(x) = f(18x^2 + 96x + 119) = (18x^2 + 96x + 119)^{10}$$

PRACTICE

Find $f \circ g \circ h(x)$.

1. $f(x) = 5x - 8$, $g(x) = 2x + 9$, and $h(x) = 6x^2 + 3x + 1$
2. $f(x) = \sqrt{3x + 4}$, $g(x) = x^2 - x - 3$, and $h(x) = x + 1$
3. $f(x) = x^{12}$, $g(x) = 3x^2 + x - 4$, and $h(x) = 9x + 2$
4. $f(x) = 1/x$, $g(x) = x^2$, and $h(x) = x^2 - 4$

SOLUTIONS

1.

$$g \circ h(x) = g(h(x)) = g(6x^2 + 3x + 1) = 2(6x^2 + 3x + 1) + 9$$
$$= 12x^2 + 6x + 11$$
$$f(12x^2 + 6x + 11) = 5(12x^2 + 6x + 11) - 8 = 60x^2 + 30x + 47$$
$$f \circ g \circ h(x) = 60x^2 + 30x + 47$$

2.

$$g \circ h(x) = g(h(x)) = g(x+1) = (x+1)^2 - (x+1) - 3$$
$$= (x+1)(x+1) - x - 1 - 3 = x^2 + x - 3$$

$$f(x^2 + x - 3) = \sqrt{3(x^2 + x - 3) + 4} = \sqrt{3x^2 + 3x - 5}$$

$$f \circ g \circ h(x) = \sqrt{3x^2 + 3x - 5}$$

3.

$$g \circ h(x) = g(h(x)) = g(9x+2) = 3(9x+2)^2 + (9x+2) - 4$$
$$= 3(9x+2)(9x+2) + 9x + 2 - 4 = 243x^2 + 117x + 10$$

$$f(243x^2 + 117x + 10) = (243x^2 + 117x + 10)^{12}$$

$$f \circ g \circ h(x) = (243x^2 + 117x + 10)^{12}$$

4.

$$g \circ h(x) = g(h(x)) = g(x^2 - 4) = (x^2 - 4)^2$$
$$= (x^2 - 4)(x^2 - 4) = x^4 - 8x^2 + 16$$

$$f(x^4 - 8x^2 + 16) = \frac{1}{x^4 - 8x^2 + 16}$$
$$f \circ g \circ h(x) = \frac{1}{x^4 - 8x^2 + 16}$$

Chapter 8 Review

1. The graph of $f(x+3) - 6$ is the graph of $f(x)$
   a) shifted to the left 3 units and down 6 units.
   b) shifted to the right 3 units and up 6 units.
   c) shifted to the left 3 units and up 6 units.
   d) shifted to the right 3 units and down 6 units.

2. The solid graph in Fig. 8-86 is the graph of $f(x) = |x|$. The dashed graph is the graph of which function?
a) $y = |x + 2| - 1$ b) $y = |x - 2| - 1$ c) $y = |x + 2| + 1$
d) $y = |x - 2| + 1$

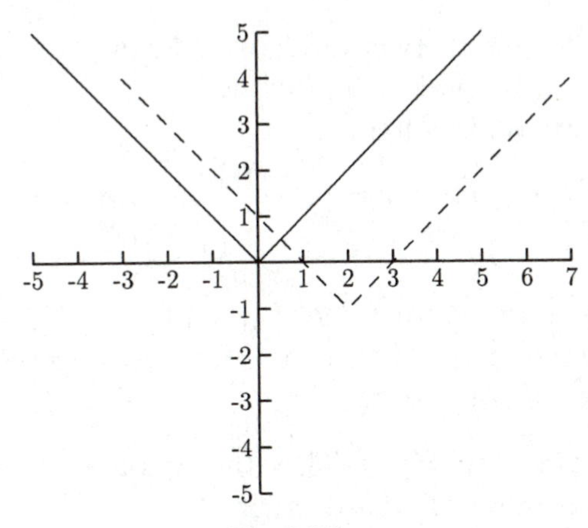

Fig. 8-86.

3. Is $f(x) = -3x^2 + 5$ an even function, odd function, or neither?
a) Even b) Odd c) Neither d) Cannot be determined

4. For $f(x) = x - 7$ and $g(x) = x^2 + 3x$, find $f \circ g(x)$.
a) $x^2 + 4x - 7$ b) $x^2 - 11x + 28$ c) $x^3 - 4x^2 - 21x$
d) $x^2 + 3x - 7$

5. The solid graph in Fig. 8-87 is the graph of $f(x)$. The dashed graph is the graph of $af(x)$. What is a?
a) $-\frac{1}{2}$ b) $\frac{1}{2}$ c) -2 d) 2

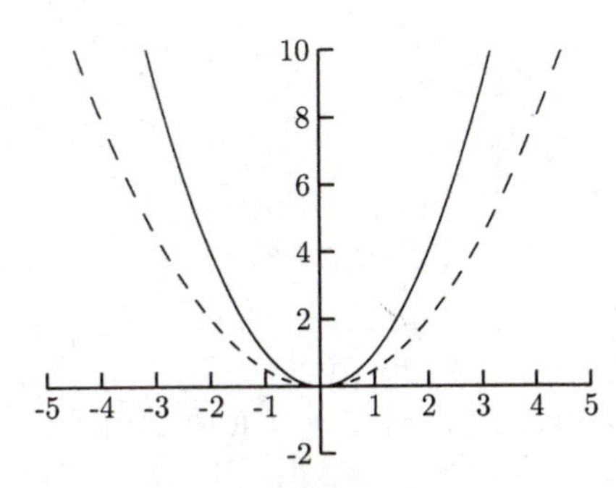

Fig. 8-87.

6. The graph of $-3f(x)$ is the graph of $f(x)$
   a) reflected about the x-axis and vertically stretched.
   b) reflected about the x-axis and vertically flattened.
   c) reflected about the y-axis and vertically stretched.
   d) reflected about the y-axis and vertically flattened.

7. Is $g(x) = 2x^3 - x$ an even function, odd function, or neither?
   a) Even b) Odd c) Neither
   d) Cannot be determined

8. Evaluate $f \circ g(x)$ for $f(x) = \sqrt{3x+5}$ and $g(x) = x^2$.
   a) $\sqrt{3x^2+5}$ b) $x^2\sqrt{3x+5}$ c) $\sqrt{x^2+5}$ d) $3x+5$

9. What is the domain for $f \circ g(x)$ when $f(x) = 1/x$ and $g(x) = x - 6$?
   a) $(-\infty, 0) \cup (0, \infty)$ b) $(0, 6) \cup (6, \infty)$ c) $(-\infty, 6) \cup (6, \infty)$
   d) $(-\infty, 0) \cup (0, 6) \cup (6, \infty)$

10. The solid graph in Fig. 8-88 is the graph of $y = \sqrt{x}$. The dashed graph is the graph of what function?
    a) $y = -\sqrt{x} + 1$ b) $y = \sqrt{-x} + 1$ c) $y = \sqrt{1-x}$
    d) $y = -\sqrt{x-1}$

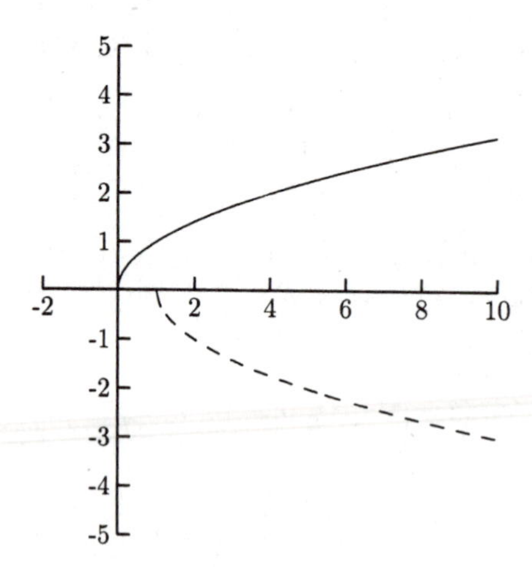

Fig. 8-88.

11. Find $f \circ g(2)$ when $f(x) = x^2 - x$ and $g(x) = 4x$.
    a) 16 b) 8 c) 56 d) 10

12. Find $f \circ g \circ h(x)$ for $f(x) = \sqrt[3]{x}$, $g(x) = x^2 + 1$, and $h(x) = 2x + 5$.
    a) $\sqrt[3]{4x^2 + 20x + 26}$ b) $(2x + 5)\sqrt[3]{x^2 + 1}$ c) $2x^{2/3} + 7$
    d) $\sqrt[3]{2x^2 + 7}$

SOLUTIONS

1. a) 2. b) 3. a) 4. d) 5. b) 6. a)
7. b) 8. a) 9. c) 10. d) 11. c) 12. a)

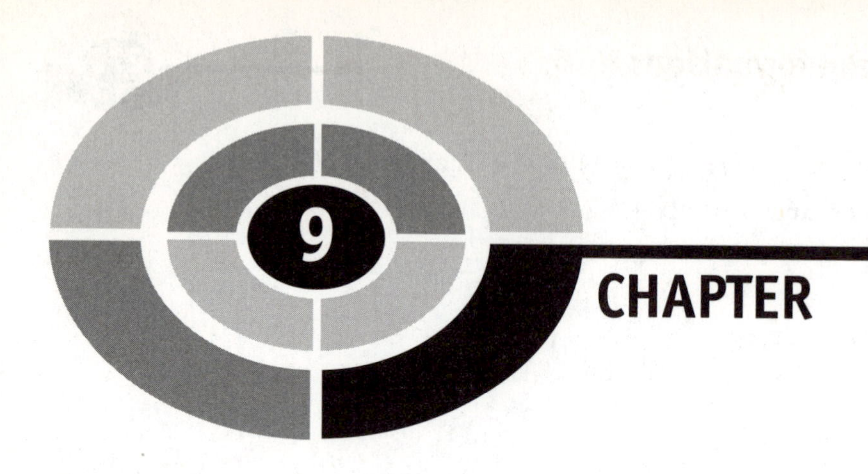

Polynomial Functions

A polynomial function is a function in the form $f(x) = a_n x^n + a_{n-1}x^{n-1} + \cdots + a_1 x + a_0$, where each a_i is a real number and the powers on x are whole numbers. There is no x under a root sign and no x in any denominator. The number a_i is called a *coefficient*. For example, in the polynomial function $f(x) = -2x^3 + 5x^2 - 4x + 8$, the coefficients are -2, 5, -4, and 8. The *constant* term (the term with no variable) is 8. The powers on x are 3, 2, and 1. The *degree* of the polynomial (and polynomial function) is the highest power on x. In this example, the degree is 3. Quadratic functions are of degree 2. Linear functions of the form $f(x) = mx + b$ (if $m \neq 0$) are of degree 1. Constant functions of the form $f(x) = b$ are of degree 0 (this is because $x^0 = 1$, making $f(x) = bx^0$).

The *leading term* of a polynomial (and polynomial function) is the term having x to the highest power. Usually, but not always, the leading term is written first. The *leading coefficient* is the coefficient on the leading term. In our example, the leading term is $-2x^3$, and the leading coefficient is -2. By looking at the leading term only, we can tell roughly what the graph looks like. The graph of any polynomial will go up on both ends, go down on both ends, or go up on one end and down on the other. This is called the *end behavior* of the

graph. The figures below illustrate the end behavior. The shape of the dashed part of each graph depends on the individual function.

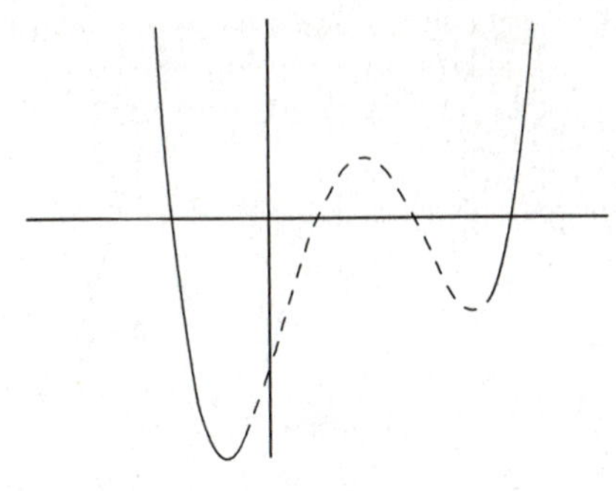

Fig. 9-1.

This graph goes up on both ends.

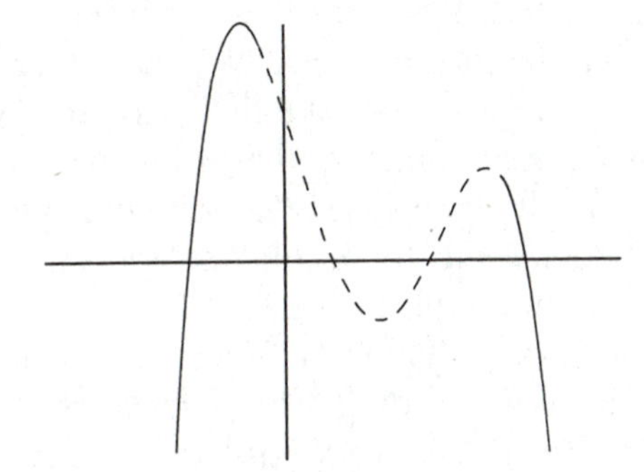

Fig. 9-2.

This graph goes down on both ends.

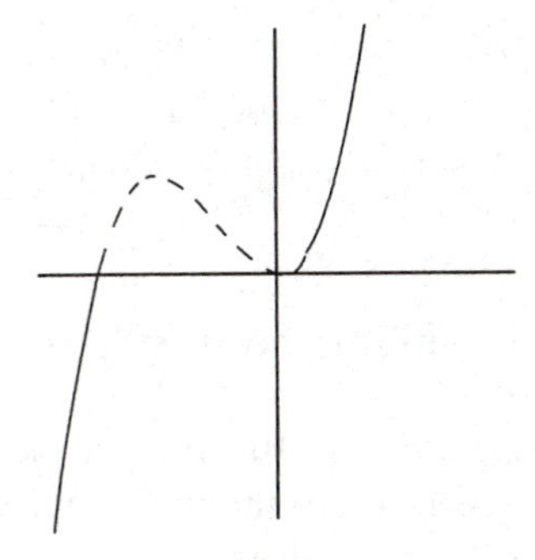

Fig. 9-3.

This graph goes down on the left and up on the right.

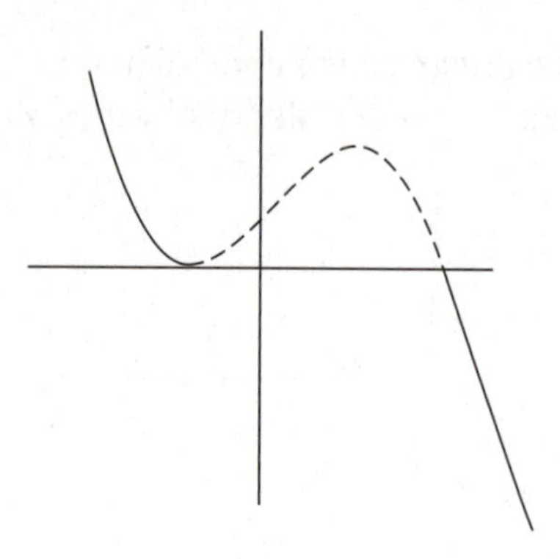

Fig. 9-4.

This graph goes up on the left and down on the right.

If the degree of the polynomial is an even number, the graph will look like the graph in Figs. 9-1 or 9-2. If the leading coefficient is a positive number, the graph will look like the graph in Fig. 9-1. If the leading coefficient is a negative number, the graph will look like the graph in Fig. 9-2. If the degree of the polynomial is an odd number, the graph will look like the one in Figs. 9-3 or 9-4. If the leading coefficient is a positive number, the graph will look like the graph in Fig. 9-3. If the leading coefficient is a negative number, the graph will look like the graph in Fig. 9-4.

How can one term in a polynomial function give us this information? For polynomial functions, the leading term dominates all of the other terms. For x-values large enough (both large positive numbers and large negative numbers), the other terms do not contribute much to the size of the y-values. "Large enough" x-values depend on the polynomial. For example, $x = 100$ is "large" for the function $f(x) = x^2 - x + 1$ but not "large" for the function $f(x) = 9718x^{10} + 30{,}162x^8 - 10{,}956x^7$.

We can avoid memorizing the effect of the leading coefficient on the graph of the function by reasoning our way to remembering. Suppose the leading term is $5x^4$. What kind of y-value will we get for a "large" positive x-value?

$$5(\text{large positive})^4$$

A "large" positive number raised to the fourth power then multiplied by 5 will be another "large" positive number. The same is true for a "large" negative number.

$$5(\text{large negative})^4$$

This is another "large" positive number. This means that the graph will look like the graph in Fig. 9-1.

If we change the leading coefficient from 5 to -5, then $-5(\text{large positive})^4$ and $-5(\text{large negative})^4$ are large negative numbers. The graph of the function will look like the graph in Fig. 9-2.

Now suppose the leading term is $5x^3$. What kind of y-value will we get by letting x be a "large" positive number?

$$5(\text{large positive})^3$$

A large positive number raised to the third power is also a large positive number. When multiplied by 5, it becomes a larger positive number.

$$5(\text{large negative})^3$$

A large negative number raised to the third power is a large negative number. When multiplied by 5, it becomes a larger negative number. The graph of the function will look like the graph in Fig. 9-3.

Finally, if the leading coefficient is -5 instead of 5, $-5(\text{large positive})^3$ is a large negative number, and $-5(\text{large negative})^3$ is a large positive number. The graph of the function will look like the graph in Fig. 9-4.

EXAMPLES

Match the graph of the given function with one of the graphs in Figs. 9-1–9-4.

- $f(x) = 4x^5 + 6x^3 - 2x^2 + 8x + 11$
  We only need to look at the leading term, $4x^5$. The degree, 5, is odd, and the leading coefficient, 4, is positive. The graph of this function looks like the one in Fig. 9-3.
- $P(x) = 5 + 2x - 6x^2$
  The leading term is $-6x^2$. The degree, 2, is even, and the leading coefficient, -6, is negative. The graph of this function looks like the one in Fig. 9-2.
- $h(x) = -2x^3 + 4x^2 - 7x + 9$
  The leading term is $-2x^3$. The degree, 3, is odd, and the leading coefficient, -2, is negative. The graph of this function looks like the one in Fig. 9-4.

- $g(x) = x^4 + 4x^3 - 8x^2 + 3x - 5$
  The leading term is x^4. The degree, 4, is even, and the leading coefficient, 1, is positive. The graph of this function looks like the one in Fig. 9-1.

PRACTICE

Match the graph of the given function with one of the graphs in Figs. 9-1–9-4.

1. $f(x) = -8x^3 + 4x^2 - 9x + 3$
2. $f(x) = 4x^5 + 10x^4 - 3x^3 + x^2$
3. $P(x) = -x^2 + x - 6$
4. $g(x) = 1 + x + x^2 + x^3$
5. $P(x) = -x^3$

SOLUTIONS

1. Figure 9-4
2. Figure 9-3
3. Figure 9-2
4. Figure 9-3
5. Figure 9-4

Finding the x-intercepts (if any) for the graph of a polynomial function is very important. The x-intercept of any graph is where the graph touches the x-axis. This happens when the y-coordinate of the point is 0. We found the x-intercepts for many quadratic functions—by factoring and setting each factor equal to zero. This is how we will find the x-intercepts for any polynomial function. This is not always easy to do. In fact, some polynomials are so hard to factor that the best we can do is approximate the x-intercepts (using graphing calculators or calculus manipulations). This will not be the case for the polynomials in this book, however. Every polynomial here will be factorable using the techniques covered later in the book.

Because an x-intercept for $f(x) = a_n x^n + a_{n-1}x^{n-1} + \cdots + a_1 x + a_0$ is a solution to the equation $0 = a_n x^n + a_{n-1}x^{n-1} + \cdots + a_1 x + a_0$, x-intercepts are also called *zeros* of the polynomial. All of the following statements have the same meaning for a polynomial. Let c be a real number, and let $P(x)$ be a polynomial function.

1. c is an x-intercept of the graph of $P(x)$.
2. c is a zero for $P(x)$.
3. $x - c$ is a factor of $P(x)$.

EXAMPLES

- $x - 1$ is a factor means that 1 is an x-intercept and a zero.
- $x + 5$ is a factor means that -5 is an x-intercept and a zero.
- x is a factor means that 0 is an x-intercept and a zero.
- 3 is a zero means that $x - 3$ is a factor and 3 is an x-intercept.

We can find the zeros of a function (or at least the approximate zeros) by looking at its graph.

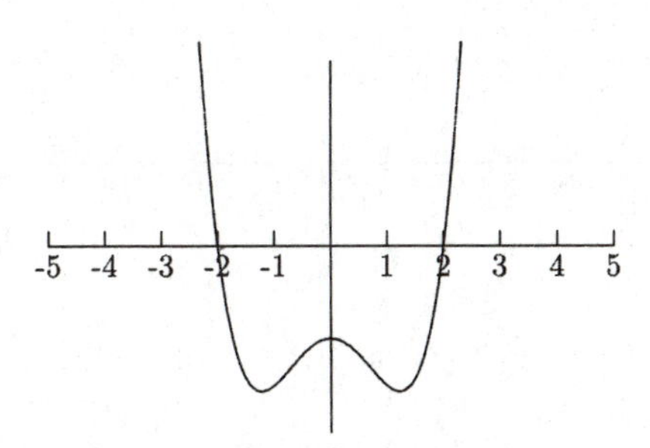

Fig. 9-5.

The x-intercepts of the graph in Fig. 9-5 are 2 and -2. Now we know that $x - 2$ and $x + 2$ (which is $x - (-2)$) are factors of the polynomial.

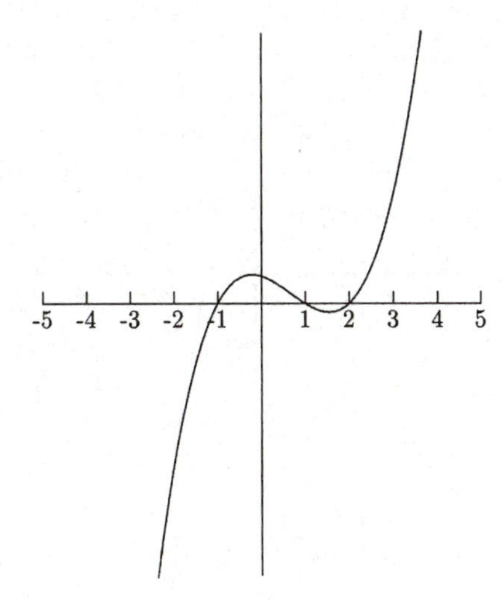

Fig. 9-6.

The graph of the polynomial function in Fig. 9-6 has x-intercepts of -1, 1, and 2. This means that $x-1$, $x-2$, and $x+1$ (as $x-(-1)$) are factors of the polynomial.

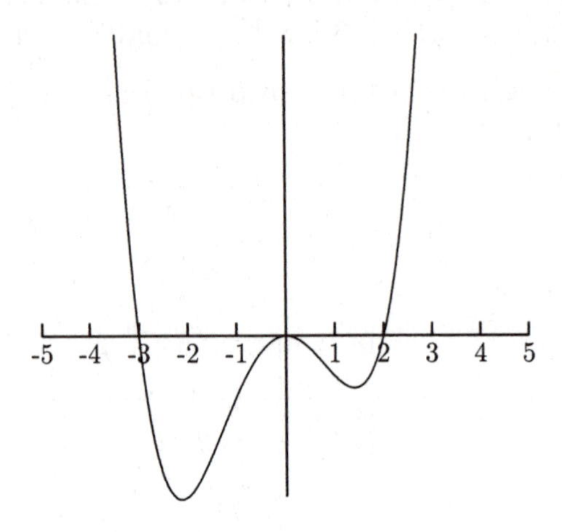

Fig. 9-7.

The x-intercepts for the graph in Fig. 9-7 are -3, 0, and 2, making $x+3$, x (as $x-0$), and $x-2$ factors of the polynomial.

PRACTICE

Identify the x-intercepts and factors for the polynomial functions whose graphs are given.

1.

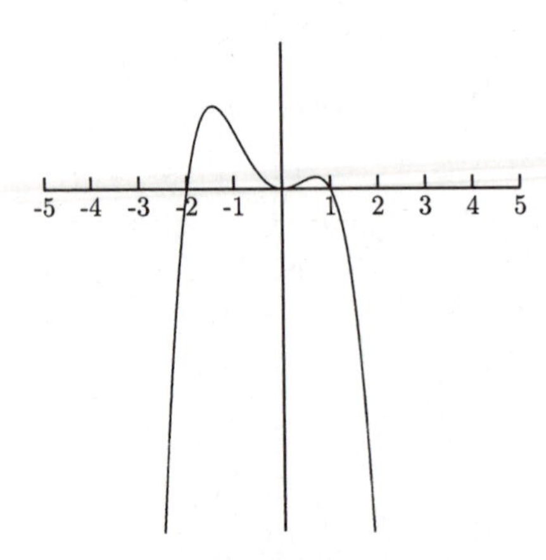

Fig. 9-8.

2.

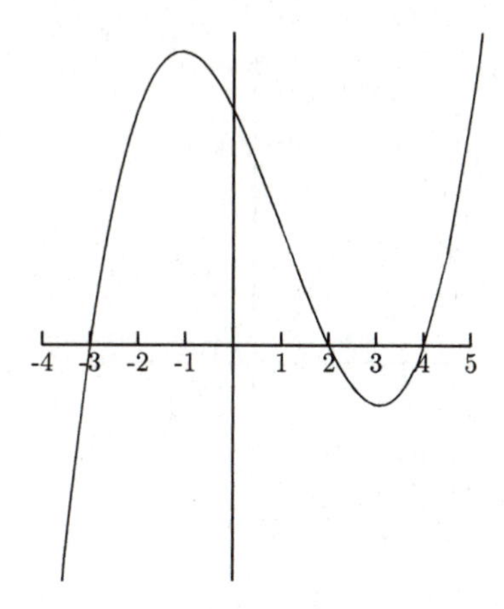

Fig. 9-9.

3.

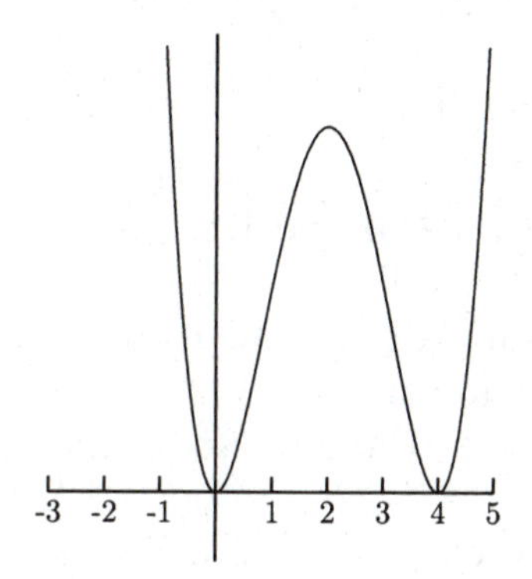

Fig. 9-10.

4.

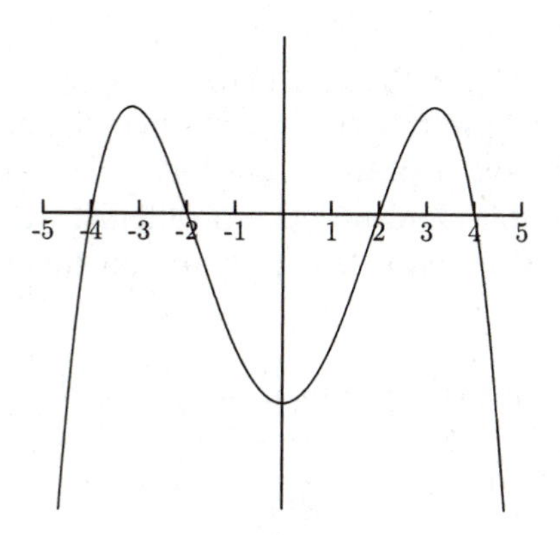

Fig. 9-11.

5.

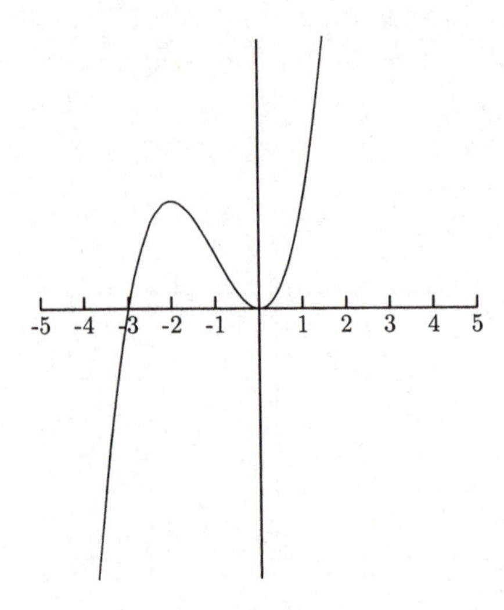

Fig. 9-12.

SOLUTIONS

1. The x-intercepts are -2, 0, and 1, so $x+2$, x, and $x-1$ are factors of the polynomial.
2. The x-intercepts are -3, 2, and 4, so $x+3$, $x-2$, and $x-4$ are factors of the polynomial.
3. The x-intercepts are 0 and 4, so x and $x-4$ are factors of the polynomial.
4. The x-intercepts are -4, -2, 2 and 4, so $x+4$, $x+2$, $x-2$, and $x-4$ are factors of the polynomial.
5. The x-intercepts are -3 and 0, so $x+3$ and x are factors of the polynomial.

Now that we know about the end behavior of the graphs of polynomial functions and the relationship between x-intercepts and factors, we can look at a polynomial and have a pretty good idea of what its graph looks like. In the next set of examples and practice problems, we will match the graphs from the above discussion with their polynomial functions.

EXAMPLES

Match the functions with the graphs in Figs. 9-5–9-7.

- $f(x) = \frac{1}{10}x^2(x+3)(x-2) = \frac{1}{10}x^4 + \frac{1}{10}x^3 - \frac{3}{5}x^2$

Because $f(x)$ is a polynomial whose degree is even and whose leading coefficient is positive, we will look for a graph that goes up on the left and up on the right. Because the factors are x^2, $x+3$, and $x-2$, we will also look for a graph with x-intercepts of 0, -3, and 2. The graph in Fig. 9-7 satisfies these conditions.

- $g(x) = \frac{1}{2}(x-1)(x-2)(x+1) = \frac{1}{2}x^3 - x^2 - \frac{1}{2}x + 1$

  Because $g(x)$ is a polynomial whose degree is odd and whose leading coefficient is positive, we will look for a graph that goes down on the left and up on the right. The factors are $x-1$, $x-2$, and $x+1$, so we will also look for a graph with 1, 2, and -1 as x-intercepts. The graph in Fig. 9-6 is the only one that satisfies all of these conditions.

- $P(x) = (x^2+2)(x-2)(x+2) = x^4 - 2x^2 - 8$

  Because $P(x)$ is a polynomial whose degree is even and whose leading term is positive, we will look for a graph that goes up on both ends. Although x^2+2 is a factor, there is no x-intercept from this factor (this is because $x^2+2=0$ has no real number solution). The x-intercepts are 2 and -2. The graph in Fig. 9-5 satisfies these conditions.

PRACTICE

Match the polynomial function with one of the graphs in Figs. 9-8–9-12.

1.
$$f(x) = -\frac{1}{8}(x+4)(x+2)(x-2)(x-4) = -\frac{1}{8}x^4 + \frac{5}{2}x^2 - 8$$

2.
$$g(x) = x^2(x-4)^2 = x^4 - 8x^3 + 16x^2$$

3.
$$P(x) = -\frac{1}{2}x^2(x+2)(x-1) = -\frac{1}{2}x^4 - \frac{1}{2}x^3 + x^2$$

4.
$$Q(x) = x^2(x+3) = x^3 + 3x^2$$

5.
$$R(x) = \frac{1}{2}(x+3)(x-2)(x-4) = \frac{1}{2}x^3 - \frac{3}{2}x^2 - 5x + 12$$

SOLUTIONS

1. Figure 9-11
2. Figure 9-10

3. Figure 9-8
4. Figure 9-12
5. Figure 9-9

Sketching Graphs of Polynomials

To sketch the graph of most polynomial functions accurately, we need to use calculus (do not let that scare you—the calculus part is easier than the algebra part!). We can still get a pretty good graph using algebra alone. The general method is to plot x-intercepts (if there are any), a point to the left of the smallest x-intercept, a point between any two x-intercepts, and a point to the right of the largest x-intercept. Because y-intercepts are easy to find, these can also be plotted.

EXAMPLES

- $f(x) = \frac{1}{2}x(x+1)(x-3)$

  The x-intercepts are 0, -1, and 3. We will use $x = -2$ for the point to the left of the smallest x-intercept, $x = -0.5$ for the point between the x-intercepts -1 and 0, $x = 1.5$ for the point between the x-intercepts 0 and 3, and $x = 4$ for the point to the right of the x-intercept 3 (Table 9-1).

Table 9-1

x	$f(x)$
−2	−5
−1	0
−0.5	0.4375
0	0
1.5	−2.8125
3	0
4	10

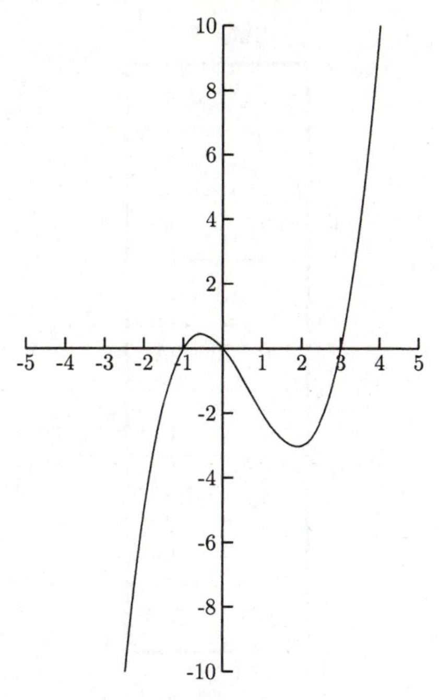

Fig. 9-13.

- $f(x) = -(2x - 1)(x + 2)(x - 3)$
  The x-intercept from the factor $2x - 1$ is $\frac{1}{2}$.

$$2x - 1 = 0$$

$$2x = 1$$

$$x = \frac{1}{2}$$

The other x-intercepts are -2 and 3. In addition to the x-intercepts, we will plot the points for $x = -2.5$ (to the left of $x = -2$), $x = -1$ (between $x = -2$ and $x = \frac{1}{2}$), $x = 2$ (between $x = \frac{1}{2}$ and $x = 3$), and $x = 3.5$ (to the right of $x = 3$) (Table 9-2). The reason we used $x = -2.5$ instead of $x = -3$ and $x = 3.5$ instead of $x = 4$ is that their y-values were too large for our graph.

Table 9-2

x	$f(x)$
-2.5	16.5
-2	0
-1	-12
0	-6
$\frac{1}{2}$	0
2	12
3	0
3.5	-16.5

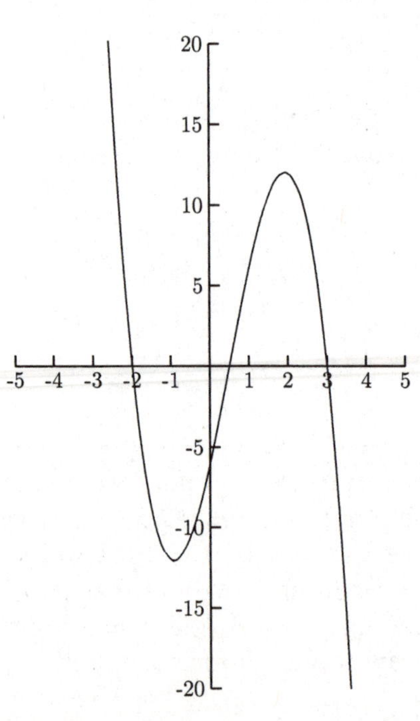

Fig. 9-14.

PRACTICE

Sketch the graph of the function.

1. $f(x) = \frac{1}{2}x(x-2)(x+2)$
2. $g(x) = -\frac{1}{2}(x+3)(x-1)(x-3)$
3. $h(x) = -\frac{1}{10}(x+4)(x+1)(x-2)(x-3)$

SOLUTIONS

1.

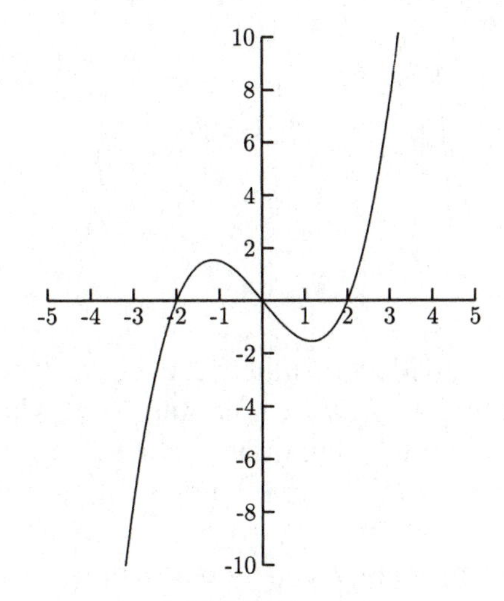

Fig. 9-15.

2.

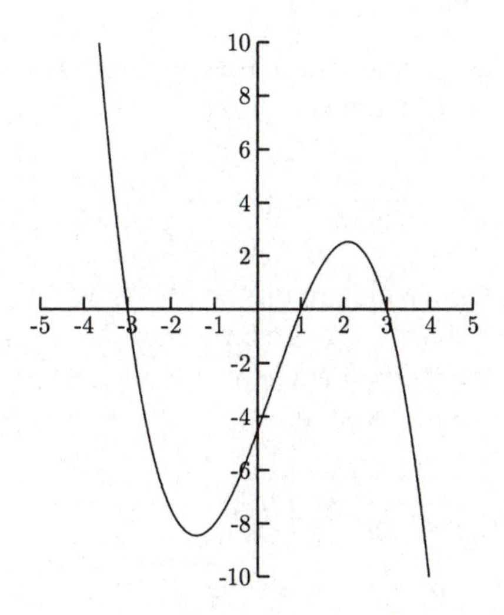

Fig. 9-16.

3.

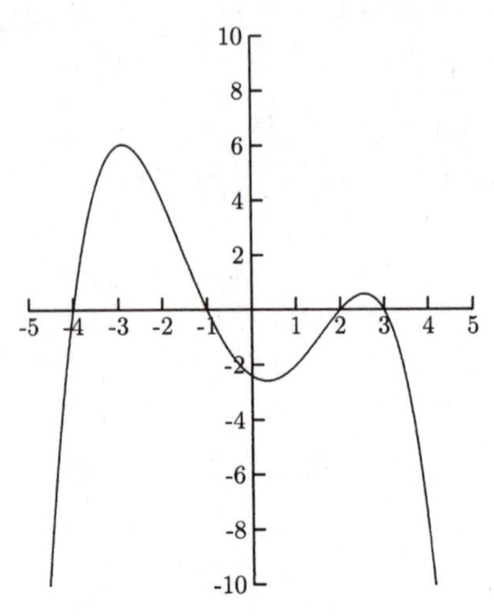

Fig. 9-17.

Polynomials can be divided in much the same way as whole numbers. When we take the quotient of two whole numbers (where the divisor is not zero), we get a quotient and a remainder. The same happens when we take the quotient of two polynomials. Polynomial division is useful in factoring polynomials.

Polynomial division problems usually come in one of two forms.

$$\frac{\text{dividend polynomial}}{\text{divisor polynomial}} \text{ or dividend polynomial} \div \text{divisor polynomial}$$

According to the division algorithm for polynomials, for any polynomials $f(x)$ and $g(x)$ (with $g(x)$ not the zero function)

$$\frac{f(x)}{g(x)} = q(x) + \frac{r(x)}{g(x)},$$

where $q(x)$ is the quotient (which might be 0) and $r(x)$ is the remainder, which has degree *strictly* less than the degree of $g(x)$. Multiplying by $g(x)$ to clear the fraction, we also get $f(x) = g(x)q(x) + r(x)$. First we will perform polynomial division using long division.

$$\begin{array}{r} q(x) \\ g(x)\overline{)\,f(x)\qquad} \\ \underline{\qquad\quad} \\ r(x) \end{array}$$

EXAMPLES

Find the quotient and remainder using long division.

- $\dfrac{4x^2 + 3x - 5}{x + 2}$

$$x + 2\overline{)4x^2 + 3x - 5}$$

We will begin by dividing the leading term of the dividend by the leading term of the divisor. For the first step in this example, we will divide $4x^2$ by x. You might see right away that $4x^2 \div x$ is $4x$. If not, write $4x^2 \div x$ as a fraction then reduce: $4x^2/x = 4x$. This will be the first term of the quotient.

$$\begin{array}{r}4x \\ x + 2\overline{)4x^2 + 3x - 5}\end{array}$$

Multiply $4x$ by the divisor: $4x(x + 2) = 4x^2 + 8x$. Subtract this from the first two terms of the dividend. Be careful to subtract all of $4x^2 + 8x$, not just $4x^2$.

$$\begin{array}{rl} & 4x \\ x+2 & \overline{)4x^2 + 3x - 5} \\ & \underline{-(4x^2 + 8x)} \\ & -5x \end{array}$$

Bring down the next term.

$$\begin{array}{rl} & 4x \\ x+2 & \overline{)4x^2 + 3x - 5} \\ & \underline{-(4x^2 + 8x)} \\ & -5x - 5 \end{array}$$

Start the process again with $-5x \div x = -5$. Multiply $x + 2$ by -5: $-5(x + 2) = -5x - 10$. Subtract this from $-5x - 5$.

$$\begin{array}{rl} & 4x \; - 5 \\ x+2 & \overline{)4x^2 + 3x - 5} \\ & \underline{-(4x^2 + 8x)} \\ & - 5x - 5 \\ & \underline{-(-5x - 10)} \\ & 5 \end{array}$$

We are finished because $5 \div x = 5/x$ cannot be a term in a polynomial. The remainder is 5 and the quotient is $4x - 5$.

- $\dfrac{3x^4 + 5x^3 - 4x^2 + 7x - 1}{x^2 + 2x - 3}$

$$x^2 + 2x - 3 \overline{)\,3x^4 + 5x^3 - 4x^2 + 7x - 1}$$

Divide $3x^4$ by x^2 to get the first term of the quotient: $3x^4/x^2 = 3x^2$. Multiply $x^2 + 2x - 3$ by $3x^2$: $3x^2(x^2 + 2x - 3) = 3x^4 + 6x^3 - 9x^2$. Subtract this from the first three terms in the dividend.

$$\begin{array}{r} 3x^2 \qquad\qquad\qquad\quad \\ x^2 + 2x - 3 \overline{)\,3x^4 + 5x^3 - 4x^2 + 7x - 1} \\ \underline{-(3x^4 + 6x^3 - 9x^2)} \qquad\quad\;\; \\ -x^3 + 5x^2 \qquad\qquad \end{array}$$

Divide $-x^3$ by x^2 to get the second term in the quotient: $-x^3/x^2 = -x$. Multiply $x^2 + 2x - 3$ by $-x$: $-x(x^2 + 2x - 3) = -x^3 - 2x^2 + 3x$. Subtract this from $-x^3 + 5x^2 + 7x$.

$$\begin{array}{r} 3x^2 - \;x \qquad\qquad\quad \\ x^2 + 2x - 3 \overline{)\,3x^4 + 5x^3 - 4x^2 + 7x - 1} \\ \underline{-(3x^4 + 6x^3 - 9x^2)} \qquad\quad\;\; \\ -x^3 + 5x^2 + 7x \qquad \\ \underline{-(-x^3 - 2x^2 + 3x)} \qquad \\ 7x^2 + 4x \qquad \end{array}$$

Divide $7x^2$ by x^2 to get the third term in the quotient: $7x^2/x^2 = 7$. Multiply $x^2 + 2x - 3$ by 7: $7(x^2 + 2x - 3) = 7x^2 + 14x - 21$. Subtract this from $7x^2 + 4x - 1$.

$$\begin{array}{r} 3x^2 - \;x + 7 \qquad\quad \\ x^2 + 2x - 3 \overline{)\,3x^4 + 5x^3 - 4x^2 + 7x - 1} \\ \underline{-(3x^4 + 6x^3 - 9x^2)} \qquad\quad\;\; \\ -x^3 + 5x^2 + 7x \qquad \\ \underline{-(-x^3 - 2x^2 + 3x)} \qquad \\ 7x^2 + \;4x - \;1 \\ \underline{-(7x^2 + 14x - 21)} \\ -10x + 20 \end{array}$$

Because $-10x/x^2$ cannot be a term in a polynomial, we are finished. The quotient is $3x^2 - x + 7$, and the remainder is $-10x + 20$.

PRACTICE

Use long division to find the quotient and remainder.

1.

$$\frac{x^3 - 6x^2 + 12x - 4}{x + 2}$$

2.

$$(6x^3 - 2x^2 + 5x - 1) \div (x^2 + 3x + 2)$$

3.

$$\frac{x^5 + 3x^4 - 3x^3 - 3x^2 + 19x - 13}{x^2 + 2x - 3}$$

SOLUTIONS

1.

$$\begin{array}{r} x^2 - 8x + 28 \\ x + 2\,\overline{)\,x^3 - 6x^2 + 12x - 4} \\ \underline{-(x^3 + 2x^2)} \\ -8x^2 + 12x \\ \underline{-(-8x^2 - 16x)} \\ 28x - 4 \\ \underline{-(28x + 56)} \\ -60 \end{array}$$

The quotient is $x^2 - 8x + 28$, and the remainder is -60.

2.

$$\begin{array}{r} 6x - 20 \\ x^2 + 3x + 2\,\overline{)\,6x^3 - 2x^2 + 5x - 1} \\ \underline{-(6x^3 + 18x^2 + 12x)} \\ -20x^2 - 7x - 1 \\ \underline{-(-20x^2 - 60x - 40)} \\ 53x + 39 \end{array}$$

The quotient is $6x - 20$, and the remainder is $53x + 39$.

3.

$$
\begin{array}{r l}
 & x^3 + x^2 - 2x + 4 \\
x^2 + 2x - 3 \,\big) & x^5 + 3x^4 - 3x^3 - 3x^2 + 19x - 13 \\
 & -(x^5 + 2x^4 - 3x^3) \\ \hline
 & x^4 + 0x^3 - 3x^2 \\
 & -(x^4 + 2x^3 - 3x^2) \\ \hline
 & -2x^3 + 0x^2 + 19x \\
 & -(-2x^3 - 4x^2 + 6x) \\ \hline
 & 4x^2 + 13x - 13 \\
 & -(4x^2 + 8x - 12) \\ \hline
 & 5x - 1
\end{array}
$$

The quotient is $x^3 + x^2 - 2x + 4$, and the remainder is $5x - 1$.

It is important that every power of x, from the highest power to the constant term, be represented in the polynomial. Although it is possible to perform long division without all powers represented, it is very easy to make an error. Also, it is not possible to perform synthetic division (discussed later in this chapter) without a coefficient for *every* term. If a power of x is not written, we need to rewrite the polynomial (either the dividend, divisor, or both) using a coefficient of 0 on the missing powers. For example, we would write $x^3 - 1$ as $x^3 + 0x^2 + 0x - 1$.

EXAMPLE

- $(x^3 - 8) \div (x + 1)$
  Rewrite as $(x^3 + 0x^2 + 0x - 8) \div (x + 1)$

$$
\begin{array}{r l}
 & x^2 - x + 1 \\
x + 1 \,\big) & x^3 + 0x^2 + 0x - 8 \\
 & -(x^3 + x^2) \\ \hline
 & -x^2 + 0x \\
 & -(-x^2 - x) \\ \hline
 & x - 8 \\
 & -(x + 1) \\ \hline
 & -9
\end{array}
$$

The quotient is $x^2 - x + 1$, and the remainder is -9.

PRACTICE

Find the quotient and remainder.

1.

$$\frac{x^3 - 1}{x - 1}$$

2. $(3x^4 - 5x + 2) \div (x - 4)$

3.

$$\frac{3x^4 - x^2 + 1}{x^2 - 2}$$

SOLUTIONS

1.

$$\begin{array}{r|l} & x^2 + x + 1 \\ \hline x - 1 & x^3 + 0x^2 + 0x - 1 \\ & -(x^3 - x^2) \\ \hline & x^2 + 0x \\ & -(x^2 - x) \\ \hline & x - 1 \\ & -(x - 1) \\ \hline & 0 \end{array}$$

The quotient is $x^2 + x + 1$, and the remainder is 0.

2.

$$\begin{array}{r|l} & 3x^3 + 12x^2 + 48x + 187 \\ \hline x - 4 & 3x^4 + 0x^3 + 0x^2 - 5x + 2 \\ & -(3x^4 - 12x^3) \\ \hline & 12x^3 + 0x^2 \\ & -(12x^3 - 48x^2) \\ \hline & 48x^2 - 5x \\ & -(48x^2 - 192x) \\ \hline & 187x + 2 \\ & -(187x - 748) \\ \hline & 750 \end{array}$$

The quotient is $3x^3 + 12x^2 + 48x + 187$, and the remainder is 750.

3.

$$
\begin{array}{r|l}
 & 3x^2 + 5 \\
\hline
x^2 + 0x - 2 & 3x^4 + 0x^3 - x^2 + 0x + 1 \\
 & -(3x^4 + 0x^3 - 6x^2) \\
\hline
 & \qquad\qquad 5x^2 + 0x + 1 \\
 & \qquad\qquad -(5x^2 + 0x - 10) \\
\hline
 & \qquad\qquad\qquad\qquad 11
\end{array}
$$

The quotient is $3x^2 + 5$, and the remainder is 11.

Polynomial division is a little more difficult when the leading coefficient of the divisor is not 1. One reason is that the terms of the quotient are harder to find and are likely to be fractions.

EXAMPLES

Find the quotient and remainder using long division.

- $\dfrac{x^2 - x + 2}{2x - 1}$

  Find the first term in the quotient by dividing the first term of the dividend by the first term of the divisor: $x^2/2x = x/2 = (1/2)x$.

$$
\begin{array}{r|l}
 & \frac{1}{2}x \\
\hline
2x - 1 & x^2 - x + 2 \\
 & -(x^2 - \frac{1}{2}x) \\
\hline
 & \qquad -\frac{1}{2}x + 2
\end{array}
$$

  The second term in the quotient is

$$\frac{-\frac{1}{2}x}{2x} = \frac{-\frac{1}{2}}{2} = -\frac{1}{2} \div 2 = -\frac{1}{2} \cdot \frac{1}{2} = -\frac{1}{4}$$

  Multiply $2x - 1$ by $-\frac{1}{4}$: $-\frac{1}{4}(2x - 1) = -\frac{1}{2}x + \frac{1}{4}$.

$$
\begin{array}{r|l}
 & \frac{1}{2}x - \frac{1}{4} \\
\hline
2x - 1 & x^2 - x + 2 \\
 & -(x^2 - \frac{1}{2}x) \\
\hline
 & \qquad -\frac{1}{2}x + 2 \\
 & \qquad -(-\frac{1}{2}x + \frac{1}{4}) \\
\hline
 & \qquad\qquad\qquad \frac{7}{4}
\end{array}
$$

  The quotient is $\frac{1}{2}x - \frac{1}{4}$, and the remainder is $\frac{7}{4}$.

- $(4x^2 + 5x - 6) \div \left(\frac{2}{3}x - 1\right)$

  Find the first term in the quotient by dividing the leading term of the dividend by the first term of the divisor.

$$\frac{4x^2}{(2/3)x} = \frac{4x}{2/3} = 4x \div \frac{2}{3} = 4x \cdot \frac{3}{2} = 6x$$

$$6x\left(\frac{2}{3}x - 1\right) = \frac{12}{3}x^2 - 6x = 4x^2 - 6x$$

$$\begin{array}{r|l} & \quad 6x \\ \hline \frac{2}{3}x - 1 & \quad 4x^2 + \ 5x - 6 \\ & -(4x^2 - \ 6x) \\ \hline & \qquad\quad 11x - 6 \end{array}$$

$$\frac{11x}{(2/3)x} = \frac{11}{2/3} = 11 \div \frac{2}{3} = 11 \cdot \frac{3}{2} = \frac{33}{2}$$

$$\frac{33}{2}\left(\frac{2}{3}x - 1\right) = \frac{66}{6}x - \frac{33}{2} = 11x - \frac{33}{2}$$

$$\begin{array}{r|l} & \quad 6x + \frac{33}{2} \\ \hline \frac{2}{3}x - 1 & \quad 4x^2 + \ 5x - 6 \\ & -(4x^2 - \ 6x) \\ \hline & \qquad\quad 11x - 6 \\ & \qquad -(11x - \frac{33}{2}) \\ \hline & \qquad\qquad\quad \frac{21}{2} \end{array}$$

The quotient is $6x + \frac{33}{2}$, and the remainder is $\frac{21}{2}$.

PRACTICE

Use long division to find the quotient and remainder.

1. $(x^3 - x^2 + 2x + 5) \div (3x - 4)$
2.

$$\frac{3x^3 - x^2 + 4x + 2}{-(1/2)x^2 + 1}$$

SOLUTIONS

1.

$$\begin{array}{r|l} & \tfrac{1}{3}x^2 + \tfrac{1}{9}x \\ \hline 3x-4 & x^3 - x^2 + 2x + 5 \\ & -(x^3 - \tfrac{4}{3}x^2) \\ \hline & \tfrac{1}{3}x^2 + 2x \\ & -(\tfrac{1}{3}x^2 - \tfrac{4}{9}x) \\ \hline & \tfrac{22}{9}x + 5 \end{array}$$

$$\frac{(22/9)x}{3x} = \frac{22/9}{3} = \frac{22}{9} \cdot \frac{1}{3} = \frac{22}{27}$$

$$\frac{22}{27}(3x-4) = \frac{66x}{27} - \frac{88}{27} = \frac{22}{9}x - \frac{88}{27}$$

$$\begin{array}{r|l} & \tfrac{1}{3}x^2 + \tfrac{1}{9}x + \tfrac{22}{27} \\ \hline 3x-4 & x^3 - x^2 + 2x + 5 \\ & -(x^3 - \tfrac{4}{3}x) \\ \hline & \tfrac{1}{3}x^2 + 2x \\ & -(\tfrac{1}{3}x^2 - \tfrac{4}{9}x) \\ \hline & \tfrac{22}{9}x + 5 \\ & -(\tfrac{22}{9}x - \tfrac{88}{27}) \\ \hline & \tfrac{223}{27} \end{array}$$

The quotient is $\frac{1}{3}x^2 + \frac{1}{9}x + \frac{22}{27}$, and the remainder is $\frac{223}{27}$.

2.

$$\frac{3x^3}{-(1/2)x^2} = \frac{3x}{-(1/2)} = 3x \div -\frac{1}{2} = 3x \cdot -2 = -6x$$

$$-6x\left(-\frac{1}{2}x^2 + 0x + 1\right) = 3x^3 + 0x^2 - 6x$$

$$\begin{array}{r|l} & -6x \\ \hline -\tfrac{1}{2}x^2 + 0x + 1 & 3x^3 - x^2 + 4x + 2 \\ & -(3x^3 - 0x^2 - 6x) \\ \hline & -x^2 + 10x + 2 \end{array}$$

$$\frac{-x^2}{-(1/2)x^2} = \frac{1}{1/2} = 1 \div \frac{1}{2} = 1 \cdot 2 = 2$$

$$2\left(-\frac{1}{2}x^2 + 0x + 1\right) = -x^2 + 0x + 2$$

$$\begin{array}{r|l} & \quad -6x \;+\; 2 \\ \hline -\frac{1}{2}x^2 + 0x + 1 & \quad 3x^3 - \;\; x^2 + 4x + 2 \\ & \underline{-(3x^3 - 0x^2 - 6x)} \\ & \qquad\quad -x^2 + 10x + 2 \\ & \qquad \underline{-(-x^2 + \;\; 0x + 2)} \\ & \qquad\qquad\qquad 10x + 0 \end{array}$$

The quotient is $-6x + 2$, and the remainder is $10x$.

Synthetic division of polynomials is much easier than long division. It only works when the divisor is of a certain form, though. Here, we will use synthetic division when the divisor is of the form $x -$ number or $x +$ number. First we will learn how to set up the problems.

For a problem of the form

$$\frac{a_n x^n + a_{n-1}x^{n-1} + \cdots + a_1 x + a_0}{x - c}$$

$$\text{or} \quad (a_n x^n + a_{n-1}x^{n-1} + \cdots + a_1 x + a_0) \div (x - c),$$

write

$$c \overline{\big| \; a_n \quad a_{n-1} \quad \cdots \quad a_1 \quad a_0}$$

Every power of x must be represented.

EXAMPLES

Set up the division problems for synthetic division.

- $\dfrac{4x^3 - 5x^2 + x - 8}{x - 2}$

  The coefficients of the dividend are 4, -5, 1, and -8. Because the divisor is $x - 2$, $c = 2$.

$$2 \overline{\big| \; 4 \quad -5 \quad 1 \quad -8}$$

- $\dfrac{x^3 - 2x + 1}{x - 4}$

  We need to think of $x^3 - 2x + 1$ as $x^3 + 0x^2 - 2x + 1$. The coefficients are 1, 0, -2, and 1. The divisor is $x - 4$, so $c = 4$.

  $$4\,\overline{)\,1 \quad 0 \quad -2 \quad 1}$$

- $\dfrac{3x^4 - x^2 + 2x + 9}{x + 5}$

  Think of $3x^4 - x^2 + 2x + 9$ as $3x^4 + 0x^3 - x^2 + 2x + 9$ and $x + 5$ as $x - (-5)$. The coefficients are 3, 0, -1, 2, and 9, and $c = -5$.

  $$-5\,\overline{)\,3 \quad 0 \quad -1 \quad 2 \quad 9}$$

PRACTICE

Set up the problems for synthetic division.

1. $$\frac{5x^3 + x^2 - 3x + 4}{x - 2}$$

2. $$\frac{x^4 - x^3 + 3x - 10}{x - 6}$$

3. $$\frac{x^3 + 2x^2 + x - 8}{x + 3}$$

4. $(x^3 + 8) \div (x + 2)$

SOLUTIONS

1. $2\,\overline{)\,5 \quad 1 \quad -3 \quad 4}$
2. $6\,\overline{)\,1 \quad -1 \quad 0 \quad 3 \quad -10}$
3. $-3\,\overline{)\,1 \quad 2 \quad 1 \quad -8}$
4. $-2\,\overline{)\,1 \quad 0 \quad 0 \quad 8}$

We are ready to learn the steps in synthetic division. The tedious work in long division is reduced to a few steps.

Find the quotient and remainder using synthetic division

$$\frac{4x^3 - 5x^2 + x - 8}{x - 2}$$

$$2 \,\big\lfloor\, 4 \quad -5 \quad 1 \quad -8$$

Bring down the first coefficient.

$$\begin{array}{r|rrrr} 2 & 4 & -5 & 1 & -8 \\ \hline & 4 & & & \end{array}$$

Multiply this coefficient by 2 (the c) and put the product under -5, the next coefficient.

$$\begin{array}{r|rrrr} 2 & 4 & -5 & 1 & -8 \\ & & 8 & & \\ \hline & 4 & & & \end{array}$$

Add -5 and 8. Put the sum under 8.

$$\begin{array}{r|rrrr} 2 & 4 & -5 & 1 & -8 \\ & & 8 & & \\ \hline & 4 & 3 & & \end{array}$$

Multiply 3 by 2 and put the product under 1, the next coefficient.

$$\begin{array}{r|rrrr} 2 & 4 & -5 & 1 & -8 \\ & & 8 & 6 & \\ \hline & 4 & 3 & & \end{array}$$

Add 1 and 6. Put the sum under 6.

$$\begin{array}{r|rrrr} 2 & 4 & -5 & 1 & -8 \\ & & 8 & 6 & \\ \hline & 4 & 3 & 7 & \end{array}$$

Multiply 7 by 2. Put the product under -8, the last coefficient.

$$\begin{array}{r|rrrr} 2 & 4 & -5 & 1 & -8 \\ & & 8 & 6 & 14 \\ \hline & 4 & 3 & 7 & \end{array}$$

Add −8 and 14. Put the sum under 14. This is the last step.

$$\begin{array}{r|rrrr} 2 & 4 & -5 & 1 & -8 \\ & & 8 & 6 & 14 \\ \hline & 4 & 3 & 7 & 6 \end{array}$$

The numbers on the last row are the coefficients of the quotient and the remainder. The degree of the remainder is smaller than the degree of the divisor. Because the divisor is $x - 2$, its degree is 1. This means that the remainder has to be a constant (which is a term of degree 0). It also means that the degree of the quotient is exactly one less than the degree of the dividend. In this example, the degree of the dividend is 3, so the degree of the quotient is 2. The last number on the bottom row is the remainder. The numbers before it are the coefficients of the quotient, in order from the highest degree to the lowest. The remainder in this example is 6. The coefficients of the quotient are 4, 3, and 7. The quotient is $4x^2 + 3x + 7$.

EXAMPLE

- $(3x^4 - x^2 + 2x + 9) \div (x + 5)$

$$\begin{array}{r|rrrrr} -5 & 3 & 0 & -1 & 2 & 9 \end{array}$$

Bring down 3, the first coefficient. Multiply it by −5. Put $3(-5) = -15$ under 0.

$$\begin{array}{r|rrrrr} -5 & 3 & 0 & -1 & 2 & 9 \\ & & -15 & & & \\ \hline & 3 & & & & \end{array}$$

Add $0 + (-15) = -15$. Multiply −15 by −5 and put $(-15)(-5) = 75$ under −1.

$$\begin{array}{r|rrrrr} -5 & 3 & 0 & -1 & 2 & 9 \\ & & -15 & 75 & & \\ \hline & 3 & -15 & & & \end{array}$$

Add −1 and 75. Multiply $-1 + 75 = 74$ by −5 and put $(74)(-5) = -370$ under 2.

$$\begin{array}{r|rrrrr} -5 & 3 & 0 & -1 & 2 & 9 \\ & & -15 & 75 & -370 & \\ \hline & 3 & -15 & 74 & & \end{array}$$

Add 2 to -370. Multiply $2+(-370)=-368$ by -5 and put $(-368)(-5)=1840$ under 9.

$$\begin{array}{r|rrrrr} -5 & 3 & 0 & -1 & 2 & 9 \\ & & -15 & 75 & -370 & 1840 \\ \hline & 3 & -15 & 74 & -368 & \end{array}$$

Add 9 to 1840. Put $9+1840=1849$ under 1840. This is the last step.

$$\begin{array}{r|rrrrr} -5 & 3 & 0 & -1 & 2 & 9 \\ & & -15 & 75 & -370 & 1840 \\ \hline & 3 & -15 & 74 & -368 & 1849 \end{array}$$

The dividend has degree 4, so the quotient has degree 3. The quotient is $3x^3-15x^2+74x-368$ and the remainder is 1849.

PRACTICE

Find the quotient and remainder using synthetic division.

1.

$$\frac{5x^3+x^2-3x+4}{x-2}$$

2.

$$\frac{x^3+2x^2+x-8}{x+3}$$

3. $(x^3+8)\div(x+2)$

4.

$$\frac{-3x^4+6x^3+4x^2+9x-11}{x+1}$$

5.

$$\frac{2x^3+x^2-4x-12}{x+(1/2)}$$

SOLUTIONS

1.

$$\begin{array}{r|rrrr} 2 & 5 & 1 & -3 & 4 \\ & & 10 & 22 & 38 \\ \hline & 5 & 11 & 19 & 42 \end{array}$$

The quotient is $5x^2 + 11x + 19$, and the remainder is 42.

2.

$$\begin{array}{r|rrrr} -3 & 1 & 2 & 1 & -8 \\ & & -3 & 3 & -12 \\ \hline & 1 & -1 & 4 & -20 \end{array}$$

The quotient is $x^2 - x + 4$, and the remainder is -20.

3.

$$\begin{array}{r|rrrr} -2 & 1 & 0 & 0 & 8 \\ & & -2 & 4 & -8 \\ \hline & 1 & -2 & 4 & 0 \end{array}$$

The quotient is $x^2 - 2x + 4$, and the remainder is 0.

4.

$$\begin{array}{r|rrrrr} -1 & -3 & 6 & 4 & 9 & -11 \\ & & 3 & -9 & 5 & -14 \\ \hline & -3 & 9 & -5 & 14 & -25 \end{array}$$

The quotient is $-3x^3 + 9x^2 - 5x + 14$, and the remainder is -25.

5.

$$\begin{array}{r|rrrr} -\frac{1}{2} & 2 & 1 & -4 & -12 \\ & & -1 & 0 & 2 \\ \hline & 2 & 0 & -4 & -10 \end{array}$$

The quotient is $2x^2 - 4$, and the remainder is -10.

When dividing a polynomial $f(x)$ by $x - c$, the remainder tells us two things. If we get a remainder of 0, then both the divisor $(x - c)$ and quotient are factors of $f(x)$. In practice problem 3 above, we had $(x^3 + 8) \div (x + 2) = x^2 - 2x + 4$, with a remainder of 0. This means that

$x^3 + 8 = (x+2)(x^2 - 2x + 4)$. Another fact we get from the remainder is that $f(c) =$ remainder.

$$f(x) = (x-c)q(x) + \text{remainder}$$
$$f(c) = (c-c)q(c) + \text{remainder}$$
$$f(c) = 0q(c) + \text{remainder}$$
$$f(c) = \text{remainder}$$

The fact that $f(c)$ is the remainder is called the *Remainder Theorem*. It is useful when trying to evaluate complicated polynomials. We can also use this fact to check our work in synthetic division and long division (providing the divisor is $x - c$).

EXAMPLE

- $(x^3 - 6x^2 + 4x - 5) \div (x - 3)$

  By the Remainder Theorem, we should get the remainder to be $3^3 - 6(3^2) + 4(3) - 5 = -20$.

3	1	−6	4	−5
		3	−9	−15
	1	−3	−5	−20

EXAMPLE

Use synthetic division and the Remainder Theorem to evaluate $f(c)$.

- $f(x) = 14x^3 - 16x^2 + 10x + 8$; $c = 1$.

  We will first perform synthetic division with $x - 1$.

1	14	−16	10	8
		14	−2	8
	14	−2	8	16

  The remainder is 16, so $f(1) = 16$.

PRACTICE

Use synthetic division and the Remainder Theorem to evaluate $f(c)$.

1. $f(x) = 6x^4 - 8x^3 + x^2 + 2x - 5$; $c = -2$
2. $f(x) = 3x^3 + 7x^2 - 3x + 4$; $c = \frac{2}{3}$
3. $f(x) = -4x^3 + 5x^2 - 3x + 4$; $c = -\frac{1}{2}$

SOLUTIONS

1.

-2	6	-8	1	2	-5
		-12	40	-82	160
	6	-20	41	-80	155

The remainder is 155, so $f(-2) = 155$.

2.

$\frac{2}{3}$	3	7	-3	4
		2	6	2
	3	9	3	6

The remainder is 6, so $f(\frac{2}{3}) = 6$.

3.

$-\frac{1}{2}$	-4	5	-3	4
		2	$-\frac{7}{2}$	$\frac{13}{4}$
	-4	7	$-\frac{13}{2}$	$\frac{29}{4}$

The remainder is $\frac{29}{4}$, so $f(-\frac{1}{2}) = \frac{29}{4}$.

Now we will use synthetic division and the Remainder Theorem to help factor polynomials. Suppose $x = c$ is a zero for a polynomial $f(x)$. Let us see what happens when we divide $f(x)$ by $x - c$.

$$f(x) = (x - c)q(x) + r(x)$$

Because $x = c$ is a zero, the remainder is 0, so $f(x) = (x - c)q(x) + 0$, which means $f(x) = (x - c)q(x)$. The next step in completely factoring $f(x)$ is factoring $q(x)$, if necessary.

EXAMPLES

Completely factor the polynomials.

- $f(x) = x^3 - 4x^2 - 7x + 10$, $c = 1$ is a zero.
  We will use the fact that $c = 1$ is a zero to get started. We will use synthetic division to divide $f(x)$ by $x - 1$.

1	1	-4	-7	10
		1	-3	-10
	1	-3	-10	0

The quotient is $x^2 - 3x - 10$. We now have $f(x)$ partially factored.

$$f(x) = x^3 - 4x^2 - 7x + 10$$
$$= (x - 1)(x^2 - 3x - 10)$$

Because the quotient is quadratic, we can factor it directly or by using the quadratic formula.

$$x^2 - 3x - 10 = (x - 5)(x + 2)$$

Now we have the complete factorization of $f(x)$:

$$f(x) = x^3 - 4x^2 - 7x + 10$$
$$= (x - 1)(x - 5)(x + 2)$$

- $R(x) = x^3 - 2x + 1$, $c = 1$ is a factor.

$$\begin{array}{r|rrrr} 1 & 1 & 0 & -2 & 1 \\ & & 1 & 1 & -1 \\ \hline & 1 & 1 & -1 & 0 \end{array}$$

$$R(x) = x^3 - 2x + 1 = (x - 1)(x^2 + x - 1)$$

We need to use the quadratic formula to find the two zeros of $x^2 + x - 1$.

$$x = \frac{-1 \pm \sqrt{1^2 - 4(1)(-1)}}{2(1)}$$
$$\frac{-1 \pm \sqrt{5}}{2} = \frac{-1 + \sqrt{5}}{2}, \frac{-1 - \sqrt{5}}{2}$$

The factors for these zeros are $x - ((-1 + \sqrt{5})/2)$ and $x - ((-1 - \sqrt{5})/2)$.

$$R(x) = (x - 1)\left(x - \frac{-1 + \sqrt{5}}{2}\right)\left(x - \frac{-1 - \sqrt{5}}{2}\right)$$

PRACTICE

Completely factor the polynomials.

1. $f(x) = x^3 + 2x^2 - x - 2$; $c = 1$ is a zero.
2. $h(x) = x^3 + x^2 - 30x - 72$; $c = -4$ is a zero.
3. $P(x) = x^3 - 5x^2 + 5x + 3$; $c = 3$ is a zero.

SOLUTIONS

1.

$$\begin{array}{r|rrrr} 1 & 1 & 2 & -1 & -2 \\ & & 1 & 3 & 2 \\ \hline & 1 & 3 & 2 & 0 \end{array}$$

$$\begin{aligned} f(x) &= (x-1)(x^2+3x+2) \\ &= (x-1)(x+1)(x+2) \end{aligned}$$

2.

$$\begin{array}{r|rrrr} -4 & 1 & 1 & -30 & -72 \\ & & -4 & 12 & 72 \\ \hline & 1 & -3 & -18 & 0 \end{array}$$

$$\begin{aligned} h(x) &= (x+4)(x^2-3x-18) \\ &= (x+4)(x-6)(x+3) \end{aligned}$$

3.

$$\begin{array}{r|rrrr} 3 & 1 & -5 & 5 & 3 \\ & & 3 & -6 & -3 \\ \hline & 1 & -2 & -1 & 0 \end{array}$$

$$P(x) = (x-3)(x^2-2x-1)$$

In order to factor $x^2 - 2x - 1$, we must first find its zeros.

$$\begin{aligned} x &= \frac{-(-2) \pm \sqrt{(-2)^2 - 4(1)(-1)}}{2(1)} \\ &= \frac{2 \pm \sqrt{8}}{2} = \frac{2 \pm 2\sqrt{2}}{2} \\ &= \frac{2(1 \pm \sqrt{2})}{2} = 1 \pm \sqrt{2} \\ &= 1 + \sqrt{2},\ 1 - \sqrt{2} \end{aligned}$$

Because $x = 1 + \sqrt{2}$ is a zero, $x - (1 + \sqrt{2}) = x - 1 - \sqrt{2}$ is a factor.
Because $x = 1 - \sqrt{2}$ is a zero, $x - (1 - \sqrt{2}) = x - 1 + \sqrt{2}$ is a factor.

$$P(x) = (x - 3)(x - 1 - \sqrt{2})(x - 1 + \sqrt{2})$$

In the above examples and practice problems, a zero was given to help us get started. Usually, we have to find a starting point ourselves. The *Rational Zero Theorem* gives us a place to start. The Rational Zero Theorem says that if a polynomial function $f(x)$, with integer coefficients, has a rational number p/q as a zero, then p is a divisor of the constant term and q is a divisor of the leading coefficient. Not all polynomials have rational zeros, but most of those in algebra courses do.

The Rational Zero Theorem is used to find a list of candidates for zeros. These candidates are rational numbers whose numerators divide the polynomial's constant term and whose denominators divide its leading coefficient. Once we have this list, we try each number in the list to see which, if any, are zeros. Once we have found a zero, we can begin to factor the polynomial.

EXAMPLES

List the possible rational zeros.

- $f(x) = 4x^3 + 6x^2 - 2x + 9$

  The numerators in our list will be the divisors of 9: 1, 3, and 9 as well as their negatives, -1, -3, and -9. The denominators will be the divisors of 4: 1, 2, and 4. The list of possible rational zeros is

$$\frac{1}{1}, \frac{3}{1}, \frac{9}{1}, -\frac{1}{1}, -\frac{3}{1}, -\frac{9}{1}, \frac{1}{2}, \frac{3}{2}, \frac{9}{2}, -\frac{1}{2}, -\frac{3}{2},$$
$$-\frac{9}{2}, \frac{1}{4}, \frac{3}{4}, \frac{9}{4}, -\frac{1}{4}, -\frac{3}{4}, \text{ and } -\frac{9}{4}.$$

Normally, we would not write a fraction with 1 as a denominator. This list could be written with a little less effort as ± 1, ± 3, ± 9, $\pm \frac{1}{2}$, $\pm \frac{3}{2}$, $\pm \frac{9}{2}$, $\pm \frac{1}{4}$, $\pm \frac{3}{4}$, $\pm \frac{9}{4}$. We only need to list the numerators with negative numbers and not the denominators. The reason is that no new numbers are added to the list, only duplicates of numbers already there. For example, $\frac{-1}{2}$ and $\frac{1}{-2}$ are the same number.

- $g(x) = 6x^4 - 5x^3 + 2x - 8$
  The possible numerators are the divisors of 8: ±1, ±2, ±4, and ±8. The possible denominators are the divisors of 6: 1, 2, 3, and 6. The list of possible rational zeros is

  $$\pm1,\ \pm2,\ \pm4,\ \pm8,\ \pm\frac{1}{2},\ \pm\frac{2}{2},\ \pm\frac{4}{2},\ \pm\frac{8}{2},\ \pm\frac{1}{3},\ \pm\frac{2}{3},\ \pm\frac{4}{3},\ \pm\frac{8}{3},\ \pm\frac{1}{6},\ \pm\frac{2}{6},\ \pm\frac{4}{6},\ \pm\frac{8}{6}$$

  There are several duplicates on this list. There will be duplicates when the constant term and leading coefficient have common factors. The duplicates do not really matter, but they could waste time when checking the list for zeros.

PRACTICE

List the candidates for rational zeros. Do not try to find the zeros.

1. $f(x) = 3x^4 + 8x^3 - 11x^2 + 3x + 4$
2. $f(x) = x^3 - 1$
3. $g(x) = x^5 - x^3 + x - 10$
4. $P(x) = 6x^4 - 24$

SOLUTIONS

1. Possible numerators: ±1, ±2, ±4
   Possible denominators: 1 and 3
   Possible rational zeros: ±1, ±2, ±4, $\pm\frac{1}{3}$, $\pm\frac{2}{3}$, $\pm\frac{4}{3}$
2. Possible numerators: ±1
   Possible denominator: 1
   Possible rational zeros: ±1
3. Possible numerators: ±1, ±2, ±5, ±10
   Possible denominator: 1
   Possible rational zeros: ±1, ±2, ±5, ±10
4. Possible numerators: ±1, ±2, ±3, ±4, ±6, ±8, ±12, ±24
   Possible denominators: 1, 2, 3, 6
   Possible rational zeros (with duplicates omitted): ±1, ±2, ±3, ±4, ±6, ±8, ±12, ±24, $\pm\frac{1}{2}$, $\pm\frac{3}{2}$, $\pm\frac{1}{3}$, $\pm\frac{2}{3}$, $\pm\frac{4}{3}$, $\pm\frac{8}{3}$, $\pm\frac{1}{6}$

Now that we have a starting place, we can factor many polynomials. Here is the strategy. First we will see if the polynomial can be factored directly. If not, we need to list the possible rational zeros. Then we will try the numbers in this list, one at a time, until we find a zero. Once we have found a zero, we will use polynomial division (long division or synthetic division) to find the quotient. Next, we will factor the quotient. If the quotient is a quadratic factor, we will either factor it directly or use the quadratic formula to find its zeros. If the quotient is a polynomial of degree 3 or higher, we will need to start over to factor the quotient. Eventually, the quotient will be a quadratic factor.

EXAMPLES

Completely factor each polynomial.

- $P(x) = x^3 + 5x^2 - x - 5$

  While this polynomial factors using factoring by grouping, we will use the above strategy. The possible rational zeros are ± 1 and ± 5.

$$P(1) = 1^3 + 5(1)^2 - 1 - 5 = 0$$

  Now that we know that $x = 1$ is a zero, we will use synthetic division to find the quotient for $(x^3 + 5x^2 - x - 5) \div (x - 1)$.

$$\begin{array}{r|rrrr} 1 & 1 & 5 & -1 & -5 \\ & & 1 & 6 & 5 \\ \hline & 1 & 6 & 5 & 0 \end{array}$$

$$\begin{aligned} P(x) = x^3 + 5x^2 - x - 5 &= (x - 1)(x^2 + 6x + 5) \\ &= (x - 1)(x + 1)(x + 5) \end{aligned}$$

- $f(x) = 3x^4 - 2x^3 - 7x^2 - 2x$

  First we will factor x from each term: $f(x) = x(3x^3 - 2x^2 - 7x - 2)$. The possible rational zeros for $3x^3 - 2x^2 - 7x - 2$ are ± 1, ± 2, $\pm \frac{1}{3}$, $\pm \frac{2}{3}$.

$$3(1)^3 - 2(1)^2 - 7(1) - 2 \neq 0$$

$$3(-1)^3 - 2(-1)^2 - 7(-1) - 2 = 0$$

  We will use synthetic division to find the quotient for $(3x^3 - 2x^2 - 7x - 2) \div (x + 1)$.

$$\begin{array}{r|rrrr} -1 & 3 & -2 & -7 & -2 \\ & & -3 & 5 & 2 \\ \hline & 3 & -5 & -2 & 0 \end{array}$$

The quotient is $3x^2 - 5x - 2$ which factors into $(3x+1)(x-2)$.

$$\begin{aligned} f(x) &= 3x^4 - 2x^3 - 7x^2 - 2x \\ &= x(3x^3 - 2x^2 - 7x - 2) \\ &= x(x+1)(3x^2 - 5x - 2) \\ &= x(x+1)(3x+1)(x-2) \end{aligned}$$

- $h(x) = 3x^3 + 4x^2 - 18x + 5$
The possible rational zeros are ± 1, ± 5, $\pm \frac{1}{3}$, and $\pm \frac{5}{3}$.

$$\begin{aligned} h(1) &= 3(1^3) + 4(1^2) - 18(1) + 5 \neq 0 \\ h(-1) &= 3(-1)^3 + 4(-1)^2 - 18(-1) + 5 \neq 0 \\ h(5) &= 3(5^3) + 4(5^2) - 18(5) + 5 \neq 0 \end{aligned}$$

Continuing in this way, we see that $h(-5) \neq 0$, $h(\frac{1}{3}) \neq 0$, $h(-\frac{1}{3}) \neq 0$ and $h(\frac{5}{3}) = 0$.

$$\begin{array}{r|rrrr} \frac{5}{3} & 3 & 4 & -18 & 5 \\ & & 5 & 15 & -5 \\ \hline & 3 & 9 & -3 & 0 \end{array}$$

$$\begin{aligned} h(x) &= \left(x - \frac{5}{3}\right)(3x^2 + 9x - 3) \\ &= \left(x - \frac{5}{3}\right)(3)(x^2 + 3x - 1) = \left[3\left(x - \frac{5}{3}\right)\right](x^2 + 3x - 1) \\ &= (3x - 5)(x^2 + 3x - 1) \end{aligned}$$

We will find the zeros of $x^2 + 3x - 1$ using the quadratic formula.

$$\begin{aligned} x &= \frac{-3 \pm \sqrt{3^2 - 4(1)(-1)}}{2(1)} \\ &= \frac{-3 \pm \sqrt{13}}{2} = \frac{-3 + \sqrt{13}}{2}, \ \frac{-3 - \sqrt{13}}{2} \end{aligned}$$

$$h(x) = (3x - 5)\left(x - \frac{-3 + \sqrt{13}}{2}\right)\left(x - \frac{-3 - \sqrt{13}}{2}\right)$$

PRACTICE

Completely factor each polynomial.

1. $f(x) = x^4 - 2x^3 - 3x^2 + 8x - 4$
2. $h(x) = 2x^3 + 5x^2 - 23x + 10$
3. $P(x) = 7x^3 + 26x^2 - 15x + 2$

SOLUTIONS

1. The possible rational zeros are ± 1, ± 2, and ± 4. $f(1) = 0$.

The (which could be

facto Ve will try $x = 1$

agair ro again.

x^3 2)$(x + 2)$

2. The possible rational zeros are ± 1, ± 2, ± 5, ± 10, $\pm \frac{1}{2}$, and $\pm \frac{5}{2}$. Because $h(2) = 0$, $x = 2$ is a zero of $h(x)$.

$$\begin{array}{r|rrrr} 2 & 2 & 5 & -23 & 10 \\ & & 4 & 18 & -10 \\ \hline & 2 & 9 & -5 & 0 \end{array}$$

$h(x) = (x - 2)(2x^2 + 9x - 5)$ $\quad (2x^2 + 9x - 5 = (2x - 1)(x + 5))$

$h(x) = (x - 2)(2x - 1)(x + 5)$

3. The possible rational zeros are ± 1, ± 2, $\pm\frac{1}{7}$, and $\pm\frac{2}{7}$. Because $P(\frac{2}{7}) = 0$, $x = \frac{2}{7}$ is a zero for $P(x)$.

$$\begin{array}{r|rrrr} \frac{2}{7} & 7 & 26 & -15 & 2 \\ & & 2 & 8 & -2 \\ \hline & 7 & 28 & -7 & 0 \end{array}$$

$$\begin{aligned} P(x) &= \left(x - \frac{2}{7}\right)(7x^2 + 28x - 7) \\ &= \left(x - \frac{2}{7}\right)(7)(x^2 + 4x - 1) = \left[7\left(x - \frac{2}{7}\right)\right](x^2 + 4x - 1) \\ &= (7x - 2)(x^2 + 4x - 1) \end{aligned}$$

We will use the quadratic formula to find the zeros for $x^2 + 4x - 1$.

$$\begin{aligned} x &= \frac{-4 \pm \sqrt{4^2 - 4(1)(-1)}}{2(1)} = \frac{-4 \pm \sqrt{20}}{2} \\ &= \frac{-4 \pm 2\sqrt{5}}{2} = \frac{2(-2 \pm \sqrt{5})}{2} \\ &= -2 \pm \sqrt{5} = -2 + \sqrt{5}, \ -2 - \sqrt{5} \end{aligned}$$

$$\begin{aligned} x^2 + 4x - 1 &= (x - (-2 + \sqrt{5}))(x - (-2 - \sqrt{5})) \\ &= (x + 2 - \sqrt{5})(x + 2 + \sqrt{5}) \end{aligned}$$

$$P(x) = (7x - 2)(x + 2 - \sqrt{5})(x + 2 + \sqrt{5})$$

For a polynomial such as $f(x) = 5x^3 + 20x^2 - 9x - 36$, the list of possible rational zeros is quite long—36! There are ways of getting around having to test every one of them. The fastest way is to use a graphing calculator to sketch the graph of $y = 5x^3 + 20x^2 - 9x - 36$. There appears to be an x-intercept at $x = -4$ (remember that x-intercepts and zeros are the same thing).

$$\begin{array}{r|rrrr} -4 & 5 & 20 & -9 & -36 \\ & & -20 & 0 & 36 \\ \hline & 5 & 0 & -9 & 0 \end{array}$$

$$f(x) = (x + 4)(5x^2 - 9)$$

We will solve $5x^2 - 9 = 0$ to find the other zeros.

$$\begin{aligned} 5x^2 - 9 &= 0 \\ 5x^2 &= 9 \\ x^2 &= \frac{9}{5} \\ x &= \pm\sqrt{\frac{9}{5}} = \pm\frac{3}{\sqrt{5}} \\ &= \pm\frac{3}{\sqrt{5}} \cdot \frac{\sqrt{5}}{\sqrt{5}} \\ &= \pm\frac{3\sqrt{5}}{5} = \frac{3\sqrt{5}}{5}, \; -\frac{3\sqrt{5}}{5} \end{aligned}$$

$$f(x) = (x+4)\left(x - \frac{3\sqrt{5}}{5}\right)\left(x + \frac{3\sqrt{5}}{5}\right)$$

There are also a couple of algebra manipulations that can help eliminate some of the possible rational zeros. The first we will learn is *Descartes' Rule of Signs*. The second is the *Upper and Lower Bounds Theorem*. Descartes' Rule of Signs counts the number of positive zeros and negative zeros. For instance, according to the rule $f(x) = x^3 + x^2 + 4x + 6$ has no positive zeros at all. This shrinks the list of possible rational zeros from ± 1, ± 2, ± 3, and ± 6 to -1, -2, -3, and -6. The Upper and Lower Bounds Theorem gives us an idea of how large (in both the positive and negative directions) the zeros can be. For example, we can use the Upper and Lower Bounds Theorem to show that all of the zeros for $f(x) = 5x^3 + 20x^2 - 9x - 36$ are between -5 and 5. This shrinks the list of possible rational zeros from ± 1, ± 2, ± 3, ± 4, ± 6, ± 9, ± 12, ± 18, ± 36, $\pm\frac{1}{5}$, $\pm\frac{2}{5}$, $\pm\frac{3}{5}$, $\pm\frac{4}{5}$, $\pm\frac{6}{5}$, $\pm\frac{9}{5}$, $\pm\frac{12}{5}$, $\pm\frac{18}{5}$, and $\pm\frac{36}{5}$ to ± 1, ± 2, ± 3, ± 4, $\pm\frac{1}{5}$, $\pm\frac{2}{5}$, $\pm\frac{3}{5}$, $\pm\frac{4}{5}$, $\pm\frac{6}{5}$, $\pm\frac{9}{5}$, $\pm\frac{12}{5}$, and $\pm\frac{18}{5}$.

Descartes' Rule of Signs counts the number of positive zeros and the number of negative zeros by counting sign changes. The maximum number of positive zeros for a polynomial function is the number of sign changes in $f(x) = a_n x^n + a_{n-1}x^{n-1} + \cdots + a_1 x + a_0$. The possible number of positive zeros is the number of sign changes minus an even whole number. For example, if there are 5 sign changes, then there are 5 or 3 or 1 positive zeros. If there are 6 sign changes, there are 6 or 4 or 2 or 0 positive zeros. The polynomial function $f(x) = 3x^4 - 2x^3 + 7x^2 + 5x - 8$ has 3 sign changes: from 3 to -2, from -2 to 7, and from 5 to -8. There are either 3 or 1 positive zeros. The maximum number of negative zeros is the number of sign changes in the polynomial $f(-x)$. The possible number of negative zeros is the number of sign changes in $f(-x)$ minus an even whole number.

EXAMPLES

Use Descartes' Rule of Signs to count the possible number of positive zeros and negative zeros for the polynomial functions.

- $f(x) = 5x^3 - 6x^2 - 10x + 4$
  There are 2 sign changes: from 5 to -6 and from -10 to 4. This means that there are either 2 or 0 positive zeros. Before we count the possible number of negative zeros, remember from earlier in the book that for a number a, $a(-x)^{\text{even power}} = ax^{\text{even power}}$ and $a(-x)^{\text{odd power}} = -ax^{\text{odd power}}$.

$$\begin{aligned} f(-x) &= 5(-x)^3 - 6(-x)^2 - 10(-x) + 4 \\ &= -5x^3 - 6x^2 + 10x + 4 \end{aligned}$$

  There is 1 sign change, from -6 to 10, so there is exactly 1 negative zero.
- $g(x) = -x^4 + 3x^2 - 9x + 1$
  There are 3 sign changes: from -1 to 3, from 3 to -9, and from -9 to 1, so there are 3 or 1 positive zeros. If we were to rewrite $g(x)$ as $g(x) = -x^4 + 0x^3 + 3x^2 - 9x + 1$, we would not consider zero coefficients as changing signs. In other words, we will ignore the zero coefficients.

$$\begin{aligned} g(-x) &= -(-x)^4 + 3(-x)^2 - 9(-x) + 1 \\ &= -x^4 + 3x^2 + 9x + 1 \end{aligned}$$

  There is 1 sign change, from -1 to 3, so there is exactly 1 negative zero.
- $P(x) = x^5 + x^3 + x + 4$
  There are no sign changes, so there are no positive zeros.

$$\begin{aligned} P(-x) &= (-x)^5 + (-x)^3 + (-x) + 4 \\ &= -x^5 - x^3 - x + 4 \end{aligned}$$

  There is 1 sign change, so there is exactly 1 negative zero.

One of the advantages of the sign test is that if we know that there are two positive zeros and we have found one of them, then we *know* that there is exactly one more.

PRACTICE

Use Descartes' Rule of Signs to count the possible number of positive zeros and the possible number of negative zeros of the polynomial functions.

1. $f(x) = 2x^4 - 6x^3 - x^2 + 4x - 8$
2. $f(x) = -x^3 - x^2 + x + 1$
3. $h(x) = -x^4 - x^2 - 6$

SOLUTIONS

1. There are 3 sign changes in $f(x)$, so there are 3 or 1 positive zeros.

$$\begin{aligned} f(-x) &= 2(-x)^4 - 6(-x)^3 - (-x)^2 + 4(-x) - 8 \\ &= 2x^4 + 6x^3 - x^2 - 4x - 8 \end{aligned}$$

There is 1 sign change in $f(-x)$, so there is exactly 1 negative zero.

2. There is 1 sign change in $f(x)$, so there is exactly 1 positive zero.

$$\begin{aligned} f(-x) &= -(-x)^3 - (-x)^2 + (-x) + 1 \\ &= x^3 - x^2 - x + 1 \end{aligned}$$

There are 2 sign changes in $f(-x)$, so there are 2 or 0 negative zeros.

3. There are no sign changes in $h(x)$, so there are no positive zeros.

$$\begin{aligned} h(-x) &= -(-x)^4 - (-x)^2 - 6 \\ &= -x^4 - x^2 - 6 \end{aligned}$$

There are no sign changes in $h(-x)$, so there are no negative zeros.

The Upper and Lower Bounds Theorem helps us to find a range of x-values that will contain all real zeros. It does *not* tell us what these bounds are. We make a guess as to what these bounds might be then check them. For a negative number $x = a$, the statement "a is a lower bound for the real zeros" means that there is no number to the left of $x = a$ on the x-axis that is a zero. For a positive number $x = b$, the statement "b is an upper bound for the real zeros" means that there is no number to the right of $x = b$ on the x-axis that is a zero.

To determine whether a negative number $x = a$ is a lower bound for a polynomial, we need to use synthetic division. If the numbers in the bottom row alternate between nonpositive and nonnegative numbers, then $x = a$ is a lower bound for the negative zeros. "Nonpositive" means 0 or a negative number, and "nonnegative" means 0 or a positive number.

To determine whether a positive number $x = b$ is an upper bound for the positive zeros, again we need to use synthetic division. If the numbers on the bottom row are all nonnegative, then $x = b$ is an upper bound for the positive zeros.

EXAMPLES

Show that the given values for a and b are lower and upper bounds, respectively, for the polynomials.

- $f(x) = x^4 + x^3 - 16x^2 - 4x + 48$; $a = -5$ and $b = 5$

−5	1	1	−16	−4	48
		−5	20	−20	120
	1	−4	4	−24	168

The bottom row alternates between positive and negative numbers, so $a = -5$ is a lower bound for the negative zeros of $f(x)$.

5	1	1	−16	−4	48
		5	30	70	330
	1	6	14	66	378

The entries on the bottom row are all positive, so $b = 5$ is an upper bound for the positive zeros of $f(x)$. All of the real zeros for $f(x)$ are between $x = -5$ and $x = 5$.

If 0 appears on the bottom row when testing for an upper bound, we can consider 0 to be positive. If 0 appears in the bottom row when testing for a lower bound, we can consider 0 to be negative if the previous entry is positive and positive if the previous entry is negative. In other words, consider it to be the opposite sign as the previous entry.

- $P(x) = 4x^4 + 20x^3 + 7x^2 + 3x - 6$ with $a = -5$

−5	4	20	7	3	−6
		−20	0	−35	160
	4	0	7	−32	154

Because 0 follows a positive number, we will consider 0 to be negative. This makes the bottom row alternate between positive and negative entries, so $a = -5$ is a lower bound for the negative zeros of $P(x)$.

- $R(x) = -x^3 + 4x^2 + 12x - 5$ with $a = -2$.

−2	−1	4	12	−5
		2	−12	0
	−1	6	0	−5

Because 0 follows a positive number, we will consider 0 to be negative. The bottom row does not alternate between negative and positive entries, so $a = -2$ is not a lower bound for the negative zeros of $R(x)$.

PRACTICE

Show that the given values for a are lower bounds and for b are upper bounds for the zeros of the polynomials.

1. $f(x) = x^3 - 6x^2 + x + 5$; $a = -3$, $b = 7$
2. $f(x) = x^4 - x^2 - 2$; $a = -2$, $b = 2$
3. $g(x) = 3x^4 + 6x^3 + 2x^2 + x - 5$; $a = -2$, $b = 1$

SOLUTIONS

1.

$$\begin{array}{r|rrrr} -3 & 1 & -6 & 1 & 5 \\ & & -3 & 27 & -84 \\ \hline & 1 & -9 & 28 & -79 \end{array}$$

The entries on the bottom row alternate between positive and negative (or nonnegative and nonpositive), so $a = -3$ is a lower bound for the zeros of $f(x)$.

$$\begin{array}{r|rrrr} 7 & 1 & -6 & 1 & 5 \\ & & 7 & 7 & 56 \\ \hline & 1 & 1 & 8 & 61 \end{array}$$

The entries on the bottom row are positive (nonnegative), so $b = 7$ is an upper bound for the positive zeros of $f(x)$.

2.

$$\begin{array}{r|rrrrr} -2 & 1 & 0 & -1 & 0 & -2 \\ & & -2 & 4 & -6 & 12 \\ \hline & 1 & -2 & 3 & -6 & 10 \end{array}$$

The entries on the bottom row alternate between positive and negative, so $a = -2$ is a lower bound for the negative zeros of $f(x)$.

$$\begin{array}{r|rrrrr} 2 & 1 & 0 & -1 & 0 & -2 \\ & & 2 & 4 & 6 & 12 \\ \hline & 1 & 2 & 3 & 6 & 10 \end{array}$$

The entries on the bottom row are all positive, so $b = 2$ is an upper bound for the positive zeros of $f(x)$.

3.

−2	3	6	2	1	−5
		−6	0	−4	6
	3	0	2	−3	1

The entries on the bottom row alternate between nonnegative and nonpositive (because 0 follows a positive number, consider it nonpositive), so $a = -2$ is a lower bound for the negative zeros of $g(x)$.

1	3	6	2	1	−5
		3	9	11	12
	3	9	11	12	7

The entries on the bottom row are all positive, so $b = 1$ is an upper bound for the positive zeros of $g(x)$.

The Upper and Lower Bounds Theorem has some limitations. For instance, it does not tell us *how* to find upper and lower bounds for the zeros of a polynomial. For any polynomial, there are infinitely many upper and lower bounds. For instance, if $x = 5$ is an upper bound, then any number larger than 5 is also an upper bound. For many polynomials, a starting place is the quotient of the constant term and the leading coefficient and its negative: $\pm$(constant term)/(leading coefficient). First show that these are bounds for the zeros, then work your way inward. For example, if $f(x) = 2x^3 - 7x^2 + x + 50$, let $a = -\frac{50}{2} = -25$ and $b = \frac{50}{2} = 25$. Then, let a and b get closer together, say $a = -10$ and $b = 10$.

The Upper and Lower Bounds Theorem does not work well when the leading coefficient is negative. For example, $f(x) = -2x^4 + x^2 + 6$ has zeros $\sqrt{2}$ and $-\sqrt{2}$ but no matter what positive number we choose for b, the bottom row will always have -2 as its first entry. The way around this is to multiply the polynomial by -1. This works because every polynomial and its "negative" have the same zeros.

We are ready to sketch the graph of polynomial functions. The general scheme of graphing most polynomials is to find the x-intercepts (the zeros), plot a point to the left of the smallest x-intercept, a point between consecutive x-intercepts, and a point to the right of the largest x-intercept. The tools we have learned will be useful. The Rational Zero Theorem and polynomial division will help us to find the x-intercepts (if there are any). Descartes' Rule of Signs and the Upper and Lower Bounds Theorem can help narrow down the list of possible rational zeros. Finally, if a function is even or odd, the work of computing points is cut in half.

EXAMPLES

Sketch the graph for the polynomial functions.

- $g(x) = x^3 - x^2 - 17x - 15$

  The possible rational zeros are ± 1, ± 3, ± 5, and ± 15; $g(-1) = 0$.

-1	1	-1	-17	-15
		-1	2	15
	1	-2	-15	0

$$g(x) = (x+1)(x^2 - 2x - 15)$$
$$= (x+1)(x+3)(x-5)$$

  The x-intercepts are -3, -1, and 5. We will plot points for $x = -3.5$, $x = -2$, $x = 0$, $x = 3$, and $x = 5.5$.

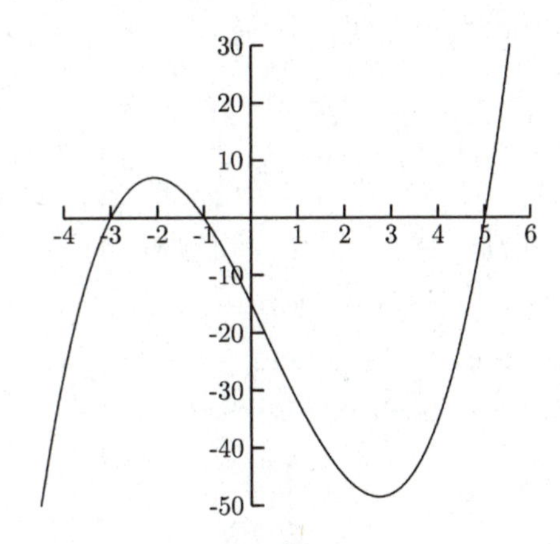

Fig. 9-18.

- $h(x) = x^3 - 4x$

  This polynomial factors easily without having to use synthetic division: $h(x) = x^3 - 4x = x(x^2 - 4) = x(x-2)(x+2)$. The x-intercepts are -2, 0, and 2. We will plot points for $x = -3$, -1, 1, and 3. Because $h(x)$ is an odd function, the y-value for $x = -3$ will be the opposite of the y-value for $x = 3$, and the y-value for $x = -1$ will be the opposite of the y-value for $x = 1$.

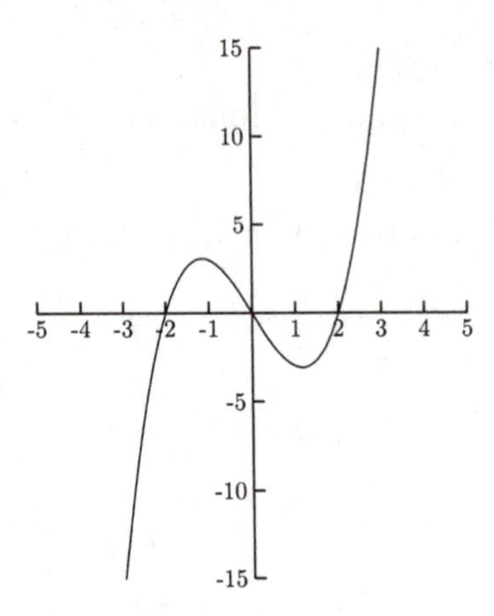

Fig. 9-19.

- $f(x) = 2x^3 + 15x^2 + 31x + 12$
  The list of possible rational zeros is $\pm1, \pm2, \pm3, \pm4, \pm6, \pm12, \pm\frac{1}{2}$, and $\pm\frac{3}{2}$. Because $f(x)$ has no sign changes, there are no positive zeros, so we only need to check $-1, -2, -3, -4, -6, -12, -\frac{1}{2}, -\frac{3}{2}$; $f(-3) = 0$.

$$\begin{array}{r|rrrr} -3 & 2 & 15 & 31 & 12 \\ & & -6 & -27 & -12 \\ \hline & 2 & 9 & 4 & 0 \end{array}$$

$$f(x) = (x+3)(2x^2 + 9x + 4)$$
$$= (x+3)(2x+1)(x+4)$$

The x-intercepts are $-4, -3$, and $-\frac{1}{2}$ (from $2x + 1 = 0$). We will plot points for $x = -5, -3.5, -2$, and 0.

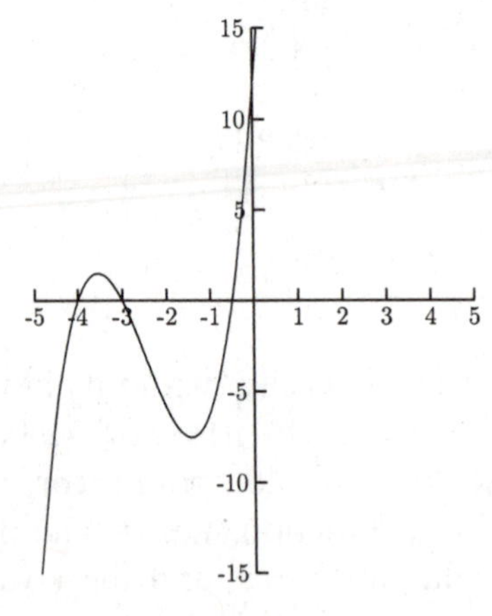

Fig. 9-20.

PRACTICE

Sketch the graph of the polynomials.

1. $f(x) = -x^3 + x^2 + 2x$
2. $g(x) = x^3 - x^2 - 10x - 8$
3. $h(x) = x^4 + 6x^3 + 11x^2 + 6x$
4. $P(x) = x^3 - 5x^2 + 18$

SOLUTIONS

1. $f(x) = -x^3 + x^2 + 2x = -x(x^2 - x - 2) = -x(x - 2)(x + 1)$

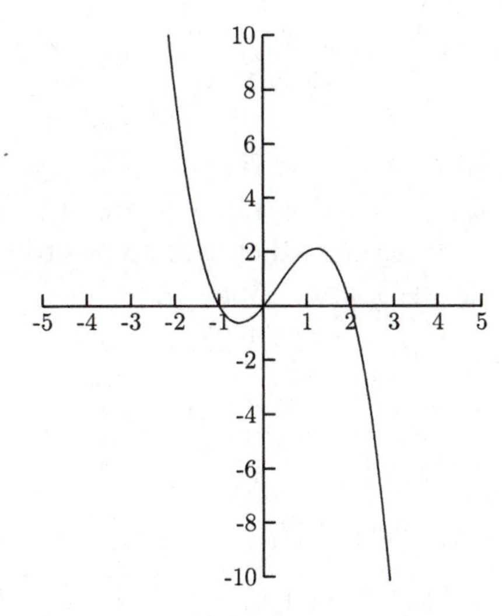

Fig. 9-21.

2. The possible rational zeros are ± 1, ± 2, ± 4, and ± 8; $g(-1) = 0$.

$$\begin{array}{r|rrrr} -1 & 1 & -1 & -10 & -8 \\ & & -1 & 2 & 8 \\ \hline & 1 & -2 & -8 & 0 \end{array}$$

$$\begin{aligned} g(x) &= x^3 - x^2 - 10x - 8 = (x + 1)(x^2 - 2x - 8) \\ &= (x + 1)(x - 4)(x + 2) \end{aligned}$$

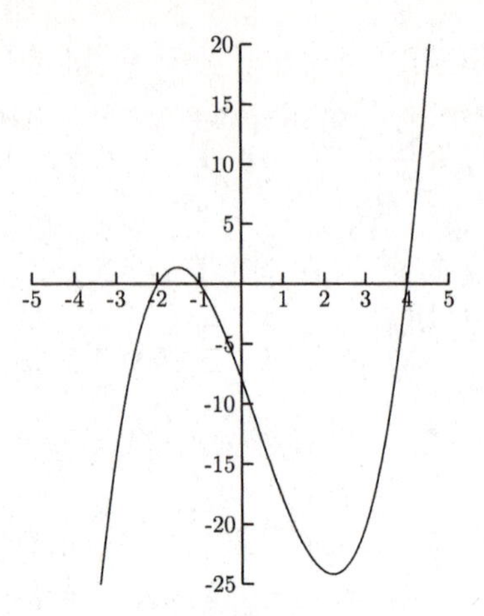

Fig. 9-22.

3. $h(x) = x^4 + 6x^3 + 11x^2 + 6x = x(x^3 + 6x^2 + 11x + 6)$. The possible rational zeros for $x^3 + 6x^2 + 11x + 6$ are ± 1, ± 2, ± 3, and ± 6. There are no sign changes, so there are no positive zeros; we only need to check -1, -2, -3, and -6: $(-1)^3 + 6(-1)^2 + 11(-1) + 6 = 0$.

$$\begin{array}{r|rrrr} -1 & 1 & 6 & 11 & 6 \\ & & -1 & -5 & -6 \\ \hline & 1 & 5 & 6 & 0 \end{array}$$

$$\begin{aligned} h(x) &= x^4 + 6x^3 + 11x^2 + 6x = x(x^3 + 6x^2 + 11x + 6) \\ &= x(x+1)(x^2 + 5x + 6) = x(x+1)(x+2)(x+3) \end{aligned}$$

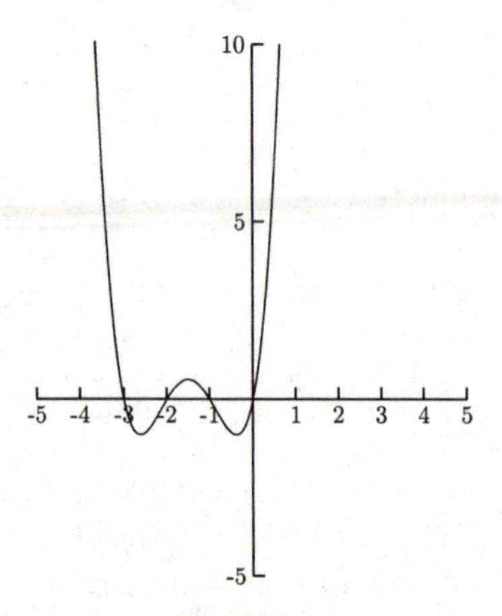

Fig. 9-23.

4. The possible rational zeros are ± 1, ± 2, ± 3, ± 6, ± 9, and ± 18. Because $a = -4$ is a lower bound for the negative zeros and $x = 6$ is an upper bound for the positive zeros, we only need to check ± 1, ± 2, and ± 3; $P(3) = 0$.

$$\begin{array}{r|rrrr} 3 & 1 & -5 & 0 & 18 \\ & & 3 & -6 & -18 \\ \hline & 1 & -2 & -6 & 0 \end{array}$$

$$P(x) = x^3 - 5x^2 + 18 = (x - 3)(x^2 - 2x - 6)$$

The zeros for $x^2 - 2x - 6$ are

$$x = \frac{-(-2) \pm \sqrt{(-2)^2 - 4(1)(-6)}}{2(1)}$$

$$= \frac{2 \pm \sqrt{28}}{2} \approx -1.65, 3.65$$

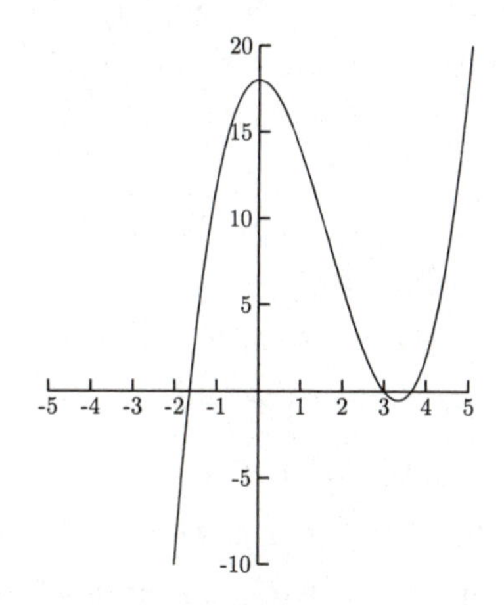

Fig. 9-24.

Complex Numbers

Until now, the zeros of polynomials have been real numbers. The next topic involves *complex* zeros. These zeros come from even roots of negative numbers like $\sqrt{-1}$. Before working with complex zeros of polynomials, we will first learn some complex number arithmetic.

Complex numbers are normally written in the form $a + bi$, where a and b are real numbers and $i = \sqrt{-1}$. Technically, real numbers are complex numbers where $b = 0$. A number such as $4 + \sqrt{-9}$ would be written as $4 + 3i$ because $\sqrt{-9} = \sqrt{9}\sqrt{-1} = 3i$.

EXAMPLES

Write the complex numbers in the form $a + bi$, where a and b are real numbers.

- $\sqrt{-64} = \sqrt{64}\sqrt{-1} = 8i$
- $\sqrt{-27} = \sqrt{27}\sqrt{-1} = \sqrt{27}i = \sqrt{9 \cdot 3}i = 3\sqrt{3}i$
  Be careful, $\sqrt{3i} \neq \sqrt{3}\,i$
- $6 + \sqrt{-8} = 6 + \sqrt{8}i = 6 + \sqrt{4 \cdot 2}i = 6 + 2\sqrt{2}i$
- $2 - \sqrt{-50} = 2 - \sqrt{50}i = 2 - \sqrt{25 \cdot 2}i = 2 - 5\sqrt{2}i$

PRACTICE

Write the complex numbers in the form $a + bi$, where a and b are real numbers.

1. $\sqrt{-25}$
2. $\sqrt{-10}$
3. $\sqrt{-24}$
4. $14 - \sqrt{-36}$
5. $8 + \sqrt{-12}$

SOLUTIONS

1. $\sqrt{-25} = \sqrt{25}\,i = 5i$
2. $\sqrt{-10} = \sqrt{10}\,i$
3. $\sqrt{-24} = \sqrt{24}\,i = \sqrt{4 \cdot 6}\,i = 2\sqrt{6}\,i$
4. $14 - \sqrt{-36} = 14 - \sqrt{36}\,i = 14 - 6i$
5. $8 + \sqrt{-12} = 8 + \sqrt{12}\,i = 8 + \sqrt{4 \cdot 3}\,i = 8 + 2\sqrt{3}\,i$

Adding complex numbers is a matter of adding like terms. Add the real parts, a and c, and the imaginary parts, b and d.

$$(a + bi) + (c + di) = (a + c) + (b + d)i$$

EXAMPLES

Perform the addition. Write the sum in the form $a + bi$, where a and b are real numbers.

- $(3 - 5i) + (4 + 8i) = (3 + 4) + (-5 + 8)i = 7 + 3i$
- $2i - 6 + 9i = -6 + 11i$
- $4 + i - 3 - i = (4 - 3) + (1 - 1)i = 1$

- $$7-\sqrt{-18}+3+5\sqrt{-2}=7-\sqrt{18}i+3+5\sqrt{2}i$$
$$=7-\sqrt{9\cdot 2}i+3+5\sqrt{2}i=7-3\sqrt{2}i+3+5\sqrt{2}i$$
$$=10+2\sqrt{2}i$$

PRACTICE

Perform the addition. Write the sum in the form $a+bi$, where a and b are real numbers.

1. $18-4i+(-15)+2i$
2. $8-2+5i$
3. $5+i+5-i$
4. $7+i+12+i$
5. $1+\sqrt{-15}-6+2\sqrt{-15}$
6. $-5+\sqrt{-12}+7+4\sqrt{-12}$

SOLUTIONS

1. $18-4i+(-15)+2i=3-2i$
2. $8-2+5i=6+5i$
3. $5+i+5-i=10+0i=10$
4. $7+i+12+i=19+2i$
5. $1+\sqrt{-15}+(-6)+2\sqrt{-15}=1+\sqrt{15}\,i-6+2\sqrt{15}\,i=-5+3\sqrt{15}\,i$
6. $$-5+\sqrt{-12}+7+4\sqrt{-12}=-5+\sqrt{12}i+7+4\sqrt{12}i$$
$$=-5+\sqrt{4\cdot 3}i+7+4\sqrt{4\cdot 3}i$$
$$=-5+2\sqrt{3}i+7+4\cdot 2\sqrt{3}i$$
$$=-5+2\sqrt{3}i+7+8\sqrt{3}i$$
$$=2+10\sqrt{3}i$$

Two complex numbers can be subtracted by distributing the minus sign in the parentheses then combining the like terms.

$$a+bi-(c+di)=a+bi-c-di=(a-c)+(b-d)i$$

EXAMPLES

Perform the subtraction and write the difference in the form $a+bi$, where a and b are real numbers.

- $11-3i-(7+6i)=11-3i-7-6i=4-9i$
- $i-(1+i)=i-1-i=-1$

- $9-(4-i)=9-4+i=5+i$
- $7+\sqrt{-8}-(1-\sqrt{-18})=7+\sqrt{8}i-1+\sqrt{18}i=7+2\sqrt{2}i-1+3\sqrt{2}i=6+5\sqrt{2}i$

PRACTICE

Perform the subtraction and write the difference in the form $a+bi$, where a and b are real numbers.

1. $2+3i-(8+7i)$
2. $4+5i-(4-5i)$
3. $\sqrt{-48}-(-1-\sqrt{-75})$

SOLUTIONS

1. $2+3i-(8+7i)=2+3i-8-7i=-6-4i$
2. $4+5i-(4-5i)=4+5i-4+5i=10i$
3.

$$\begin{aligned}\sqrt{-48}-(-1-\sqrt{-75}) &= \sqrt{48}i+1+\sqrt{75}i\\ &= \sqrt{16\cdot 3}i+1+\sqrt{25\cdot 3}i\\ &= 4\sqrt{3}i+1+5\sqrt{3}i=1+9\sqrt{3}i\end{aligned}$$

Multiplying complex numbers is not as straightforward as are adding and subtracting them. First we will take the product of two purely imaginary numbers (numbers whose real parts are 0). Remember that $i=\sqrt{-1}$, which makes $i^2=-1$. In most complex number multiplication problems, we will have a term with i^2. Replace i^2 with -1.

EXAMPLES

Write the product in the form $a+bi$, where a and b are real numbers.

- $(5i)(6i)=30i^2=30(-1)=-30$
- $(2i)(-9i)=-18i^2=-18(-1)=18$
- $(\sqrt{-6})(\sqrt{-9})=(\sqrt{6}i)(\sqrt{9}i)=(\sqrt{6})(3)i^2=3\sqrt{6}(-1)=-3\sqrt{6}$

PRACTICE

Write the product in the form $a+bi$, where a and b are real numbers.

1. $(2i)(10i)$
2. $(4\sqrt{-25})(2\sqrt{-25})$

3. $\sqrt{-3} \cdot \sqrt{-12}$
4. $\sqrt{-6} \cdot \sqrt{-15}$

SOLUTIONS

1. $(2i)(10i) = 20i^2 = 20(-1) = -20$
2. $(4\sqrt{-25})(2\sqrt{-25}) = 4(5i)[2(5i)] = 200i^2 = 200(-1) = -200$
3. $\sqrt{-3} \cdot \sqrt{-12} = \sqrt{3}\,i \cdot \sqrt{12}\,i = \sqrt{3 \cdot 12}\,i^2 = \sqrt{36}\,i^2 = 6(-1) = -6$
4. $\sqrt{-6} \cdot \sqrt{-15} = \sqrt{6}\,i \cdot \sqrt{15}\,i = \sqrt{6 \cdot 15}\,i^2 = \sqrt{90}\,i^2 = 3\sqrt{10}(-1) = -3\sqrt{10}$

Two complex numbers in the form $a + bi$ can be multiplied using the FOIL method, substituting -1 for i^2 and combining like terms.

EXAMPLES

Perform the multiplication. Write the product in the form $a + bi$, where a and b are real numbers.

- $(4 + 2i)(5 + 3i) = 20 + 12i + 10i + 6i^2 = 20 + 22i + 6(-1) = 14 + 22i$
- $(1 - i)(2 + i) = 2 + i - 2i - i^2 = 2 - i - (-1) = 3 - i$
- $(8 - 2i)(8 + 2i) = 64 + 16i - 16i - 4i^2 = 64 - 4(-1) = 68$

PRACTICE

1. $(15 + 3i)(-2 + i)$
2. $(-1 + 3i)(4 - 2i)$
3. $(3 + 2i)(3 - 2i)$
4. $(2 - i)(2 + i)$

SOLUTIONS

1. $(15 + 3i)(-2 + i) = -30 + 15i - 6i + 3i^2 = -30 + 9i + 3(-1) = -33 + 9i$
2. $(-1 + 3i)(4 - 2i) = -4 + 2i + 12i - 6i^2 = -4 + 14i - 6(-1) = 2 + 14i$
3. $(3 + 2i)(3 - 2i) = 9 - 6i + 6i - 4i^2 = 9 - 4(-1) = 13$
4. $(2 - i)(2 + i) = 4 + 2i - 2i - i^2 = 4 - (-1) = 5$

The two complex numbers $a + bi$ and $a - bi$ are called *complex conjugates.* The only difference between a complex number and its conjugate is the sign between the real part and the imaginary part.

EXAMPLES

- The complex conjugate of $3 + 2i$ is $3 - 2i$.
- The complex conjugate of $-7 - i$ is $-7 + i$.
- The complex conjugate of $10i$ is $-10i$.

PRACTICE
Identify the complex conjugate.

1. $15+7i$
2. $-3+i$
3. $-9i$

SOLUTIONS
1. The complex conjugate of $15+7i$ is $15-7i$.
2. The complex conjugate of $-3+i$ is $-3-i$.
3. The complex conjugate of $-9i$ is $9i$.

The product of any complex number and its conjugate is a real number.

$$\begin{aligned}(a+bi)(a-bi) &= a^2 - abi + abi - b^2i^2 \\ &= a^2 - b^2(-1) \\ &= a^2 + b^2\end{aligned}$$

EXAMPLES
- $(7-2i)(7+2i)$. Here, $a=7$ and $b=2$, so $a^2=49$ and $b^2=4$, making $(7-2i)(7+2i)=49+4=53$.
- $(1-i)(1+i)$. Here $a=1$ and $b=1$, so $a^2=1$ and $b^2=1$, making $(1-i)\,(1+i)=1+1=2$.
- $(-6+3i)(-6-3i)=36+9=45$

PRACTICE
Perform the multiplication.

1. $(8-10i)(8+10i)$
2. $(1-9i)(1+9i)$
3. $(5-2i)(5+2i)$

SOLUTIONS
1. $(8-10i)(8+10i)=64+100=164$
2. $(1-9i)(1+9i)=1+81=82$
3. $(5-2i)(5+2i)=25+4=29$

Dividing two complex numbers can be complicated. These problems are normally written in fraction form. If the denominator is purely imaginary, we can simply multiply the fraction by i/i and simplify.

EXAMPLES

Perform the division. Write the quotient in the form $a + bi$, where a and b are real numbers.

- $\dfrac{2+3i}{i}$

$$\begin{aligned} &= \frac{2+3i}{i} \cdot \frac{i}{i} = \frac{(2+3i)i}{i^2} \\ &= \frac{2i+3i^2}{i^2} = \frac{2i+3(-1)}{-1} \\ &= \frac{-3+2i}{-1} = -(-3+2i) \\ &= 3-2i \end{aligned}$$

- $\dfrac{4+5i}{2i}$

$$\begin{aligned} &= \frac{4+5i}{2i} \cdot \frac{i}{i} = \frac{4i+5i^2}{2i^2} \\ &= \frac{4i+5(-1)}{2(-1)} = \frac{4i-5}{-2} \\ &= \frac{-5+4i}{-2} = \frac{-(-5+4i)}{2} \\ &= \frac{5-4i}{2} = \frac{5}{2} - 2i \end{aligned}$$

PRACTICE

Write the quotient in the form $a + bi$, where a and b are real numbers.

1. $\dfrac{12+5i}{2i}$
2. $\dfrac{4-9i}{-3i}$
3. $\dfrac{1+i}{i}$

SOLUTIONS

1.

$$\begin{aligned} \frac{12+5i}{2i} &= \frac{12+5i}{2i} \cdot \frac{i}{i} = \frac{12i+5i^2}{2i^2} \\ &= \frac{12i+5(-1)}{2(-1)} = \frac{-5+12i}{-2} \\ &= \frac{-(-5+12i)}{2} = \frac{5-12i}{2} = \frac{5}{2} - 6i \end{aligned}$$

2.

$$\frac{4-9i}{-3i}=\frac{4-9i}{-3i}\cdot\frac{i}{i}=\frac{4i-9i^2}{-3i^2}$$
$$=\frac{4i-9(-1)}{-3(-1)}=\frac{9+4i}{3}=3+\frac{4}{3}i$$

3.

$$\frac{1+i}{i}=\frac{1+i}{i}\cdot\frac{i}{i}=\frac{i+i^2}{i^2}$$
$$=\frac{i+(-1)}{-1}=\frac{-1+i}{-1}$$
$$=\frac{-(-1+i)}{1}=1-i$$

When the divisor (denominator) is in the form $a+bi$, multiplying the fraction by i/i will not work.

$$\frac{2-5i}{3+6i}\cdot\frac{i}{i}=\frac{2i-5i^2}{3i+6i^2}=\frac{5+2i}{-6+3i}$$

What *does* work is to multiply the fraction by the denominator's conjugate over itself. This works because the product of any complex number and its conjugate is a real number. We will use the FOIL method in the numerator (if necessary) and the fact that $(a+bi)(a-bi)=a^2+b^2$ in the denominator.

EXAMPLES

Write the quotient in the form $a+bi$, where a and b are real numbers.

- $\dfrac{2+7i}{6+i}$

$$=\frac{2+7i}{6+i}\cdot\frac{6-i}{6-i}=\frac{12-2i+42i-7i^2}{6^2+1^2}$$
$$=\frac{12+40i-7(-1)}{37}=\frac{12+40i+7}{37}$$
$$=\frac{19+40i}{37}=\frac{19}{37}+\frac{40}{37}i$$

- $\dfrac{4-9i}{5-2i}$

$$= \frac{4-9i}{5-2i} \cdot \frac{5+2i}{5+2i} = \frac{20+8i-45i-18i^2}{5^2+2^2}$$
$$= \frac{20-37i-18(-1)}{25+4} = \frac{20-37i+18}{29}$$
$$= \frac{38-37i}{29} = \frac{38}{29} - \frac{37}{29}i$$

PRACTICE

Write the quotient in the form $a+bi$, where a and b are real numbers.

1. $\dfrac{1-2i}{1-i}$
2. $\dfrac{4+2i}{1-3i}$
3. $\dfrac{8-i}{2-5i}$
4. $\dfrac{6+4i}{6-4i}$

SOLUTIONS

1.

$$\frac{1-2i}{1-i} = \frac{1-2i}{1-i} \cdot \frac{1+i}{1+i} = \frac{1+i-2i-2i^2}{1^2+1^2}$$
$$= \frac{1-i-2(-1)}{2} = \frac{3-i}{2} = \frac{3}{2} - \frac{1}{2}i$$

2.

$$\frac{4+2i}{1-3i} = \frac{4+2i}{1-3i} \cdot \frac{1+3i}{1+3i} = \frac{4+12i+2i+6i^2}{1^2+3^2}$$
$$= \frac{4+14i+6(-1)}{10} = \frac{-2+14i}{10} = -\frac{1}{5} + \frac{7}{5}i$$

3.

$$\frac{8-i}{2-5i} = \frac{8-i}{2-5i} \cdot \frac{2+5i}{2+5i} = \frac{16+40i-2i-5i^2}{2^2+5^2}$$
$$= \frac{16+38i-5(-1)}{29} = \frac{21+38i}{29} = \frac{21}{29} + \frac{38}{29}i$$

4.

$$\frac{6+4i}{6-4i}=\frac{6+4i}{6-4i}\cdot\frac{6+4i}{6+4i}=\frac{36+24i+24i+16i^2}{6^2+4^2}$$
$$=\frac{36+48i+16(-1)}{36+16}=\frac{20+48i}{52}=\frac{5}{13}+\frac{12}{13}i$$

There are reasons to write complex numbers in the form $a+bi$. One is that complex numbers are plotted in the plane (real numbers are plotted on the number line), where the x-axis becomes the *real* axis and the y-axis becomes the *imaginary* axis. The number $3-4i$ is plotted in Fig. 9-25.

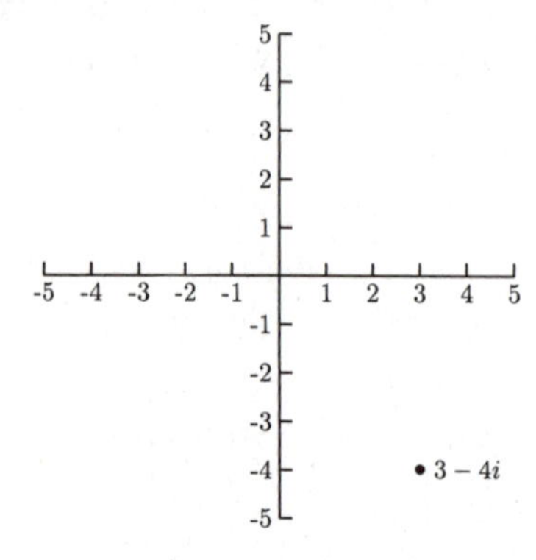

Fig. 9-25.

Complex Solutions to Quadratic Equations

Every quadratic equation has a solution, real or complex. The real solution, or solutions, for a quadratic equation is, or are, the x-intercept, or intercepts, for the graph of the quadratic function. The graph for $f(x)=x^2+1$ has no real solutions and no x-intercepts.

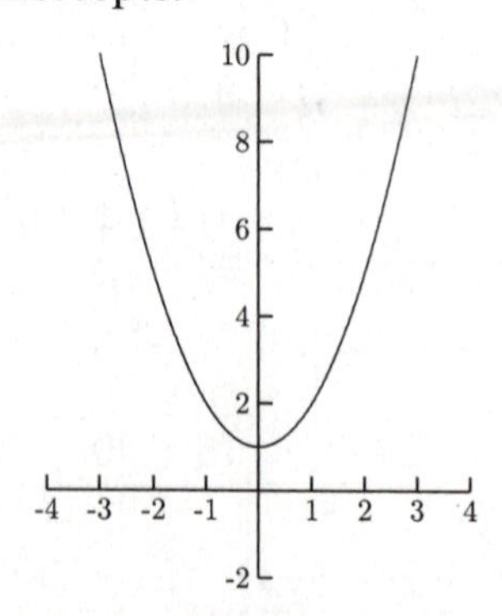

Fig. 9-26.

The equation $x^2 + 1 = 0$ does have two complex solutions.

$$\begin{aligned} x^2 + 1 &= 0 \\ x^2 &= -1 \\ x &= \pm\sqrt{-1} \\ &= \pm i \end{aligned}$$

EXAMPLES

Solve the equations and write the solutions in the form $a + bi$, where a and b are real numbers.

- $3x^2 + 8x + 14 = 0$

$$\begin{aligned} x &= \frac{-8 \pm \sqrt{8^2 - 4(3)(14)}}{2(3)} = \frac{-8 \pm \sqrt{-104}}{6} \\ &= \frac{-8 \pm 2\sqrt{26}\,i}{6} = \frac{2(-4 \pm \sqrt{26}\,i)}{6} \\ &= \frac{-4 \pm \sqrt{26}\,i}{3} = -\frac{4}{3} \pm \frac{\sqrt{26}}{3}i \\ &= -\frac{4}{3} + \frac{\sqrt{26}}{3}i,\ -\frac{4}{3} - \frac{\sqrt{26}}{3}i \end{aligned}$$

- $9x^2 + 25 = 0$

$$\begin{aligned} 9x^2 + 25 &= 0 \\ 9x^2 &= -25 \\ x^2 &= -\frac{25}{9} \\ x &= \pm\sqrt{-\frac{25}{9}} = \pm\sqrt{\frac{25}{9}}\,i \\ x &= \pm\frac{5}{3}i = \frac{5}{3}i,\ -\frac{5}{3}i \end{aligned}$$

PRACTICE

Solve the equations and write the solutions in the form $a + bi$, where a and b are real numbers.

1. $x^2 + 2x + 4 = 0$
2. $x^2 + 25 = 0$
3. $9x^2 + 4 = 0$
4. $6x^2 + 8x + 9 = 0$

SOLUTIONS

1.

$$\begin{aligned} x^2+2x+4&=0\\ x^2+2x+\left(\frac{2}{2}\right)^2&=-4+\left(\frac{2}{2}\right)^2\\ (x+1)^2&=-3\\ x+1&=\pm\sqrt{-3}\\ x&=-1\pm\sqrt{3}\,i=-1+\sqrt{3}\,i,\ -1-\sqrt{3}\,i \end{aligned}$$

2.

$$\begin{aligned} x^2+25&=0\\ x^2&=-25\\ x&=\pm\sqrt{-25}=\pm 5i=5i,\ -5i \end{aligned}$$

3.

$$\begin{aligned} 9x^2+4&=0\\ 9x^2&=-4\\ x^2&=-\frac{4}{9}\\ x&=\pm\sqrt{-\frac{4}{9}}=\pm\frac{2}{3}i=\frac{2}{3}i,\ -\frac{2}{3}i \end{aligned}$$

4.

$$\begin{aligned} x&=\frac{-8\pm\sqrt{8^2-4(6)(9)}}{2(6)}\\ &=\frac{-8\pm\sqrt{-152}}{12}=\frac{-8\pm 2\sqrt{38}\,i}{12}\\ &=\frac{2(-4\pm\sqrt{38}\,i)}{12}=\frac{-4\pm\sqrt{38}\,i}{6}\\ &=-\frac{4}{6}\pm\frac{\sqrt{38}}{6}i=-\frac{2}{3}\pm\frac{\sqrt{38}}{6}i\\ &=-\frac{2}{3}+\frac{\sqrt{38}}{6}i,\ -\frac{2}{3}-\frac{\sqrt{38}}{6}i \end{aligned}$$

In all the previous examples and practice problems, complex solutions to quadratic equations came in conjugate pairs. This always happens when the solutions are complex numbers. A quadratic expression that has complex zeros are called *irreducible* (over the reals) because they cannot be factored using real numbers. For example, the polynomial function $f(x) = x^4 - 1$ can be factored using real numbers as $(x^2 - 1)(x^2 + 1) = (x - 1)(x + 1)(x^2 + 1)$. The factor $x^2 + 1$ is irreducible because it is factored as $(x - i)(x + i)$. We can tell which quadratic factors are irreducible without having to use the quadratic formula.

We only need part of the quadratic formula, the part under the square root sign, $b^2 - 4ac$. When this number is negative, the quadratic factor has two complex zeros, $(-b \pm \sqrt{\text{negative number}})/2a$. When this number is positive, there are two real number solutions, $(-b \pm \sqrt{\text{positive number}})/2a$. When this number is zero, there is one real zero, $(-b \pm \sqrt{0})/2a = -b/2a$. For this reason, $b^2 - 4ac$ is called the *discriminant*.

The graphs of some polynomials having irreducible quadratic factors need extra points plotted to get a more accurate graph. The graph in Fig. 9-27 shows the graph of $f(x) = x^4 - 3x^2 - 4$ plotted using our usual method—plotting the x-intercepts, a point to the left of the smallest x-intercept, a point between each consecutive pair of x-intercepts, and a point to the right of the largest x-intercept.

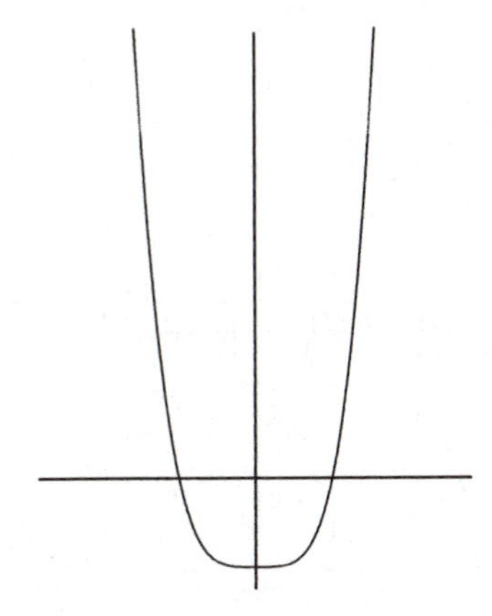

Fig. 9-27.

See what happens to the graph when we plot the points for $x = 1$ and $x = -1$.

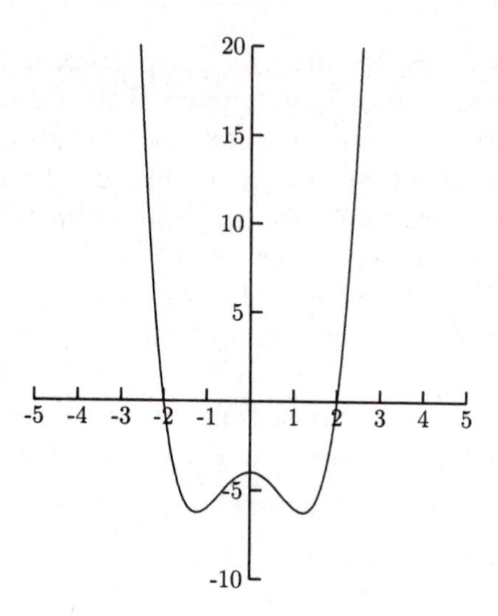

Fig. 9-28.

The graph of $f(x) = (x - 2)(x^2 + 6x + 10)$ is sketched in Fig. 9-29. The graphs we have sketched have several vertices between x-intercepts. When this happens, we need calculus to find them.

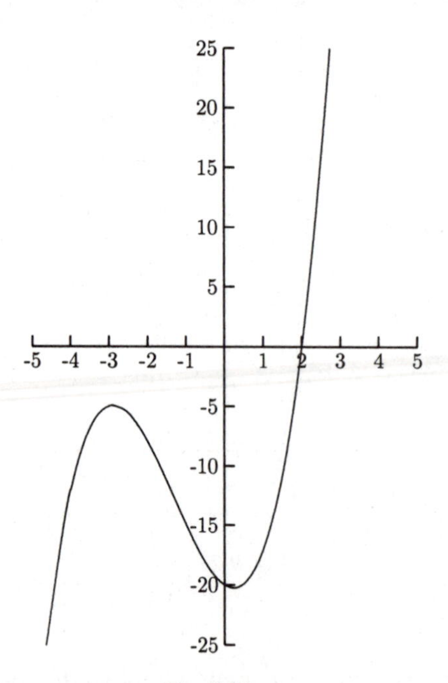

Fig. 9-29.

The Fundamental Theorem of Algebra

By the *Fundamental Theorem of Algebra*, every polynomial of degree n has exactly n zeros (some might be counted more than once). Because $x = c$ is a zero implies $x - c$ is a factor, every polynomial can be completely factored in the form $a(x - c_n)(x - c_{n-1}) \cdots (x - c_1)$, where a is a real number and c_i is real or complex. Factors in the form $x - c$ are called *linear factors*. Factors such as $2x + 1$ can be written in the form $x - c$ by factoring 2: $2(x + \frac{1}{2})$ or $2(x - (-\frac{1}{2}))$.

To completely factor a polynomial, we usually need to first find its zeros. At times, we will use the Rational Zero Theorem, polynomial division, and the quadratic formula.

EXAMPLES

Find all zeros, real and complex.

- $h(x) = x^4 - 16$
  This is the difference of two squares.

$$x^4 - 16 = (x^2 - 4)(x^2 + 4) = (x - 2)(x + 2)(x^2 + 4)$$

The real zeros are 2 and −2. Find the complex zeros by solving $x^2 + 4 = 0$.

$$x^2 + 4 = 0$$

$$x^2 = -4$$

$$x = \pm\sqrt{-4} = \pm 2i$$

The complex zeros are $\pm 2i$.

- $x^4 + 6x^3 + 9x^2 - 6x - 10$
  The possible rational zeros are ± 1, ± 2, ± 5, and ± 10. $P(1) = 0$.

$$\begin{array}{r|rrrrr} 1 & 1 & 6 & 9 & -6 & -10 \\ & & 1 & 7 & 16 & 10 \\ \hline & 1 & 7 & 16 & 10 & 0 \end{array}$$

$$P(x) = (x - 1)(x^3 + 7x^2 + 16x + 10)$$

Because $x^3 + 7x^2 + 16x + 10$ has no sign changes, there are no positive zeros; $x = -1$ is a zero for $x^3 + 7x^2 + 16x + 10$.

$$\begin{array}{r|rrrr} -1 & 1 & 7 & 16 & 10 \\ & & -1 & -6 & -10 \\ \hline & 1 & 6 & 10 & 0 \end{array}$$

$$P(x) = (x-1)(x+1)(x^2 + 6x + 10)$$

Solve $x^2 + 6x + 10 = 0$ to find the complex zeros.

$$\begin{aligned} x &= \frac{-6 \pm \sqrt{6^2 - 4(1)(10)}}{2(1)} = \frac{-6 \pm \sqrt{-4}}{2} \\ &= \frac{-6 \pm 2i}{2} = \frac{2(-3 \pm i)}{2} = -3 \pm i \end{aligned}$$

The zeros are $\pm 1, \ -3 \pm i$.

PRACTICE

Find all zeros, real and complex.

1. $f(x) = x^4 - 81$
2. $h(x) = x^3 + 13x - 34$
3. $f(x) = x^4 + 5x^2 + 4$
4. $P(x) = x^4 - 6x^3 + 29x^2 - 76x + 68$

SOLUTIONS

1. $f(x) = (x^2 - 9)(x^2 + 9) = (x-3)(x+3)(x^2 + 9)$

$$\begin{aligned} x^2 + 9 &= 0 \\ x^2 &= -9 \\ x &= \pm\sqrt{-9} = \pm 3i \end{aligned}$$

The zeros are $\pm 3, \ \pm 3i$.

2. $h(2) = 0$

$$\begin{array}{r|rrrr} 2 & 1 & 0 & 13 & -34 \\ & & 2 & 4 & 34 \\ \hline & 1 & 2 & 17 & 0 \end{array}$$

$h(x) = (x-2)(x^2 + 2x + 17)$

$$x^2 + 2x + 17 = 0$$

$$x = \frac{-2 \pm \sqrt{2^2 - 4(1)(17)}}{2(1)} = \frac{-2 \pm 8i}{2}$$

$$= \frac{2(-1 \pm 4i)}{2} = -1 \pm 4i$$

The zeros are 2, $-1 \pm 4i$.

3. $(x^4 + 5x^2 + 4) = (x^2 + 1)(x^2 + 4)$

$$x^2 + 1 = 0 \qquad x^2 + 4 = 0$$

$$x^2 = -1 \qquad x^2 = -4$$

$$x = \pm i \qquad x = \pm 2i$$

The zeros are $\pm i$, $\pm 2i$.

4. $P(2) = 0$

2	1	−6	29	−76	68
		2	−8	42	−68
	1	−4	21	−34	0

$$P(x) = (x - 2)(x^3 - 4x^2 + 21x - 34)$$

$x = 2$ is a zero for $x^3 - 4x^2 + 21x - 34$

2	1	−4	21	−34
		2	−4	34
	1	−2	17	0

$$P(x) = (x - 2)(x - 2)(x^2 - 2x + 17)$$

$$x^2 - 2x + 17 = 0$$

$$x = \frac{-(-2) \pm \sqrt{(-2)^2 - 4(1)(17)}}{2(1)} = \frac{2 \pm \sqrt{-64}}{2}$$

$$= \frac{2 \pm 8i}{2} = \frac{2(1 \pm 4i)}{2} = 1 \pm 4i$$

The zeros are 2, $1 \pm 4i$.

If we know a complex number is a zero for a polynomial, we automatically know another zero—the complex conjugate is also a zero. This gives us a quadratic factor for the polynomial. Once we have this computed, we can use long division to find the quotient, which will be another factor of the polynomial. Each time we factor a polynomial, we are closer to finding its zeros.

EXAMPLES

Find all zeros, real and complex.

- $f(x) = 3x^4 + x^3 + 17x^2 + 4x + 20$ and $x = 2i$ is a zero.
  Because $x = 2i$ is a zero, its conjugate, $-2i$ is another zero. This tells us that two factors are $x - 2i$ and $x + 2i$.

$$(x - 2i)(x + 2i) = x^2 + 2ix - 2ix - 4i^2 = x^2 - 4(-1) = x^2 + 4$$

We will divide $f(x)$ by $x^2 + 4 = x^2 + 0x + 4$.

$$\begin{array}{r|l} & \quad 3x^2 + \quad x + \quad 5 \\ \hline x^2 + 0x + 4 & \quad 3x^4 + x^3 + 17x^2 + 4x + 20 \\ & -(3x^4 + 0x^3 + 12x^2) \\ \hline & \qquad\quad x^3 + 5x^2 + 4x \\ & \qquad -(x^3 + 0x^2 + 4x) \\ \hline & \qquad\qquad\quad 5x^2 + 0x + 20 \\ & \qquad\qquad -(5x^2 + 0x + 20) \\ \hline & \qquad\qquad\qquad\qquad 0 \end{array}$$

$f(x) = (x^2 + 4)(3x^2 + x + 5)$. Solving $3x^2 + x + 5 = 0$, we get the solutions

$$x = \frac{-1 \pm \sqrt{1^2 - 4(3)(5)}}{2(3)} = \frac{-1 \pm \sqrt{-59}}{6} = \frac{-1 \pm \sqrt{59}\, i}{6}.$$

The zeros are $\pm 2i$, $(-1 \pm \sqrt{59}\, i)/6$.

- $h(x) = 2x^3 - 7x^2 + 170x - 246$, $x = 1 + 9i$ is a zero.
  Because $x = 1 + 9i$ is a zero, we know that $x = 1 - 9i$ is also a zero. We also know that $x - (1 + 9i) = x - 1 - 9i$ and $x - (1 - 9i) = x - 1 + 9i$ are factors. We will multiply these two factors.

$$\begin{aligned} (x - 1 - 9i)(x - 1 + 9i) &= x^2 - x + 9ix - x + 1 - 9i - 9ix + 9i - 81i^2 \\ &= x^2 - 2x + 1 - 81(-1) = x^2 - 2x + 82 \end{aligned}$$

$$\begin{array}{r|l} & \quad 2x - 3 \\ \hline x^2 - 2x + 82 & \quad 2x^3 - 7x^2 + 170x - 246 \\ & -(2x^3 - 4x^2 + 164x) \\ \hline & \qquad\quad -3x^2 + 6x - 246 \\ & \qquad -(-3x^2 + 6x - 246) \\ \hline & \qquad\qquad\qquad\qquad 0 \end{array}$$

$h(x) = (2x - 3)(x^2 - 2x + 82)$.
The zeros are $1 \pm 9i$ and $\frac{3}{2}$ (from $2x - 3 = 0$).

PRACTICE

Find all zeros, real and complex.

1. $f(x) = x^4 - x^3 + 8x^2 - 9x - 9$; $x = -3i$ is a zero.
2. $g(x) = x^3 - 5x^2 + 7x + 13$; $x = 3 - 2i$ is a zero.
3. $h(x) = x^4 - 8x^3 + 21x^2 + 32x - 100$; $x = 4 + 3i$ is a zero.

SOLUTIONS

1. $x = -3i$ is a zero, so $x = 3i$ is also a zero. One factor of $f(x)$ is $(x - 3i)(x + 3i) = x^2 + 9 = x^2 + 0x + 9$.

$$\begin{array}{r|l} & x^2 - x - 1 \\ \hline x^2 + 0x + 9 & x^4 - x^3 + 8x^2 - 9x - 9 \\ & -(x^4 + 0x^3 + 9x^2) \\ \hline & -x^3 - x^2 - 9x \\ & -(-x^3 + 0x^2 - 9x) \\ \hline & -x^2 + 0x - 9 \\ & -(-x^2 + 0x - 9) \\ \hline & 0 \end{array}$$

$f(x) = (x^2 + 9)(x^2 - x - 1)$. Solve $x^2 - x - 1 = 0$.

$$x = \frac{-(-1) \pm \sqrt{(-1)^2 - 4(1)(-1)}}{2(1)} = \frac{1 \pm \sqrt{5}}{2}$$

The zeros are $\pm 3i$, $(1 \pm \sqrt{5})/2$.

2. $x = 3 - 2i$ is a zero, so $x = 3 + 2i$ is also a zero. One factor of $g(x)$ is

$$\begin{aligned} &(x - (3 - 2i))(x - (3 + 2i)) \\ &\quad = (x - 3 + 2i)(x - 3 - 2i) \\ &\quad = x^2 - 3x - 2ix - 3x + 9 + 6i + 2ix - 6i - 4i^2 \\ &\quad = x^2 - 6x + 9 - 4(-1) = x^2 - 6x + 13. \end{aligned}$$

$$\begin{array}{r|l} & x + 1 \\ \hline x^2 - 6x + 13 & x^3 - 5x^2 + 7x + 13 \\ & -(x^3 - 6x^2 + 13x) \\ \hline & x^2 - 6x + 13 \\ & -(x^2 - 6x + 13) \\ \hline & 0 \end{array}$$

$g(x) = (x + 1)(x^2 - 6x + 13)$.
The zeros are -1, $3 \pm 2i$.

3. $x = 4 + 3i$ is a zero, so $x = 4 - 3i$ is also a zero. One factor of $g(x)$ is

$$\begin{aligned}&(x-(4+3i))(x-(4-3i))\\&= (x-4-3i)(x-4+3i)\\&= x^2-4x+3ix-4x+16-12i-3ix+12i-9i^2\\&= x^2-8x+16-9(-1) = x^2-8x+25.\end{aligned}$$

$$\begin{array}{r|l} & x^2 - 4 \\ \hline x^2-8x+25 & x^4 - 8x^3 + 21x^2 + 32x - 100 \\ & -(x^4 - 8x^3 + 25x^2) \\ \hline & \qquad -4x^2 + 32x - 100 \\ & \qquad -(-4x^2 + 32x - 100) \\ \hline & \qquad\qquad 0 \end{array}$$

$h(x) = (x^2 - 4)(x^2 - 8x + 25)$.
The zeros are $4 \pm 3i$ and ± 2 (from $x^2 - 4 = 0$).

A consequence of the Fundamental Theorem of Algebra is that a polynomial of degree n will have n zeros, although not necessarily n different zeros. For example, the polynomial $f(x) = (x-2)^3 = (x-2)(x-2)(x-2)$ has $x = 2$ as a zero three times. The number of times an x-value is a zero is called its *multiplicity*. In the above example, $x = 2$ is a zero with multiplicity 3.

EXAMPLE

- $f(x) = x^4(x+3)^2(x-6)$
  $x = 0$ is a zero with multiplicity 4. (We can think of x^4 as $(x-0)^4$.)
  $x = -3$ is a zero with multiplicity 2.
  $x = 6$ is a zero with multiplicity 1.

PRACTICE

State each zero and its multiplicity.

1. $f(x) = x^2(x+4)(x+9)^6(x-5)^3$

SOLUTION

1. $x = 0$ is a zero with multiplicity 2.
   $x = -4$ is a zero with multiplicity 1.
   $x = -9$ is a zero with multiplicity 6.
   $x = 5$ is a zero with multiplicity 3.

Now, instead of finding the zeros for a given polynomial, we will find a polynomial with the given zeros. Because we will know the zeros, we will know the factors. Once we know the factors of a polynomial, we have a fairly good idea of the polynomial.

EXAMPLES

Find a polynomial with integer coefficients having the given degree and zeros.

- Degree 3 with zeros 1, 2, and 5.
  Because $x = 1$ is a zero, $x - 1$ is a factor. Because $x = 2$ is a zero, $x - 2$ is a factor. And because $x = 5$ is a zero, $x - 5$ is a factor. Such a polynomial will be of the form $a(x-1)(x-2)(x-5)$, where a is some nonzero number. We will want to choose a so that the coefficients are integers.

$$\begin{aligned} a(x-1)(x-2)(x-5) &= a(x-1)[(x-2)(x-5)] \\ &= a(x-1)(x^2-7x+10) \\ &= a(x^3-7x^2+10x-x^2+7x-10) \\ &= a(x^3-8x^2+17x-10) \end{aligned}$$

  Because the coefficients are already integers, we can let $a = 1$. One polynomial of degree 3 having integer coefficients and 1, 2, and 5 as zeros is $x^3 - 8x^2 + 17x - 10$.
- Degree 4 with zeros -3 and $2 - 5i$, with -3 a zero of multiplicity 2. Because -3 is a zero of multiplicity 2, $(x+3)^2 = x^2 + 6x + 9$ is a factor. Because $2 - 5i$ is a zero, $2 + 5i$ is another zero. Another factor of the polynomial is

$$\begin{aligned} &(x-(2-5i))(x-(2+5i)) \\ &\quad = (x-2+5i)(x-2-5i) \\ &\quad = x^2-2x-5ix-2x+4+10i+5ix-10i-25i^2 \\ &\quad = x^2-4x+4-25(-1) = x^2-4x+29. \end{aligned}$$

  The polynomial has the form $a(x^2+6x+9)(x^2-4x+29)$, where a is any real number that makes all coefficients integers.

$$\begin{aligned} &a(x^2+6x+9)(x^2-4x+29) \\ &\quad = a(x^4-4x^3+29x^2+6x^3 \\ &\qquad -24x^2+174x+9x^2-36x+261) \\ &\quad = a(x^4+2x^3+14x^2+138x+261) \end{aligned}$$

Because the coefficients are already integers, we can let $a = 1$. One polynomial that satisfies the given conditions is $x^4 + 2x^3 + 14x^2 + 138x + 261$.

PRACTICE

Find a polynomial with integer coefficients having the given degree and zeros.

1. Degree 3 with zeros 0, -4, and 6.
2. Degree 4 with zeros $-5i$ and $3i$.
3. Degree 4 with zeros -1 and $6 - 7i$, where $x = -1$ has multiplicity 2.

SOLUTIONS

1. One polynomial with integer coefficients, with degree 3 and zeros 0, -4 and 6 is

$$x(x+4)(x-6) = x(x^2 - 2x - 24) = x^3 - 2x^2 - 24x.$$

2. One polynomial with integer coefficients, with degree 4 and zeros $-5i$ and $3i$ is

$$\begin{aligned}(x+5i)(x-5i)(x-3i)(x+3i) &= (x^2+25)(x^2+9) \\ &= x^4 + 34x^2 + 225.\end{aligned}$$

3. One polynomial with integer coefficients, with degree 4 and zeros -1, $6 - 7i$, where $x = -1$ has multiplicity 2 is

$$\begin{aligned}&(x+1)^2(x-(6-7i))(x-(6+7i)) \\ &\quad = (x+1)^2(x-6+7i)(x-6-7i) \\ &\quad = [(x+1)(x+1)][(x^2 - 6x - 7ix - 6x \\ &\qquad + 36 + 42i + 7ix - 42i - 49i^2)] \\ &\quad = (x^2+2x+1)(x^2-12x+85) \\ &\quad = x^4 - 12x^3 + 85x^2 + 2x^3 - 24x^2 \\ &\qquad + 170x + x^2 - 12x + 85 \\ &\quad = x^4 - 10x^3 + 62x^2 + 158x + 85.\end{aligned}$$

In the previous problems, there were infinitely many answers because a could be any integer. In the following problems, there will be exactly one polynomial that satisfies the given conditions. This means that a will likely be a number other than 1.

EXAMPLES

Find a polynomial that satisfies the given conditions.

- Degree 3 with zeros -1, -2, and 4, where the coefficient for x is -20.

$$\begin{aligned} a(x+1)(x+2)(x-4) &= a(x+1)[(x+2)(x-4)] \\ &= a(x+1)(x^2-2x-8) \\ &= a(x^3-2x^2-8x+x^2-2x-8) \\ &= a(x^3-x^2-10x-8) \\ &= ax^3-ax^2-10ax-8a \end{aligned}$$

Because we need the coefficient of x to be -20, we need $-10ax = -20x$, so we need $a=2$ (from $-10a=-20$). The polynomial that satisfies the conditions is

$$2x^3-2x^2-20x-16.$$

- Degree 3 with zeros $\frac{2}{3}$ and $-1-5i$, where the coefficient of x^2 is -4. If $x=\frac{2}{3}$ is a zero, then $3x-2$ is a factor.

$$\begin{aligned} x-\frac{2}{3} &= 0 \\ 3\left(x-\frac{2}{3}\right) &= 3(0) \\ 3x-2 &= 0 \end{aligned}$$

The other factors are $x-(-1-5i)=x+1+5i$ and $x-(-1+5i)=x+1-5i$.

$$\begin{aligned} \bigl((x+1+5i)(x+1-5i) &= x^2+x-5ix+x+1-5i+5ix+5i-25i^2 \\ &= x^2+2x+26\bigr) \end{aligned}$$

$$\begin{aligned} a(3x-2)(x^2+2x+26) &= a(3x^3+6x^2+78x-2x^2-4x-52) \\ &= a(3x^3+4x^2+74x-52) \\ &= 3ax^3+4ax^2+74ax-52a \end{aligned}$$

We want $4ax^2=-4x^2$, so we need $a=-1$. The polynomial that satisfies the conditions is

$$-3x^3-4x^2-74x+52.$$

PRACTICE

Find the polynomial that satisfies the given conditions.

1. Degree 3, zeros 4 and ± 1, with leading coefficient 3.
2. Degree 3 with zeros $-\frac{3}{5}$ and 1, where the multiplicity of 1 is 2, and the coefficient of x is 2.
3. Degree 4 with zeros i and $4i$, with constant term 8.

SOLUTIONS

1. The factors are $x-4$, $x-1$, and $x+1$.

$$\begin{aligned} a(x-4)(x-1)(x+1) &= a(x-4)[(x-1)(x+1)] \\ &= a(x-4)(x^2-1) \\ &= a[(x-4)(x^2-1)] \\ &= a(x^3-4x^2-x+4) \\ &= ax^3-4ax^2-ax+4a \end{aligned}$$

We want the leading coefficient to be 3, so $a=3$. The polynomial that satisfies the conditions is $3x^3-12x^2-3x+12$.

2. Because $x=-\frac{3}{5}$ is a zero, $5x+3$ is a factor.

$$x-\left(-\frac{3}{5}\right)=0$$

$$5\left(x+\frac{3}{5}\right)=5(0)$$

$$5x+3=0$$

The other factor is $(x-1)^2=(x-1)(x-1)=x^2-2x+1$.

$$\begin{aligned} &a(5x+3)(x^2-2x+1) \\ &= a(5x^3-10x^2+5x+3x^2-6x+3) \\ &= a(5x^3-7x^2-x+3) \\ &= 5ax^3-7ax^2-ax+3a \end{aligned}$$

We want $-ax=2x$, so $a=-2$. The polynomial that satisfies the conditions is $-10x^3+14x^2+2x-6$.

3. The factors are $x+i$, $x-i$, $x-4i$, and $x+4i$.

$$\begin{aligned} &a(x+i)(x-i)(x-4i)(x+4i) \\ &\quad = a[(x+i)(x-i)][(x-4i)(x+4i)] \\ &\quad = a(x^2+1)(x^2+16) = a(x^4+17x^2+16) \\ &\quad = ax^4+17ax^2+16a \end{aligned}$$

We want $16a=8$, so $a=\frac{1}{2}$. The polynomial that satisfies the conditions is $\frac{1}{2}x^4+\frac{17}{2}x^2+8$.

Chapter 9 Review

1. The graph of a polynomial function whose leading term is $5x^4$
   a) goes up on the left and up on the right.
   b) goes down on the left and down on the right.
   c) goes up on the left and down on the right.
   d) goes down on the left and up on the right.
2. The zeros for the function $f(x)=x(x+1)^2(x-2)$ are
   a) $x=1, -2$ b) $x=0, 1, -2$ c) $x=-1, 2$ d) $x=0, -1, 2$
3. The graph in Fig. 9-30 is the graph of what function?
   a) $f(x)=x(x-2)(x+3)=x^3+x^2-6x$
   b) $f(x)=-x(x-2)(x+3)=-x^3-x^2+6x$
   c) $f(x)=x^2(x-2)(x+3)=x^4+x^3-6x^2$
   d) $f(x)=-x^2(x-2)(x+3)=-x^4-x^3+6x^2$

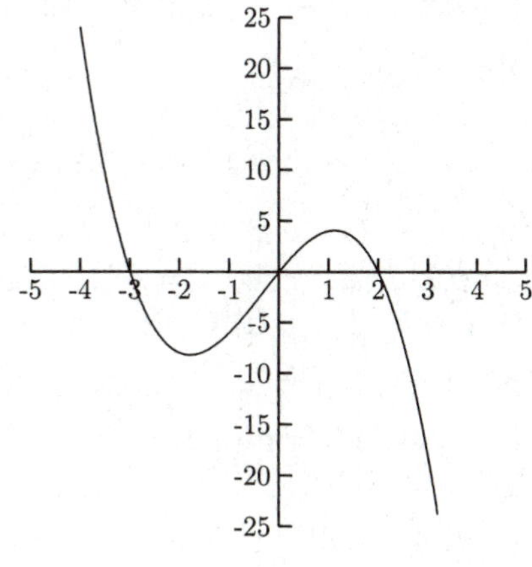

Fig. 9-30.

4. For the polynomial function $f(x) = 3x^4 + 5x^3 - x - 10$
   a) the degree is 4, and the constant term is 3.
   b) the degree is 3, and the constant term is 10.
   c) the degree is 3, and the constant term is -10.
   d) the degree is 4, and the constant term is -10.

5. For the polynomial function $f(x) = \frac{1}{5}(x+3)(x-1)(x-5)$, which set of points would be best to plot for the graph?
   a) $(4, -4.2), (-3, 0), (0, 3), (1, 0), (2, -3), (5, 0)$
   b) $(-4, -9), (-3, 0), (0, 3), (1, 0), (2, -3), (5, 0)$ $(6, 9)$
   c) $(-4, -9), (-3, 0)$ $(-2, 4.2), (1, 0), (2, -3), (4, -4.2), (5, 0)$
   d) $(-2, 4.2), (4, -4.2), (-3, 0), (-1, 4.8), (1, 0), (3, -4.8), (5, 0)$

6. Find the quotient and remainder for $(x^3 + x + 1) \div (x^2 - 1)$.
   a) The quotient is $x + 2$, and the remainder is 3.
   b) The quotient is $x^2 + x + 2$, and the remainder is 3.
   c) The quotient is x, and the remainder is $2x + 1$.
   d) The quotient is x, and the remainder is 1.

7. Find the quotient and remainder for $(5x^2 + 2x + 3) \div (2x + 1)$.
   a) The quotient is $\frac{5}{2}x - \frac{1}{4}$, and the remainder is $\frac{13}{4}$.
   b) The quotient is $\frac{5}{2}x + \frac{9}{4}$, and the remainder is $\frac{21}{4}$.
   c) The quotient is $\frac{5}{2}x - \frac{1}{2}$, and the remainder is $\frac{5}{2}$.
   d) The quotient is $\frac{5}{2}x - \frac{9}{4}$, and the remainder is $-\frac{11}{4}$.

8. Find the quotient and remainder for $(4x^3 - 2x^2 + x - 5) \div (x - 2)$.
   a) The quotient is $4x^2 - 10x + 21$, and the remainder is -47.
   b) The quotient is $4x^2 + 6x + 13$, and the remainder is 21.
   c) The quotient is $4x^2 + 6x - 13$, and the remainder is 21.
   d) The quotient is $4x^2 + 6x + 13$, and the remainder is -31.

9. What are the solutions for $2x^3 - x^2 - 5x - 2 = 0$?
   a) $x = 0, 1, 2$ b) $x = 1, \frac{1}{2}, -2$
   c) $x = -1, -\frac{1}{2}, 2$ d) Cannot be determined

10. What are the zeros for $f(x) = x^3 + 2x^2 + 4x + 3$?
    a) $x = -1, (-1 \pm \sqrt{11}\, i)/2$ b) $x = -1, (-1 \pm \sqrt{13})/2$
    c) $x = -1, (-3 \pm \sqrt{19}\, i)/2$ d) Cannot be determined

11. According to the Rational Zero Theorem, which of the following is *not* a possible rational zero for the polynomial $P(x) = 10x^4 - 6x^3 + x^2 + 6$?
    a) $-\frac{1}{2}$ b) $\frac{1}{3}$ c) 6 d) -1

12. What are the zeros for $P(x) = 3x^3 - 13x^2 - 32x + 12$?
    a) $x = -\frac{1}{3}, -6, -2$ b) $x = 3, 6, -2$
    c) $x = 1, -6, 2$ d) $x = \frac{1}{3}, 6, -2$
13. Which is a polynomial function having a zero of $2 - 3i$?
    a) $f(x) = x^3 - 5x^2 + 17x - 13$ b) $f(x) = x^3 - 5x^2 - 17x + 13$
    c) $f(x) = x^3 + 5x^2 - 17x - 13$ d) $f(x) = x^3 + 5x^2 + 17x - 13$
14. Write the quotient for $(5 + 2i)/(1 + i)$ in the form $a + bi$, where a and b are real numbers.
    a) $\frac{7}{2} + \frac{7}{2}i$ b) $\frac{7}{2} - \frac{3}{2}i$ c) $\frac{3}{2} + \frac{7}{2}i$ d) $\frac{5}{2} + i$
15. According to Descartes' Rule of Signs, how many zeros does the polynomial function $f(x) = -2x^3 + x^2 - x - 5$ have?
    a) 2 or 0 positive zeros and 0 negative zeros.
    b) 3 or 1 positive zeros and 2 or 0 negative zeros.
    c) 2 or 0 positive zeros and 1 negative zero.
    d) 3 or 1 positive zeros and 1 negative zero.

SOLUTIONS

1. a) 2. d) 3. b) 4. d) 5. b) 6. c) 7. a) 8. b)
9. c) 10. a) 11. b) 12. d) 13. a) 14. b) 15. c)

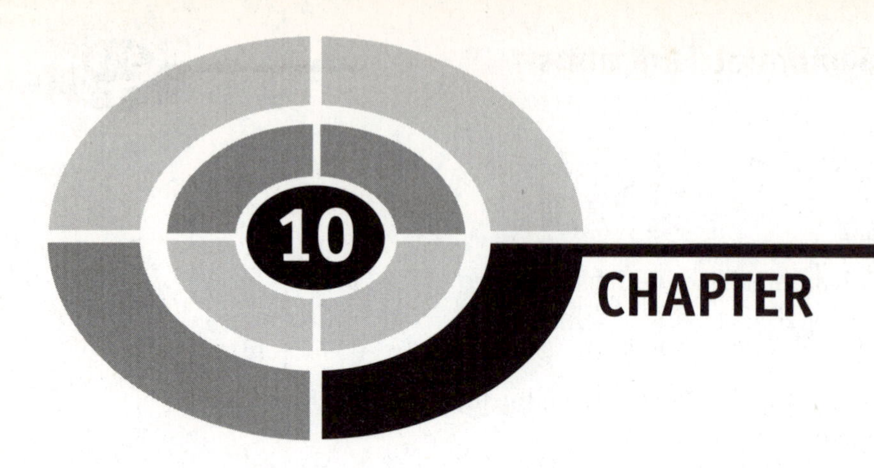

Systems of Equations and Inequalities

A system of equations is a collection of two or more equations whose graphs might or might not intersect (share a common point or points). If the graphs do intersect, then we say that the solution to the system is the point or points where the graphs intersect. For example, the solution to the system

$$\begin{cases} x + y = 4 \\ 3x - y = 0 \end{cases}$$

is (1, 3) because the graphs intersect at (1, 3). We say that (1, 3) *satisfies* the system because if we let $x = 1$ and $y = 3$ in each equation, they will both be true.

$$1 + 3 = 4 \qquad \text{This is a true statement}$$

$$3(1) - 3 = 0 \qquad \text{This is a true statement}$$

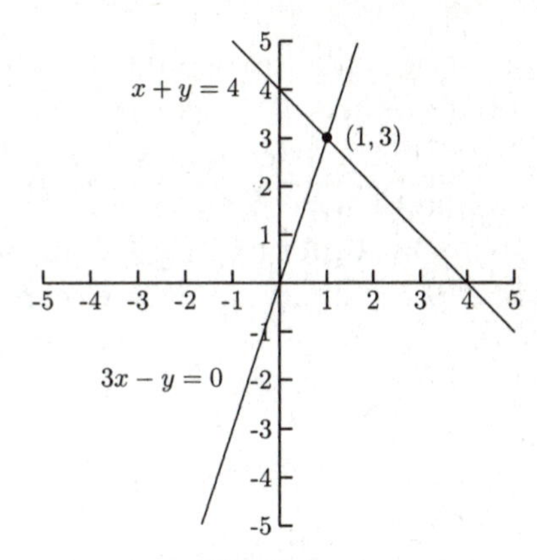

Fig. 10-1.

There are several methods for solving systems of equations. One of them is by sketching the graphs and seeing where, if anywhere, the graphs intersect. Even with a graphing calculator, though, these solutions might only be approximations. When the equations are lines, *matrices* can be used. Graphing calculators are also useful for these. We will concentrate on two methods in this book. One of them is called *substitution* and the other is called *elimination by addition*. Both methods will work with many kinds of systems of equations, but we will start with systems of linear equations.

Substitution

Substitution works by solving for one variable in one equation and making a substitution in the other equation. Technically, it does not matter which variable we use or which equation we begin with, but some choices are easier than others.

EXAMPLES

Solve the systems of equations. Put your solutions in the form of a point, (x, y).

- $\begin{cases} x + y & = 5 \\ -2x + y & = -1 \end{cases}$

  We have four places to start.

  1. Solve for x in the first equation: $x = 5 - y$
  2. Solve for y in the first equation: $y = 5 - x$
  3. Solve for x in the second equation: $x = \frac{1}{2} + \frac{1}{2}y$
  4. Solve for y in the second equation: $y = 2x - 1$

The third option looks like it would be the most trouble, so we will use one of the others. We will use the first option. Because $x = 5 - y$ came from the *first* equation, we will substitute it for x in the *second* equation. Then $-2x+y=-1$ becomes $-2(5-y)+y=-1$. This is the substitution step, substituting what x is equal to, namely $5-y$, for x. Now we can solve the equation $-2(5-y)+y=-1$.

$$-2(5-y)+y=-1$$

$$-10+2y+y=-1$$

$$3y=9$$

$$y=3$$

Now that we know $y = 3$, we could use any of the equations above to find x. We know that $x = 5 - y$, so we will use this.

$$x = 5 - 3 = 2$$

The solution is $x = 2$ and $y = 3$ or the point (2, 3). It is a good idea to check the solution.

$$2+3=5 \qquad \text{This is true.}$$

$$-2(2)+3=-1 \qquad \text{This is true.}$$

- $\begin{cases} 4x - y = 12 & \text{A} \\ 3x + y = 2 & \text{B} \end{cases}$

  We will solve for y in equation B: $y = 2 - 3x$. Next we will substitute $2 - 3x$ for y in equation A and solve for x.

$$4x - y = 12$$

$$4x - (2 - 3x) = 12$$

$$4x - 2 + 3x = 12$$

$$7x = 14$$

$$x = 2$$

Now that we know $x = 2$, we will put $x = 2$ in one of the above equations. We will use $y = 2 - 3x$: $y = 2 - 3(2) = -4$. The solution is $x = 2$, $y = -4$, or $(2, -4)$. The graphs in Fig. 10-2 verify that the solution $(2, -4)$ is on both lines.

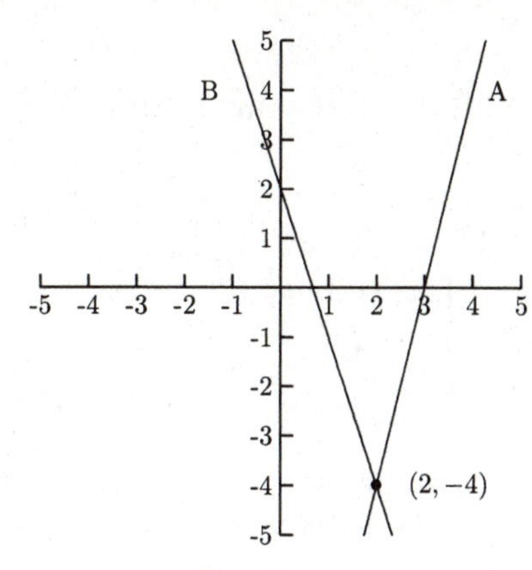

Fig. 10-2.

- $\begin{cases} y = 4x + 1 & \text{A} \\ y = 3x + 2 & \text{B} \end{cases}$

  Both equations are already solved for y, so all we need to do is to set them equal to each other.

$$4x + 1 = 3x + 2$$
$$x = 1$$

Use either equation A or equation B to find y when $x = 1$. We will use A: $y = 4x + 1 = 4(1) + 1 = 5$. The solution is $x = 1$ and $y = 5$, or $(1, 5)$. We can see from the graphs in Fig. 10-3 that $(1, 5)$ is the solution to the system.

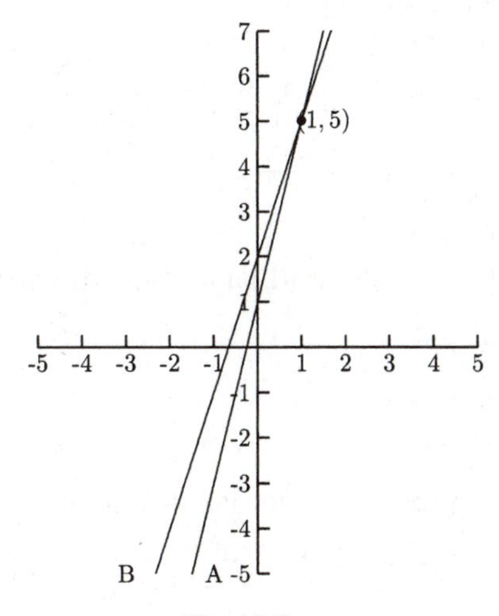

Fig. 10-3.

PRACTICE

Solve the systems of equations. Put your solutions in the form of a point, (x, y).

1.

$$\begin{cases} 2x + 3y = 1 & \text{A} \\ x - 2y = -3 & \text{B} \end{cases}$$

2.

$$\begin{cases} x + y = 3 & \text{A} \\ x + 4y = 0 & \text{B} \end{cases}$$

3.

$$\begin{cases} 2x + y = -2 & \text{A} \\ -3x + 2y = -4 & \text{B} \end{cases}$$

4.

$$\begin{cases} y = x + 1 & \text{A} \\ -3x + 2y = -2 & \text{B} \end{cases}$$

SOLUTIONS

1. Solve for x in B: $x = -3 + 2y$ and substitute this for x in A.

$$2x + 3y = 1$$
$$2(-3 + 2y) + 3y = 1$$
$$-6 + 4y + 3y = 1$$
$$7y = 7$$
$$y = 1 \quad \text{Put } y = 1 \text{ in } x = -3 + 2y$$
$$x = -3 + 2(1) = -1$$

The solution is $(-1, 1)$.

2. Solve for x in B: $x = -4y$ and substitute this for x in A.

$$x + y = 3$$
$$-4y + y = 3$$
$$-3y = 3$$
$$y = -1 \quad \text{Put } y = -1 \text{ in } x = -4y$$
$$x = -4(-1) = 4$$

The solution is $(4, -1)$.

3. Solve for y in A: $y = -2 - 2x$ and substitute for y in B.

$$-3x + 2y = -4$$
$$-3x + 2(-2 - 2x) = -4$$
$$-3x - 4 - 4x = -4$$
$$-7x = 0$$
$$x = 0 \quad \text{Put } x = 0 \text{ in } y = -2 - 2x$$
$$y = -2 - 2(0) = -2$$

The solution is $(0, -2)$

4. Equation A is already solved for y. Substitute $x + 1$ for y in B.

$$-3x + 2y = -2$$
$$-3x + 2(x + 1) = -2$$
$$-3x + 2x + 2 = -2$$
$$-x = -4$$
$$x = 4 \quad \text{Put } x = 4 \text{ in A}$$
$$y = x + 1 = 4 + 1 = 5$$

The solution is $(4, 5)$.

Elimination by Addition

Solving a system of equations by substitution can be more difficult when none of the coefficients is 1. Fortunately, there is another way. We can always *add* the two equations to eliminate one of the variables. Sometimes, though, we might need to multiply one or both equations by a number to make it work.

EXAMPLE

Solve the systems of equations. Put your solutions in the form of a point, (x, y).

- $\begin{cases} 2x - 3y = 16 & \text{A} \\ 5x + 3y = -2 & \text{B} \end{cases}$

  Add the equations by adding like terms. Because we will be adding $-3y$ to $3y$, the y-term will cancel, leaving one equation with only one variable.

$$\begin{aligned} 2x - 3y &= 16 \\ 5x + 3y &= -2 \\ \hline 7x + 0y &= 14 \\ x &= 2 \end{aligned}$$

We can put $x = 2$ into either A or B to find y. We will put $x = 2$ into A.

$$\begin{aligned} 2x - 3y &= 16 \\ 2(2) - 3y &= 16 \\ -3y &= 12 \\ y &= -4 \end{aligned}$$

The solution is $(2, -4)$.

PRACTICE

Solve the systems of equations. Put your solutions in the form of a point, (x, y).

1.

$$\begin{cases} -2x + 7y = 19 & \text{A} \\ 2x - 4y = -10 & \text{B} \end{cases}$$

2.

$$\begin{cases} 15x - y = 9 & \text{A} \\ 2x + y = 8 & \text{B} \end{cases}$$

3.

$$\begin{cases} -5x + 4y = -3 & \text{A} \\ -3x - 4y = 11 & \text{B} \end{cases}$$

SOLUTIONS

1.

$$\begin{aligned} -2x + 7y &= 19 \quad \text{A} \\ 2x - 4y &= -10 \quad +\text{B} \\ \hline 3y &= 9 \\ y &= 3 \\ -2x + 7(3) &= 19 \quad \text{Put } y = 3 \text{ in A} \\ x &= 1 \end{aligned}$$

The solution is $(1, 3)$.

2.

$$\begin{array}{rl} 15x - y = 9 & \text{A} \\ 2x + y = 8 & +\text{B} \\ \hline 17x = 17 & \\ x = 1 & \\ 15(1) - y = 9 & \text{Put } x = 1 \text{ in A} \\ y = 6 & \end{array}$$

The solution is $(1, 6)$.

3.

$$\begin{array}{rl} -5x + 4y = -3 & \text{A} \\ -3x - 4y = 11 & +\text{B} \\ \hline -8x = 8 & \\ x = -1 & \\ -5(-1) + 4y = -3 & \text{Put } x = -1 \text{ in A} \\ y = -2 & \end{array}$$

The solution is $(-1, -2)$.

Sometimes we need to multiply one or both equations by some number or numbers so that one of the variables cancels. Multiplying both sides of *any* equation by a nonzero number never changes the solution.

EXAMPLES

- $\begin{cases} 3x + 6y = -12 & \text{A} \\ 2x + 6y = -14 & \text{B} \end{cases}$

  Because the coefficients on y are the same, we only need to make one of them negative. Multiply either A or B by -1, then add.

$$\begin{array}{rl} -3x - 6y = 12 & -\text{A} \\ 2x + 6y = -14 & +\text{B} \\ \hline -x = -2 & \\ x = 2 & \\ 3(2) + 6y = -12 & \text{Put } x = 2 \text{ in A} \\ y = -3 & \end{array}$$

  The solution is $(2, -3)$.

- $$\begin{cases} 2x + 7y = 1 & \text{A} \\ 4x - 2y = 18 & \text{B} \end{cases}$$

  Several options will work. We could multiply A by -2 so that we could add $-4x$ (in -2A) to $4x$ in B. We could multiply A by 2 and multiply B by 7 so that we could add $14y$ (in 2A) to $-14y$ (in 7B). We could also divide B by -2 so that we could add $2x$ (in A) to $-2x$ (in $-\frac{1}{2}$B). We will add -2A + B.

$$\begin{aligned} -4x - 14y &= -2 \quad -2\text{A} \\ 4x - 2y &= 18 \quad +\text{B} \\ \hline -16y &= 16 \\ y &= -1 \\ 2x + 7(-1) &= 1 \quad \text{Put } y = -1 \text{ in A} \\ x &= 4 \end{aligned}$$

  The solution is $(4, -1)$.

PRACTICE

Solve the systems of equations.

1.
$$\begin{cases} -3x + 2y = 12 & \text{A} \\ 4x + 2y = -2 & \text{B} \end{cases}$$

2.
$$\begin{cases} 6x - 5y = 1 & \text{A} \\ 3x - 2y = 1 & \text{B} \end{cases}$$

3.
$$\begin{cases} 15x + 4y = -1 & \text{A} \\ 5x + 2y = -3 & \text{B} \end{cases}$$

SOLUTIONS

1. We will add $-$A + B.

$$\begin{aligned} 3x - 2y &= -12 \quad -\text{A} \\ 4x + 2y &= -2 \quad +\text{B} \\ \hline 7x &= -14 \\ x &= -2 \\ -3(-2) + 2y &= 12 \quad \text{Put } x = -2 \text{ in A} \\ y &= 3 \end{aligned}$$

   The solution is $(-2, 3)$.

2. We will compute A−2B.

$$\begin{array}{rl} 6x - 5y = 1 & \text{A} \\ \underline{-6x + 4y = -2} & -2\text{B} \\ -y = -1 & \\ y = 1 & \\ 6x - 5(1) = 1 & \text{Put } y = 1 \text{ in A} \\ x = 1 & \end{array}$$

The solution is (1, 1).

3. We will compute A−2B.

$$\begin{array}{rl} 15x + 4y = -1 & \text{A} \\ \underline{-10x - 4y = 6} & -2\text{B} \\ 5x = 5 & \\ x = 1 & \\ 15(1) + 4y = -1 & \text{Put } x = 1 \text{ in A} \\ y = -4 & \end{array}$$

The solution is (1, −4).

Both equations in each of the following systems will need to be changed to eliminate one of the variables.

EXAMPLES

- $\begin{cases} 8x - 5y = -2 & \text{A} \\ 3x + 2y = 7 & \text{B} \end{cases}$

  There are many options. Some are 3A − 8B, −3A + 8B, and 2A + 5B. We will compute 2A + 5B.

$$\begin{array}{rl} 16x - 10y = -4 & 2\text{A} \\ \underline{15x + 10y = 35} & +5\text{B} \\ 31x = 31 & \\ x = 1 & \\ 8(1) - 5y = -2 & \text{Put } x = 1 \text{ in A} \\ y = 2 & \end{array}$$

  The solution is (1, 2).

- $$\begin{cases} \frac{2}{3}x - \frac{1}{4}y &= \frac{25}{72} \quad \text{A} \\ \frac{1}{2}x + \frac{2}{5}y &= -\frac{1}{30} \quad \text{B} \end{cases}$$

  First, we will eliminate the fractions. The lowest common denominator for A is 72, and the lowest common denominator for B is 30.

  $$48x - 18y = 25 \quad 72\text{A}$$
  $$15x + 12y = -1 \quad 30\text{B}$$

  Now we will multiply the first equation by 2 and the second by 3.

  $$\begin{aligned} 96x - 36y &= 50 \\ 45x + 36y &= -3 \\ \hline 141x &= 47 \\ x &= \frac{47}{141} = \frac{1}{3} \\ 96\left(\frac{1}{3}\right) - 36y &= 50 \\ y &= -\frac{1}{2} \end{aligned}$$

  The solution is $(\frac{1}{3}, -\frac{1}{2})$.

PRACTICE

Solve the systems of equations. Put your solutions in the form of a point, (x, y).

1.
$$\begin{cases} 5x - 9y &= -26 \quad \text{A} \\ 3x + 2y &= 14 \quad \text{B} \end{cases}$$

2.
$$\begin{cases} 7x + 2y &= 1 \quad \text{A} \\ 2x + 3y &= -7 \quad \text{B} \end{cases}$$

3.
$$\begin{cases} 3x + 8y &= 12 \quad \text{A} \\ 5x + 6y &= -2 \quad \text{B} \end{cases}$$

4.
$$\begin{cases} \frac{3}{4}x + \frac{1}{5}y &= \frac{23}{60} \quad \text{A} \\ \frac{1}{6}x - \frac{1}{4}y &= -\frac{1}{9} \quad \text{B} \end{cases}$$

SOLUTIONS

1.

$$15x - 27y = -78 \quad 3A$$
$$\underline{-15x - 10y = -70} \quad -5B$$
$$-37y = -148$$
$$y = 4$$
$$5x - 9(4) = -26 \quad \text{Put } y = 4 \text{ in A}$$
$$x = 2$$

The solution is $(2, 4)$.

2.

$$21x + 6y = 3 \quad 3A$$
$$\underline{-4x - 6y = 14} \quad -2B$$
$$17x = 17$$
$$x = 1$$
$$7(1) + 2y = 1 \quad \text{Put } x = 1 \text{ in A}$$
$$y = -3$$

The solution is $(1, -3)$.

3.

$$9x + 24y = 36 \quad 3A$$
$$\underline{-20x - 24y = 8} \quad -4B$$
$$-11x = 44$$
$$x = -4$$
$$3(-4) + 8y = 12 \quad \text{Put } x = -4 \text{ in A}$$
$$y = 3$$

The solution is $(-4, 3)$.

4. First clear the fractions.

$$45x + 12y = 23 \quad 60A$$
$$6x - 9y = -4 \quad 36B$$

Add 3 times the first to 4 times the second.

$$\begin{array}{r} 135x + 36y = 69 \\ 24x - 36y = -16 \\ \hline 159x = 53 \end{array}$$

$$x = \frac{53}{159} = \frac{1}{3}$$

$$45\left(\frac{1}{3}\right) + 12y = 23$$

$$y = \frac{2}{3}$$

The solution is $(\frac{1}{3}, \frac{2}{3})$.

Application to Problems

Systems of two linear equations can be used to solve many kinds of word problems. In these problems, two facts will be given about two variables. Each pair of facts can be represented by a linear equation.

EXAMPLES

- A movie theater charges \$4 for each child's ticket and \$6.50 for each adult's ticket. One night 200 tickets were sold, amounting to \$1100 in ticket sales. How many of each type of ticket was sold?

  Let x represent the number of child tickets sold and y the number of adult tickets sold. One equation comes from the fact that a total of 200 adult and child tickets were sold, giving us $x + y = 200$. The other equation comes from the fact that the ticket revenue was \$1100. The ticket revenue from child tickets is $4x$, and the ticket revenue from adult tickets is $6.50y$. Their sum is 1100 giving us $4x + 6.50y = 1100$.

$$\begin{cases} 4x + 6.50y = 1100 & \text{A} \\ x + y = 200 & \text{B} \end{cases}$$

We could use either substitution or addition to solve this system. Substitution is a little faster. We will solve for x in B.

$$x = 200 - y$$
$$4(200 - y) + 6.50y = 1100 \quad \text{Put } 200 - y \text{ into A}$$
$$800 - 4y + 6.50y = 1100$$
$$y = 120$$
$$x = 200 - y = 200 - 120 = 80$$

Eighty child tickets were sold, and 120 adult tickets were sold.

- A farmer had a soil test performed. The farmer was told that a field needed 1080 pounds of Mineral A and 920 pounds of Mineral B. Two mixtures of fertilizers provide these minerals. Each bag of Brand I provides 25 pounds of Mineral A and 15 pounds of Mineral B. Brand II provides 20 pounds of Mineral A and 20 pounds of Mineral B. How many bags of each brand should the farmer buy?

Let x represent the number of bags of Brand I and y represent the number of bags of Brand II. Then the number of pounds of Mineral A obtained from Brand I is $25x$ and the number of pounds of Mineral B is $15x$. The number of pounds of Mineral A obtained from Brand II is $20y$ and the number of pounds of Mineral B is $20y$. The farmer needs 1080 pounds of Mineral A, $25x$ pounds will come from Brand I and $20y$ will come from Brand II. This gives us the equation $25x + 20y = 1080$. The farmer needs 920 pounds of Mineral B, $15x$ will come from Brand I and $20y$ will come from Brand II. This gives us the equation $15x + 20y = 920$.

$$\begin{cases} 25x + 20y = 1080 & \text{A} \\ 15x + 20y = 920 & \text{B} \end{cases}$$

We will compute A–B.

$$\begin{array}{rl} 25x + 20y = 1080 & \text{A} \\ \underline{-15x - 20y = -920} & -\text{B} \\ 10x = 160 & \\ x = 16 & \\ 25(16) + 20y = 1080 & \\ y = 34 & \end{array}$$

The farmer needs 16 bags of Brand I and 34 bags of Brand II.

- A furniture manufacturer has some discontinued fabric and trim in stock. It can use them on sofas and chairs. There are 160 yards of fabric and 110 yards of trim. Each sofa takes 6 yards of fabric and 4.5 yards of trim. Each chair takes 4 yards of fabric and 2 yards of trim. How many sofas and chairs should be produced in order to use all the fabric and trim?

  Let x represent the number of sofas to be produced and y the number of chairs. The manufacturer needs to use 160 yards of fabric, $6x$ will be used on sofas and $4y$ yards on chairs. This gives us the equation $6x + 4y = 160$. There are 110 yards of trim, $4.5x$ yards will be used on the sofas and $2y$ on the chairs. This gives us the equation $4.5x + 2y = 110$.

$$\begin{cases} 6x + 4y = 160 & \text{F} \\ 4.5x + 2y = 110 & \text{T} \end{cases}$$

We will compute F−2T.

$$\begin{array}{rl} 6x + 4y = 160 & \text{F} \\ -9x - 4y = -220 & -2\text{T} \\ \hline -3x = -60 & \\ x = 20 & \\ 6(20) + 4y = 160 & \\ y = 10 & \end{array}$$

The manufacturer needs to produce 20 sofas and 10 chairs.

PRACTICE

1. A grocery store sells two different brands of milk. The price for the name brand is \$3.50 per gallon, and the price for the store's brand is \$2.25 per gallon. On one Saturday, 4500 gallons of milk were sold for sales of \$12,875. How many of each brand were sold?
2. A cable company offers two services—basic cable and premium cable. It charges \$25 per month for the basic service and \$45 per month for the premium service. Last month, it had 94,000 subscribers and had \$3,030,000 in billing. How many subscribers used the premium service?
3. A gardener wants to add 39 pounds of Nutrient A and 16 pounds of Nutrient B to a garden. Each bag of Brand X provides 3 pounds of Nutrient A and 2 pounds of Nutrient B. Each bag of Brand Y

provides 4 pounds of Nutrient A and 1 pound of Nutrient B. How many bags of each brand should be bought?

4. A clothing manufacturer has 70 yards of a certain fabric and 156 buttons in stock. It manufactures jackets and slacks that use this fabric and button. Each jacket requires $1\frac{1}{3}$ yards of fabric and 4 buttons. Each pair of slacks required $1\frac{3}{4}$ yards of fabric and 3 buttons. How many jackets and pairs of slacks should the manufacturer produce to use all the available fabric and buttons?

SOLUTIONS

1. Let x represent the number of gallons of the name brand sold and y represent the number of gallons of the store brand sold. The total number of gallons sold is 4500, giving us $x + y = 4500$. Revenue from the name brand is $3.50x$ and is $2.25y$ for the store brand. Total revenue is \$12,875, giving us the equation $3.50x + 2.25y = 12{,}875$.

$$\begin{cases} x + y & = 4500 \\ 3.50x + 2.25y & = 12{,}875 \end{cases}$$

We will use substitution.

$$x = 4500 - y$$

$$3.50(4500 - y) + 2.25y = 12{,}875$$

$$y = 2300$$

$$x = 4500 - y = 4500 - 2300 = 2200$$

The store sold 2200 gallons of the name brand and 2300 gallons of the store brand.

2. Let x represent the number of basic service subscribers and y the number of premium service subscribers. The total number of subscribers is 94,000, so $x + y = 94{,}000$. Revenue from basic services is $25x$ and $45y$ from premium services. Billing was \$3,030,000, giving us $25x + 45y = 3{,}030{,}000$.

$$\begin{cases} x + y & = 94{,}000 \\ 25x + 45y & = 3{,}030{,}000 \end{cases}$$

We will use substitution.

$$x = 94{,}000 - y$$
$$25(94{,}000 - y) + 45y = 3{,}030{,}000$$
$$y = 34{,}000$$

There are 34,000 premium service subscribers.

3. Let x represent the number of bags of Brand X and y the number of bags of Brand Y. The gardener will get $3x$ pounds of Nutrient A from x bags of Brand X and $4y$ pounds from y bags of Brand Y, so we need $3x + 4y = 39$. The gardener will get $2x$ pounds of Nutrient B from x bags of Brand X and $1y$ pounds of Nutrient B from y bags of Brand Y, so we need $2x + y = 16$. We will use substitution.

$$y = 16 - 2x$$
$$3x + 4(16 - 2x) = 39$$
$$x = 5$$
$$y = 16 - 2x = 16 - 2(5) = 6$$

The gardener needs to buy 5 bags of Brand X and 6 bags of Brand Y.

4. Let x represent the number of jackets to be produced and y the number of pairs of slacks. To use 70 yards of fabric, we need $1\frac{1}{3}x + 1\frac{3}{4}y = 70$. To use 156 buttons, we need $4x + 3y = 156$.

$$\frac{4}{3}x + \frac{7}{4}y = 70 \qquad \text{F}$$
$$4x + 3y = 156 \qquad \text{B}$$
$$16x + 21y = 840 \qquad 12\text{F}$$
$$\underline{-16x - 12y = -624} \qquad -4\text{B}$$
$$9y = 216$$
$$y = 24$$
$$4x + 3(24) = 156$$
$$x = 21$$

The manufacturer should produce 21 jackets and 24 pairs of slacks.

Systems with No Solutions

Two lines in the plane either intersect in one point, are parallel, or are really the same line. Until now, our lines have intersected in one point. When solving a system of two linear equations that are parallel or are the same line, both variables will cancel and we are left with a true statement such as "3 = 3" or a false statement such as "5 = 1." We will get a true statement when the two lines are the same and a false statement when they are parallel.

EXAMPLES

- $\begin{cases} 2x - 3y = 6 & \text{A} \\ -4x + 6y = 8 & \text{B} \end{cases}$

$$\begin{array}{rl} 4x - 6y = 12 & 2\text{A} \\ -4x + 6y = 8 & +\text{B} \\ \hline 0 = 20 & \end{array}$$

This is a false statement, so the lines are parallel. They are sketched in Fig. 10-4.

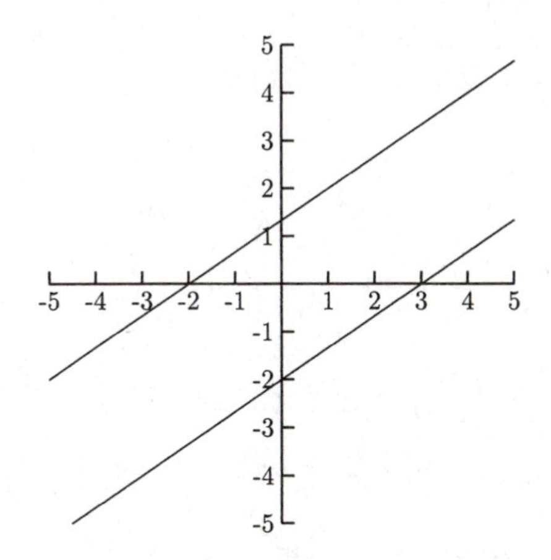

Fig. 10-4.

- $\begin{cases} y = \frac{2}{3}x - 1 \\ 2x - 3y = 3 \end{cases}$

We will use substitution.

$$2x - 3\left(\frac{2}{3}x - 1\right) = 3$$

$$2x - 2x + 3 = 3$$

$$0 = 0$$

Because $0 = 0$ is a true statement, these lines are the same.

Systems Containing Nonlinear Equations

When the system of equations is not a pair of lines, there could be no solutions, one solution, or more than one solution. The same methods used for pairs of lines will work with other kinds of systems.

EXAMPLES

- $\begin{cases} y = x^2 - 2x - 3 & \text{A} \\ 3x - y = 7 & \text{B} \end{cases}$

  Elimination by addition would not work to eliminate x^2 because B has no x^2 term to cancel x^2 in A. Solving for x in B and substituting it in for x in A would work to eliminate x. Both addition and substitution will work to eliminate y. We will use addition to eliminate y.

$$\begin{array}{rl} y = x^2 - 2x - 3 & \text{A} \\ \underline{3x - y = 7 \qquad\qquad} & \text{B} \end{array}$$

$$3x = x^2 - 2x + 4$$

$$0 = x^2 - 5x + 4$$

$$0 = (x - 1)(x - 4)$$

  The solutions occur when $x = 1$ or $x = 4$. We need to find two y-values. We will let $x = 1$ and $x = 4$ in A.

  $y = 1^2 - 2(1) - 3 = -4;$ $(1, -4)$ is one solution.

  $y = 4^2 - 2(4) - 3 = 5;$ $(4, 5)$ is the other solution.

We can see from the graphs in Fig. 10-5 that these solutions are correct.

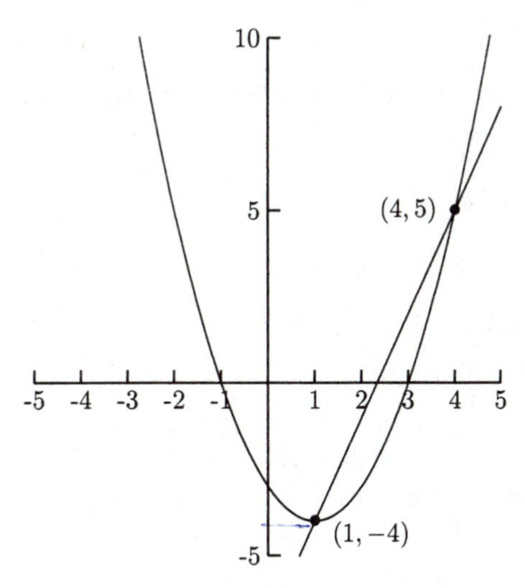

Fig. 10-5.

- $$\begin{cases} x^2 + y^2 = 25 & \text{A} \\ y = -\frac{1}{3}x^2 + 7 & \text{B} \end{cases}$$

We could solve for x^2 in A and substitute this in B. We cannot add the equations to eliminate y or y^2 because A does not have a y term to cancel y in B and B does not have a y^2 term to cancel y^2 in A. We will move $-\frac{1}{3}x^2$ to the left side of B and multiply B by -3. Then we can add this to A to eliminate x^2.

$$\frac{1}{3}x^2 + y = 7 \qquad \text{B}$$

$$x^2 + y^2 = 25 \qquad \text{A}$$

$$\underline{-x^2 - 3y = -21} \qquad -3\text{B}$$

$$y^2 - 3y = 4$$

$$y^2 - 3y - 4 = 0$$

$$(y - 4)(y + 1) = 0$$

The solutions occur when $y = 4, -1$. Put $y = 4, -1$ in A to find their x-values.

$$x^2 + 4^2 = 25$$
$$x^2 = 9$$
$$x = \pm 3; \quad (-3, 4) \text{ and } (3, 4) \text{ are solutions.}$$
$$x^2 + (-1)^2 = 25$$
$$x^2 = 24$$
$$x = \pm\sqrt{24} = \pm 2\sqrt{6}; \quad (2\sqrt{6}, -1) \text{ and } (-2\sqrt{6}, -1) \text{ are solutions.}$$

- $$\begin{cases} x^2 + y^2 = 4 & \text{A} \\ y = 2/x & \text{B} \end{cases}$$

Addition will not work on this system but substitution will. We will substitute $y = 2/x$ for y in A.

$$x^2 + \left(\frac{2}{x}\right)^2 = 4$$
$$x^2 + \frac{4}{x^2} = 4$$
$$x^2\left(x^2 + \frac{4}{x^2}\right) = x^2(4)$$
$$x^4 + 4 = 4x^2$$
$$x^4 - 4x^2 + 4 = 0$$
$$(x^2 - 2)(x^2 - 2) = 0$$
$$x^2 = 2$$
$$x = \pm\sqrt{2}$$

We will put $x = \sqrt{2}$ and $x = -\sqrt{2}$ in $y = 2/x$.

$$y = \frac{2}{\sqrt{2}} = \frac{2\sqrt{2}}{\sqrt{2}\sqrt{2}} = \frac{2\sqrt{2}}{2} = \sqrt{2}; \quad (\sqrt{2}, \sqrt{2}) \text{ is a solution.}$$

$$y = \frac{2}{-\sqrt{2}} = \frac{2\sqrt{2}}{-\sqrt{2}\sqrt{2}} = \frac{2\sqrt{2}}{-2} = -\sqrt{2}; \quad (-\sqrt{2}, -\sqrt{2}) \text{ is a solution.}$$

PRACTICE

Solve the systems of equations. Put your solutions in the form of a point, (x, y).

1.

$$\begin{cases} y = x^2 - 4 & \text{A} \\ x + y = 8 & \text{B} \end{cases}$$

2.

$$\begin{cases} x^2 + y^2 + 6x - 2y = -5 & \text{A} \\ y = -2x - 5 & \text{B} \end{cases}$$

3.

$$\begin{cases} x^2 - y^2 = 16 & \text{A} \\ x^2 + y^2 = 16 & \text{B} \end{cases}$$

4.

$$\begin{cases} 4x^2 + y^2 = 5 & \text{A} \\ y = 1/x & \text{B} \end{cases}$$

SOLUTIONS

1.

$$\begin{array}{rl} y = x^2 - 4 & \text{A} \\ -x - y = -8 & -\text{B} \\ \hline -x = x^2 - 12 & \end{array}$$

$$0 = x^2 + x - 12 = (x + 4)(x - 3)$$

There are solutions for $x = -4$ and $x = 3$. Put these in A.

$$y = (-4)^2 - 4 = 12; \quad (-4, 12) \text{ is a solution.}$$

$$y = 3^2 - 4 = 5; \quad (3, 5) \text{ is a solution.}$$

2. Substitute $-2x - 5$ for y in A.

$$x^2 + (-2x - 5)^2 + 6x - 2(-2x - 5) = -5$$

$$x^2 + 4x^2 + 20x + 25 + 6x + 4x + 10 = -5$$

$$5x^2 + 30x + 40 = 0 \quad \text{Divide by 5}$$

$$x^2 + 6x + 8 = 0$$

$$(x + 4)(x + 2) = 0$$

There are solutions for $x = -4$ and $x = -2$. We will put these in B instead of A because there is less computation to do in B.

$$y = -2(-4) - 5 = 3; \quad (-4, 3) \text{ is a solution.}$$

$$y = -2(-2) - 5 = -1; \quad (-2, -1) \text{ is a solution.}$$

3.

$$\begin{aligned} x^2 - y^2 &= 16 \qquad \text{A} \\ \underline{x^2 + y^2 = 16} & \qquad +\text{B} \\ 2x^2 &= 32 \\ x^2 &= 16 \\ x &= \pm 4 \end{aligned}$$

Put $x = 4$ and $x = -4$ in A.

$$\begin{aligned} (-4)^2 - y^2 &= 16 \qquad & 4^2 - y^2 &= 16 \\ 16 - y^2 &= 16 & 16 - y^2 &= 16 \\ y^2 &= 0 & y^2 &= 0 \\ y &= 0 & y &= 0 \end{aligned}$$

The solutions are $(-4, 0)$ and $(4, 0)$.

4. Substitute $1/x$ for y in A.

$$4x^2 + \left(\frac{1}{x}\right)^2 = 5$$

$$x^2\left(4x^2 + \frac{1}{x^2}\right) = x^2(5)$$

$$4x^4 + 1 = 5x^2$$

$$4x^4 - 5x^2 + 1 = 0$$

$$(4x^2 - 1)(x^2 - 1) = 0$$

$$(2x - 1)(2x + 1)(x - 1)(x + 1) = 0$$

The solutions are $x = \pm\frac{1}{2}$ (from $2x - 1 = 0$ and $2x + 1 = 0$) and $x = \pm 1$. Put these in B.

$$y = \frac{1}{1/2} = 2; \quad \left(\frac{1}{2}, 2\right) \text{ is a solution.}$$

$$y = \frac{1}{-(1/2)} = -2; \quad \left(-\frac{1}{2}, -2\right) \text{ is a solution.}$$

$$y = \frac{1}{1} = 1; \quad (1, 1) \text{ is a solution.}$$

$$y = \frac{1}{-1} = -1; \quad (-1, -1) \text{ is a solution.}$$

Systems of Inequalities

The solution (if any) for a system of inequalities is usually a region in the plane. The solution to a polynomial inequality (the only kind considered in this book) is the region above or below the curve. We will begin with linear inequalities.

When sketching the graph for an inequality, we will use a solid graph for "$\leq$" and "$\geq$" inequalities, and a dashed graph for "$<$" and "$>$" inequalities. We can decide which side of the graph to shade by choosing *any* point not on the graph itself. We will put this point into the inequality. If it makes the inequality true, then we will shade the side that has that point. If it makes the inequality false, we will shade the other side.

EXAMPLES

- $2x + 3y \leq 6$

  We will sketch the line $2x + 3y = 6$, using a solid line because the inequality is "$\leq$."

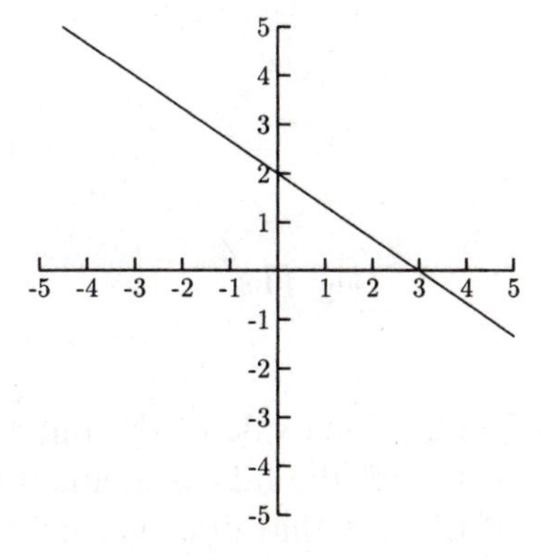

Fig. 10-6.

We will always use the origin $(0, 0)$ in our inequalities unless the graph goes through the origin. Does $(0, 0)$ make $2x + 3y \leq 6$ true? $2(0) + 3(0) \leq 6$ is a true statement, so we will shade the side that has the origin.

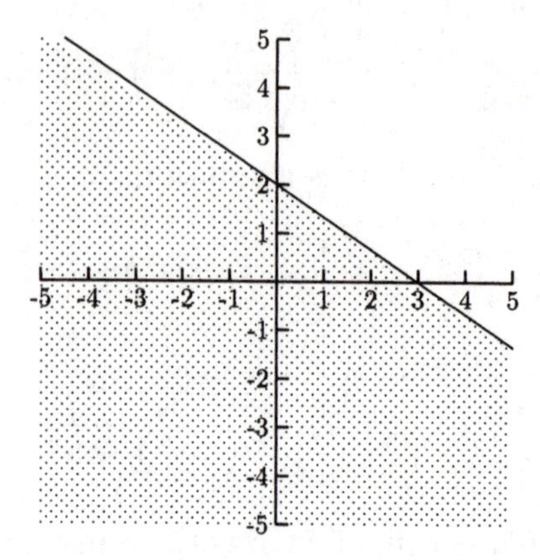

Fig. 10-7.

- $x - 2y > 4$
  We will sketch the line $x - 2y = 4$ using a dashed line because the inequality is ">."

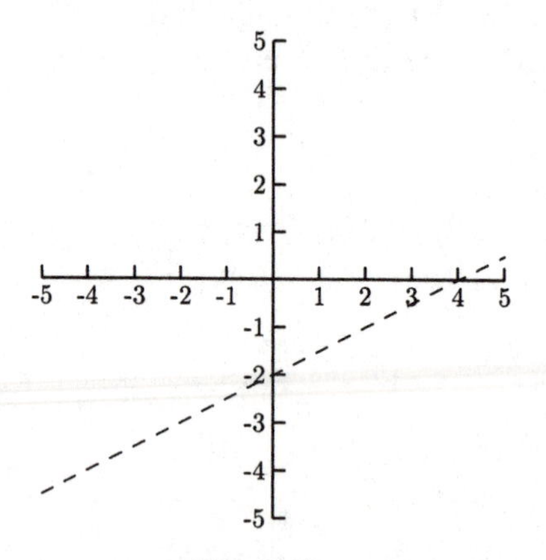

Fig. 10-8.

Now we need to decide which side of the line to shade. When we put $(0, 0)$ in $x - 2y > 4$, we get the false statement $0 - 2(0) > 4$. We need to shade the side of the line that does *not* have the origin.

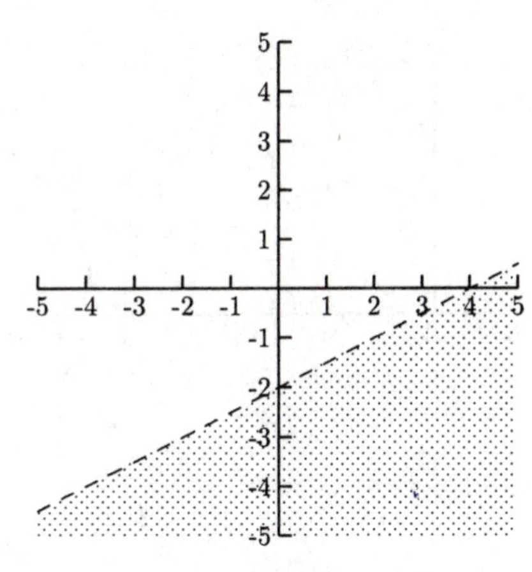

Fig. 10-9.

Every point in the shaded region is a solution to the inequality.

- $y < 3x$

  We use a dashed line to sketch the line $y = 3x$. Because the line goes through $(0, 0)$, we cannot use it to determine which side of the line to shade. This is because any point on the line makes the equality true. We want to know where the inequality is true. The point $(1, 0)$ is not on the line, so we can use it. $0 < 3(1)$ is true so we will shade the side of the line that has the point $(1, 0)$, which is the right-hand side.

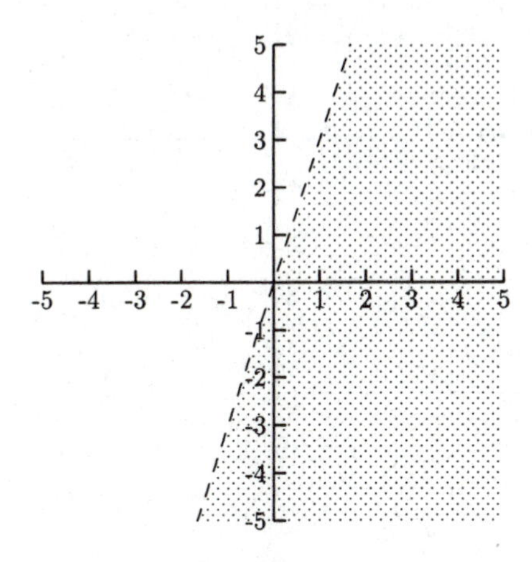

Fig. 10-10.

- $x \geq -3$

  The line $x = -3$ is a vertical line through $x = -3$. Because we want $x \geq -3$ we will shade to the right of the line.

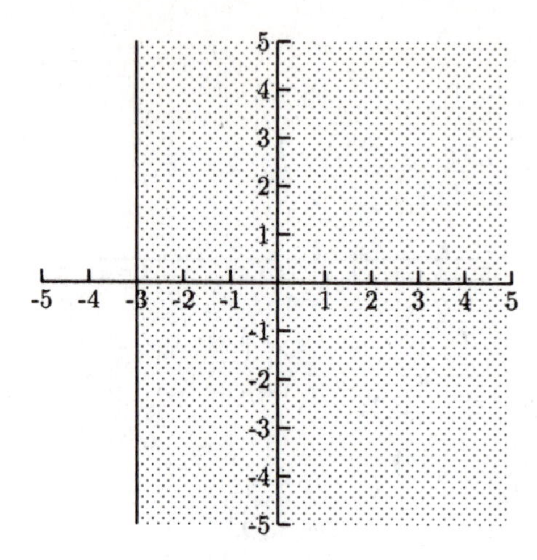

Fig. 10-11.

- $y < 2$
  The line $y = 2$ is a horizontal line at $y = 2$. Because we want $y < 2$, we will shade below the line.

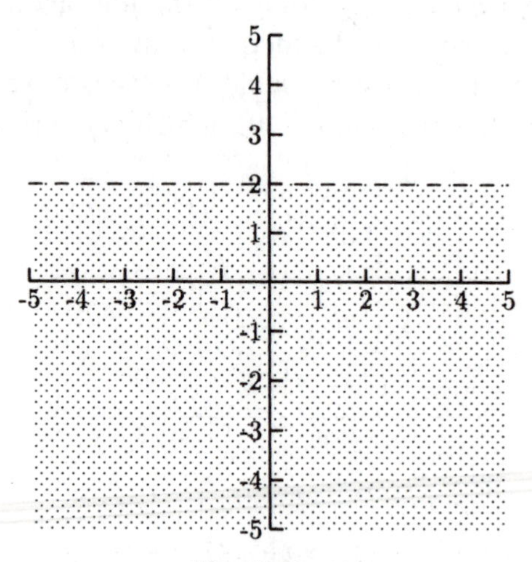

Fig. 10-12.

PRACTICE

Graph the solution.

1. $x + y \geq 2$
2. $2x - 4y < 4$
3. $y \leq \frac{2}{3}x - 1$

4. $x > 1$
5. $y \leq -1$

SOLUTIONS

1.

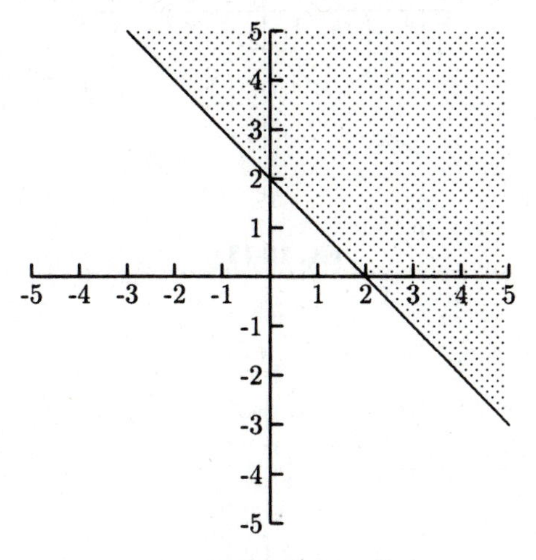

Fig. 10-13.

2.

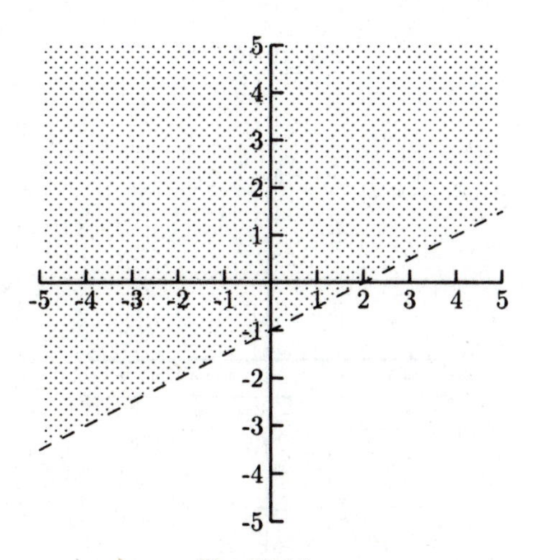

Fig. 10-14.

3.

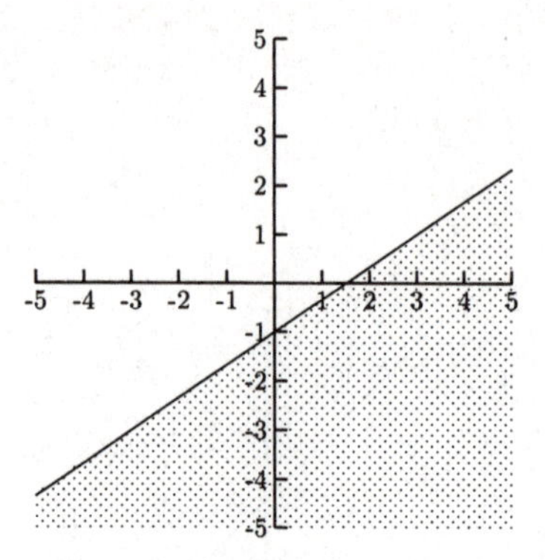

Fig. 10-15.

4.

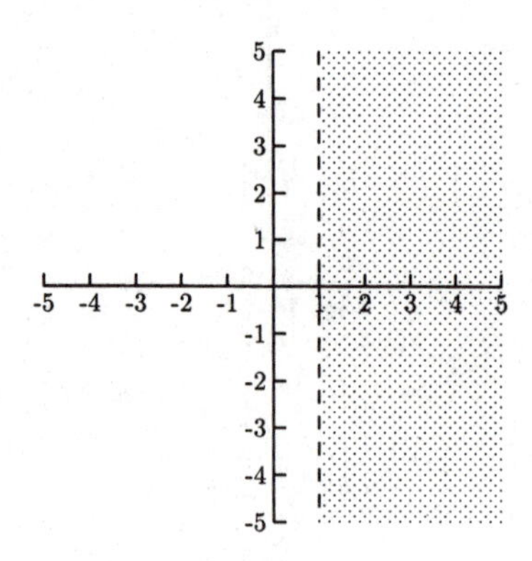

Fig. 10-16.

5.

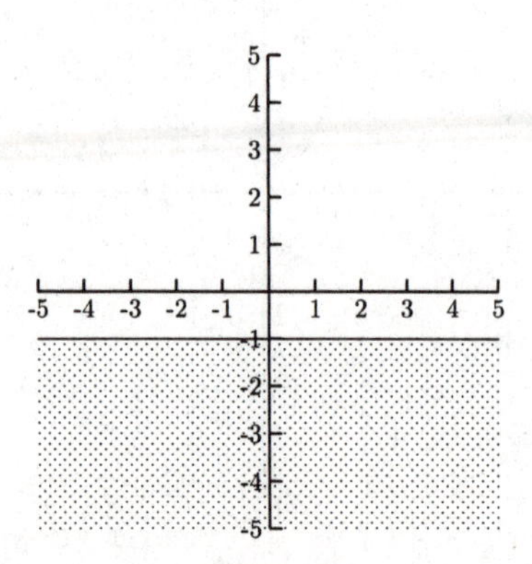

Fig. 10-17.

Graphing the solution region for nonlinear inequalities is done the same way—graph the inequality, using a solid graph for "$\leq$" and "$\geq$" inequalities and a dashed graph for "$<$" and "$>$" inequalities, then check a point to see which side of the graph to shade.

EXAMPLES

- $y \leq x^2 - x - 2$

  The equality is $y = x^2 - x - 2 = (x - 2)(x + 1)$. The graph for this equation is a parabola.

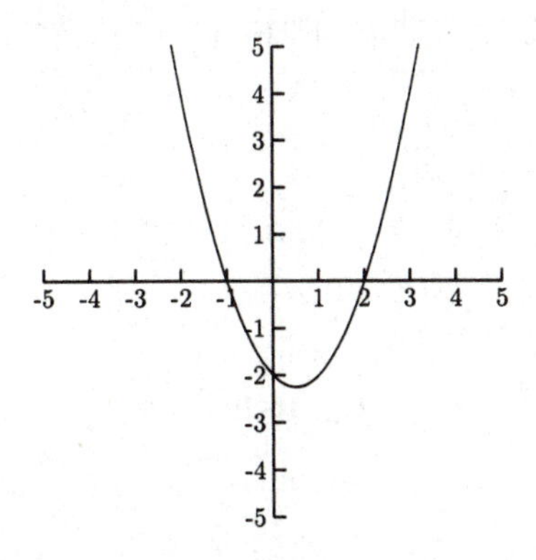

Fig. 10-18.

Because $(0, 0)$ is not on the graph, we can use it to decide which side to shade; $0 \leq 0^2 - 0 - 2$ is false, so we shade below the graph, the side that does not contain $(0, 0)$.

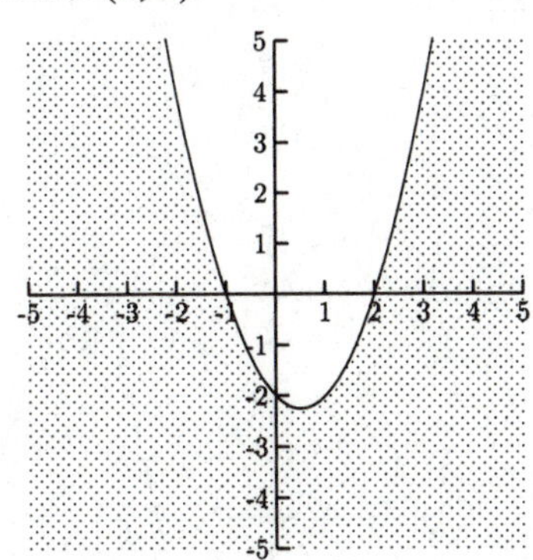

Fig. 10-19.

- $y > (x + 2)(x - 2)(x - 4)$

  When we check $(0, 0)$ in the inequality, we get the false statement $0 > (0 + 2)(0 - 2)(x - 4)$. We will shade above the graph, the region that does not contain $(0, 0)$.

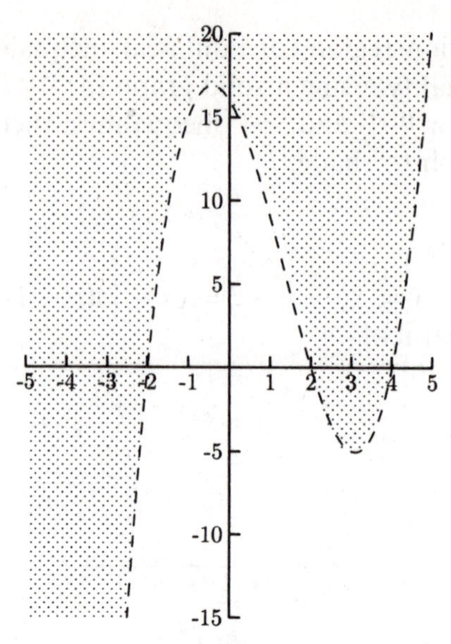

Fig. 10-20.

PRACTICE

Graph the solution.

1. $y \leq x^2 - 4$
2. $y > x^3$
3. $y < |x|$
4. $y \geq (x - 3)(x + 1)(x + 3)$

SOLUTIONS

1.

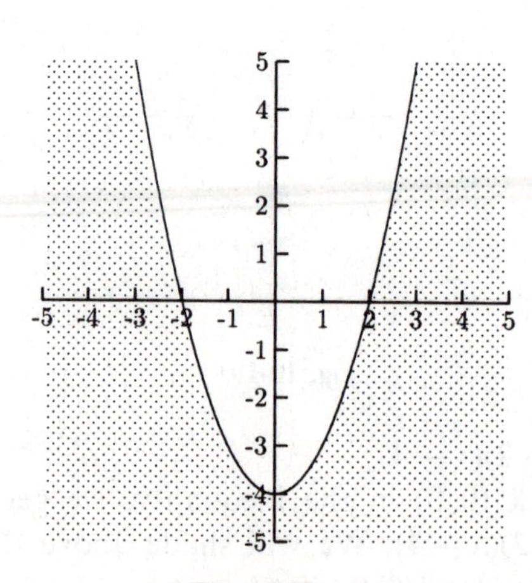

Fig. 10-21.

2.

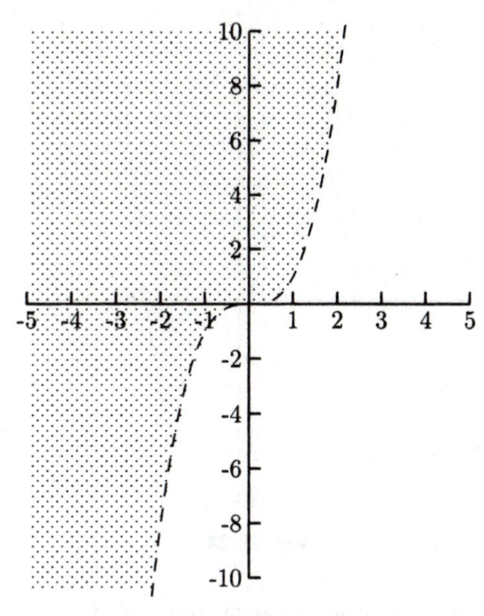

Fig. 10-22.

3.

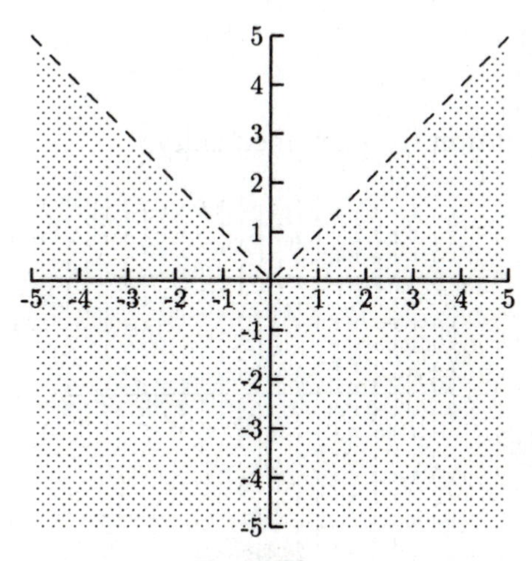

Fig. 10-23.

4.

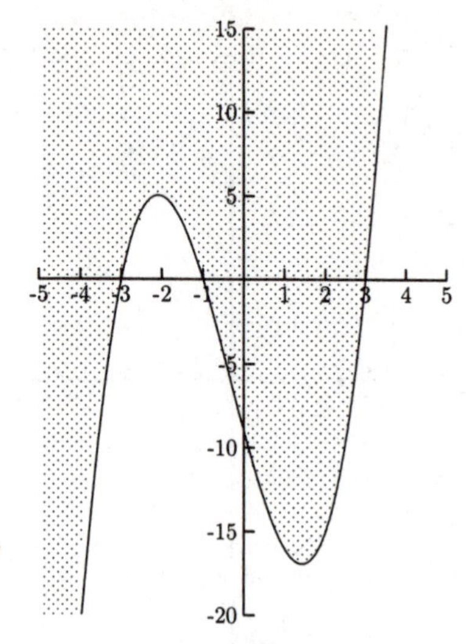

Fig. 10-24.

The solution (if there is one) to a system of two or more inequalities is the region where the shaded regions overlap. For example, if we have a system of two inequalities and shade the solution to one inequality in blue and the other in yellow, then the solution to the system would be the region in green.

EXAMPLES

- $\begin{cases} x - y & < 3 \\ x + 2y & > 1 \end{cases}$

  Sketch the solution for each inequality.

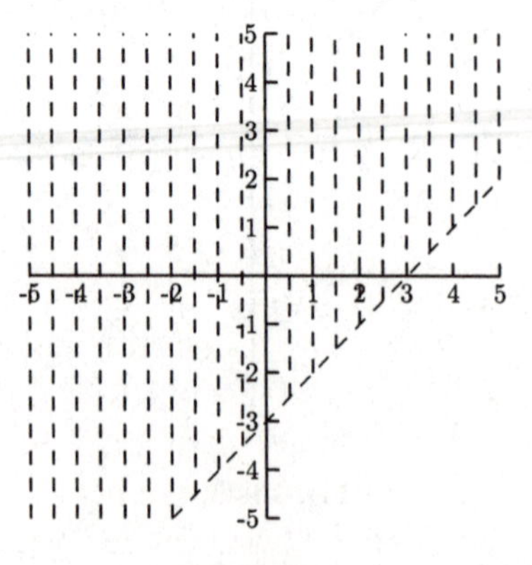

Fig. 10-25.

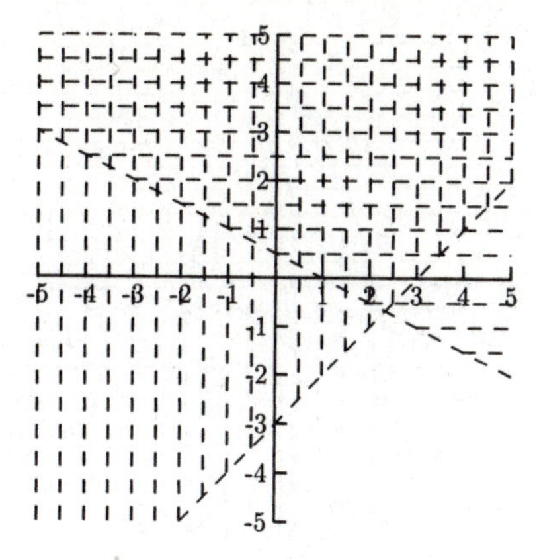

Fig. 10-26.

The region that is in both solutions is above and between the lines.

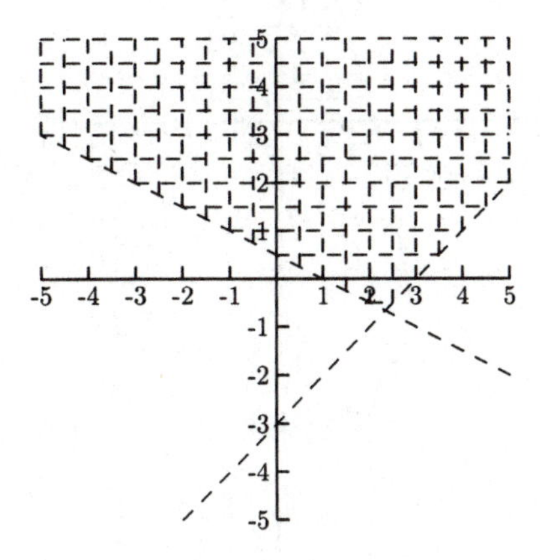

Fig. 10-27.

- $\begin{cases} y \le 4 - x^2 \\ x - 7y \le 4 \end{cases}$

The first inequality is shaded vertically and the second, horizontally (see Fig. 10-28). The region that is in both solutions is above the line and inside the parabola (see Fig. 10-29). Because a solid line indicates that the points on the graph are also solutions, to be absolutely accurate, the correct solution uses dashes for the parts of the graphs that are not on the border of the shaded region (see Fig. 10-30). We will not quibble with this technicality here.

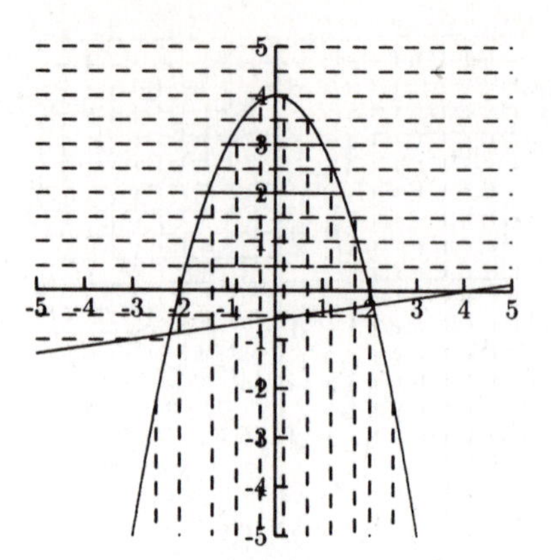

Fig. 10-28.

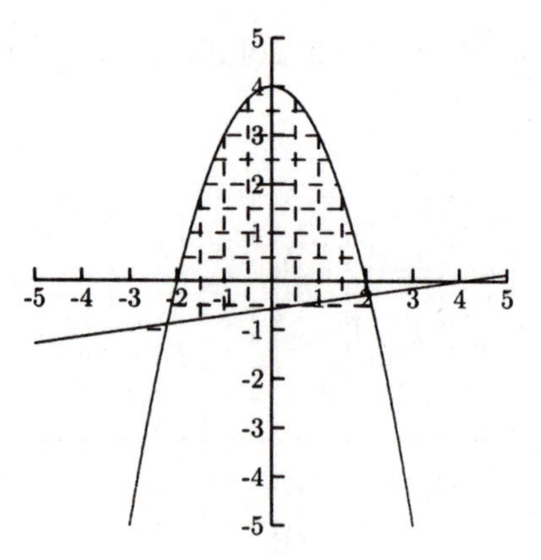

Fig. 10-29.

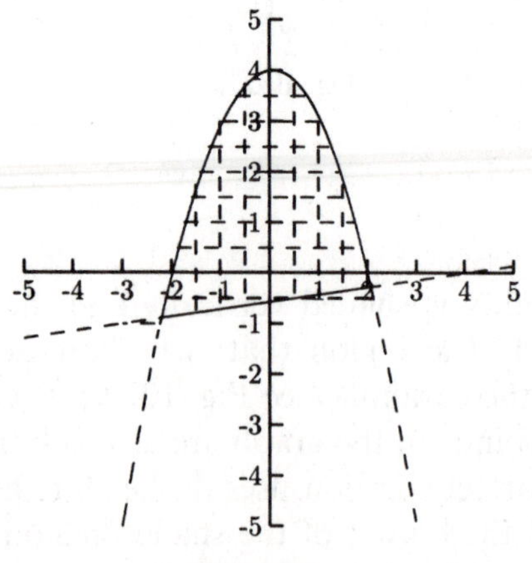

Fig. 10-30.

- $\begin{cases} 2x + y \leq 5 \\ x \geq 0 \\ y \geq 0 \end{cases}$

The inequalities $x \geq 0$ and $y \geq 0$ mean that we only need the top right corner of the graph. These inequalities are common in word problems.

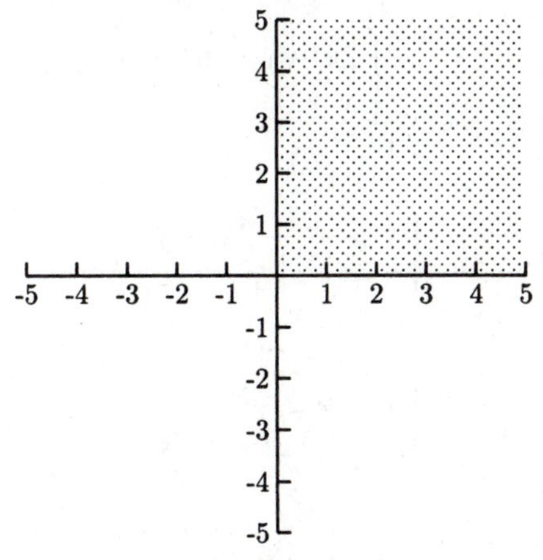

Fig. 10-31.

The solution to the system is the region in the top right corner of the graph below the line $2x + y = 5$.

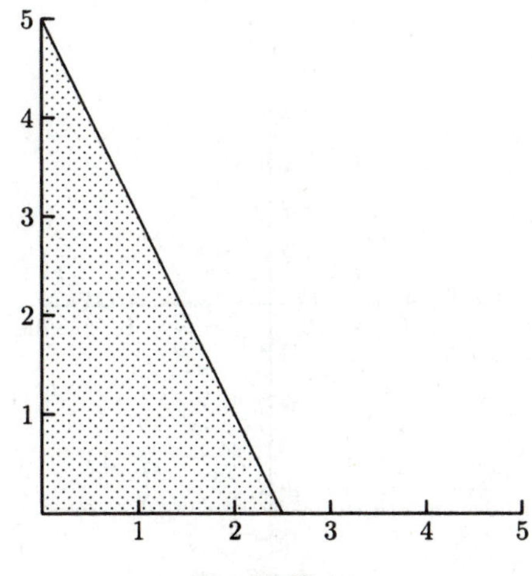

Fig. 10-32.

PRACTICE

Sketch the solutions to the inequalities.

1.

$$\begin{cases} x + y < 4 \\ x - y > 4 \end{cases}$$

2.

$$\begin{cases} 2x - y \le 6 \\ x \ge 3 \end{cases}$$

3.

$$\begin{cases} y > x^2 + 2x - 3 \\ x + y < 5 \end{cases}$$

4.

$$\begin{cases} y \le 9 - x^2 \\ y \ge x^2 + 4x - 5 \end{cases}$$

5.

$$\begin{cases} 2x + 3y \ge 6 \\ x \ge 0 \\ y \ge 0 \end{cases}$$

SOLUTIONS

1.

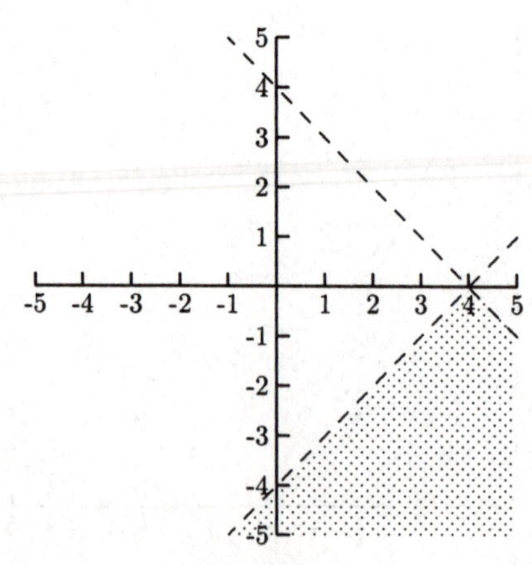

Fig. 10-33.

2.

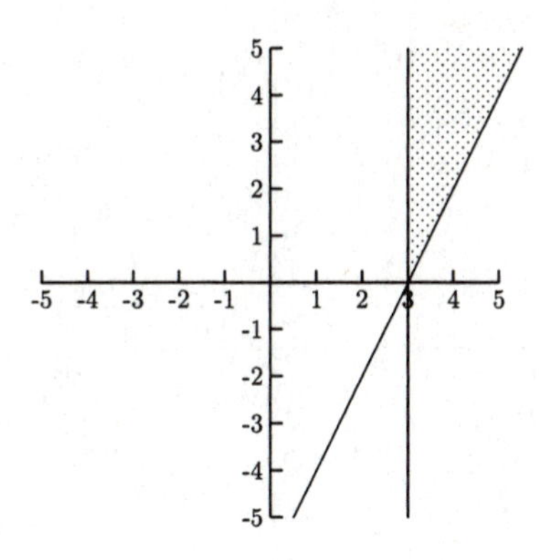

Fig. 10-34.

3.

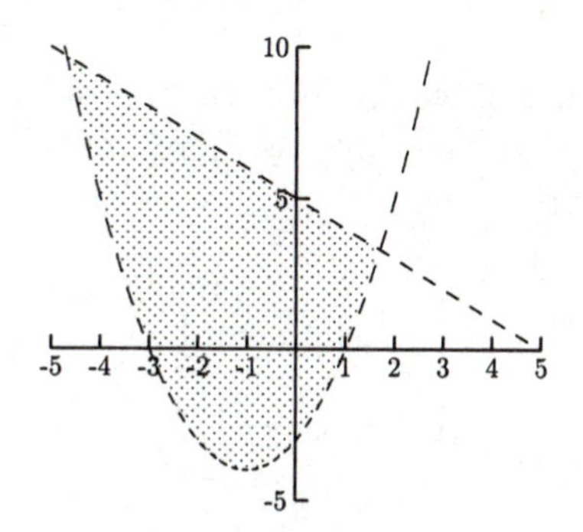

Fig. 10-35.

4.

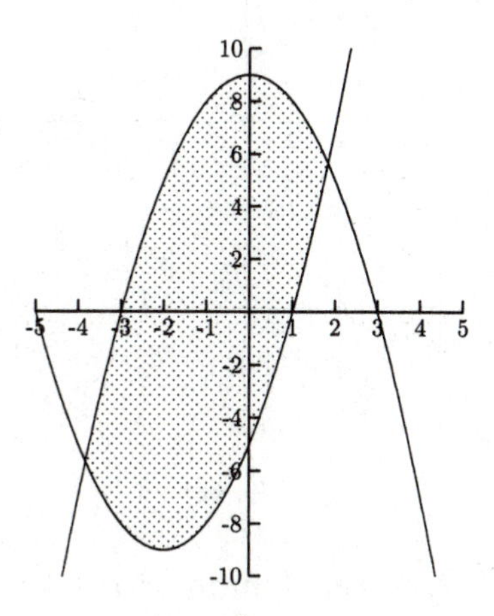

Fig. 10-36.

5.

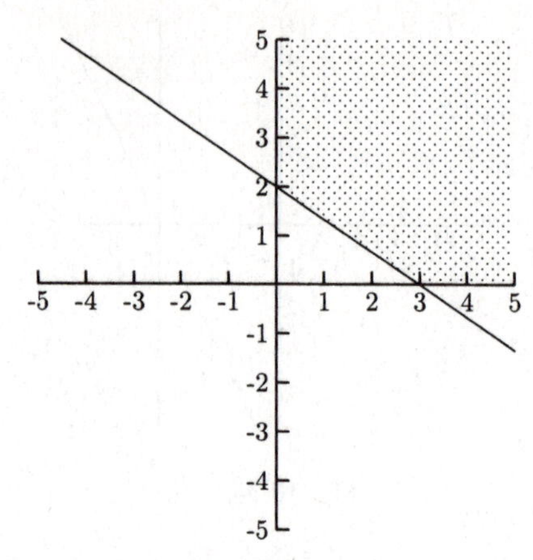

Fig. 10-37.

Some systems of inequalities do not have a solution. In the following example, the regions do not overlap, so there are no ordered pairs (points) that make both inequalities true.

- $\begin{cases} y \geq x^2 + 4 \\ x - y \geq 1 \end{cases}$

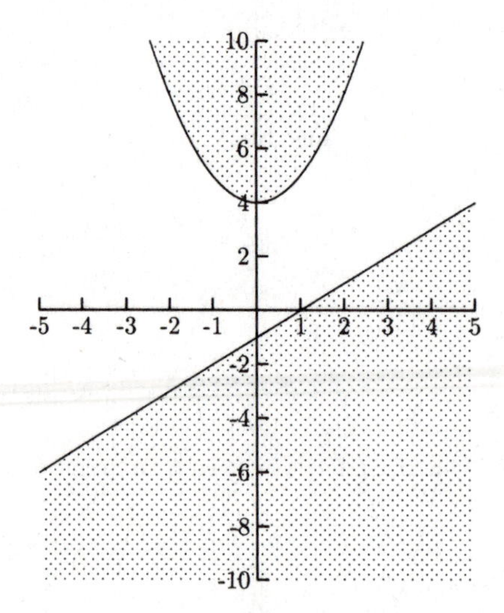

Fig. 10-38.

It is easy to lose track of the solution for a system of three or more inequalities. There are a couple of things you can do to make it easier.

First, make sure the graph is large enough, using graph paper if possible. Second, shade the solution for each inequality in a different way, with different colors or shaded with horizontal, vertical, and slanted lines. The solution (if there is one) would be shaded all different ways. You could also shade one region at a time, erasing the part of the previous region that is not part of the inequality.

EXAMPLES

- $\begin{cases} x + y \leq 4 \\ x \geq 1 \\ y \leq x \end{cases}$

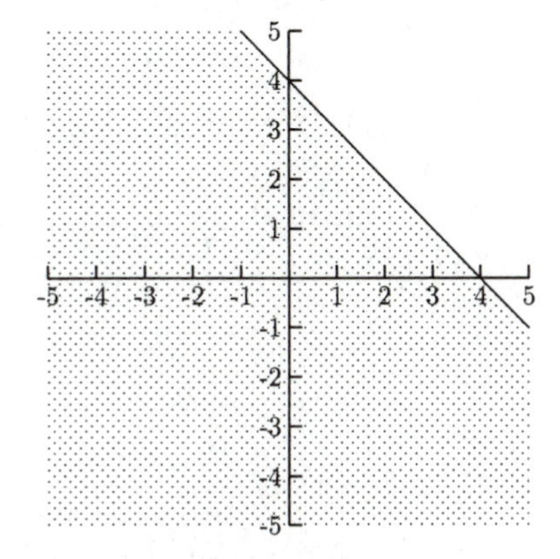

Fig. 10-39.

The region for $x \geq 1$ is to the right of the line $x = 1$, so we will erase the region to the *left* of $x = 1$.

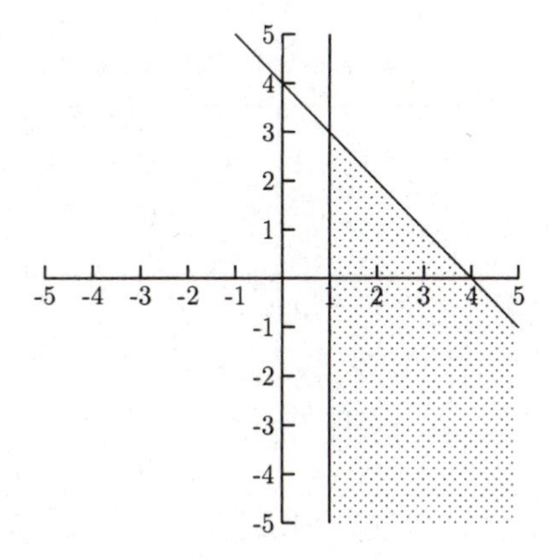

Fig. 10-40.

The solution to $y \leq x$ is the region below the line $y = x$, so we will erase the shading *above* the line $y = x$. The shaded region in Fig. 10-41 is the solution for the system.

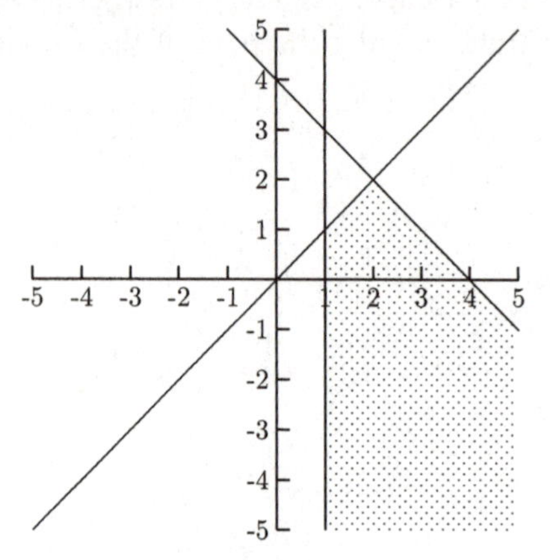

Fig. 10-41.

- $$\begin{cases} y > x^2 - 16 \\ x < 2 \\ y < -5 \\ -x + y < -8 \end{cases}$$

We will begin with $y = x^2 - 16$.

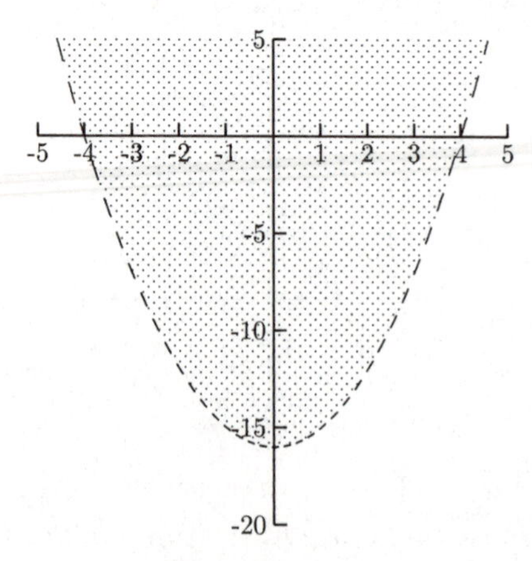

Fig. 10-42.

The solution to $x < 2$ is the region to the left of the line $x = 2$. We will erase the shading to the right of $x = 2$.

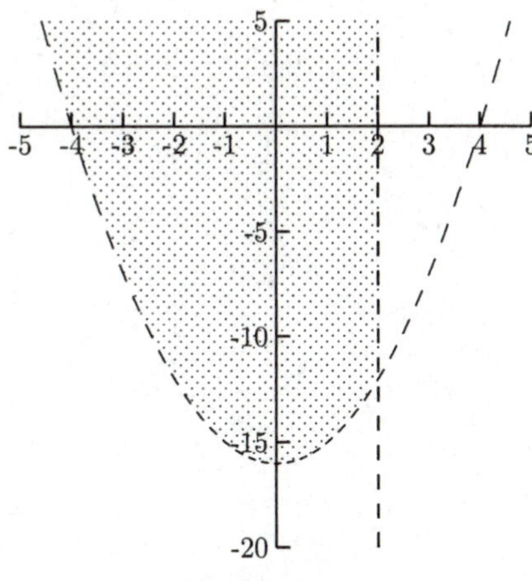

Fig. 10-43.

The solution to $y < -5$ is the region below the line $y = -5$. We will erase the shading above the line $y = -5$.

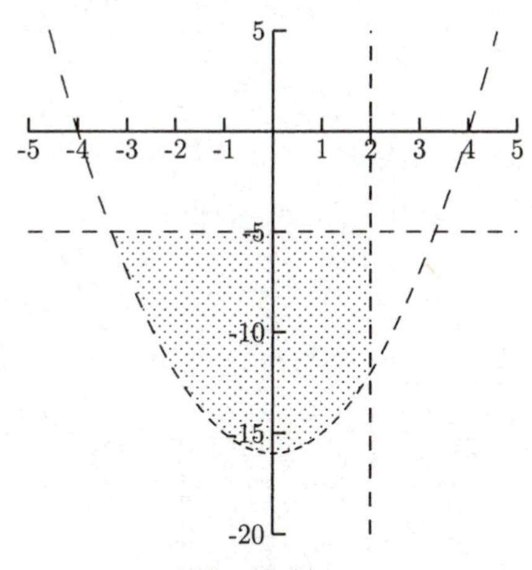

Fig. 10-44.

The solution to $-x + y < -8$ is the region below the line $-x + y = -8$, so we will erase the shading above the line. The solution to the system is shown in Fig. 10-45.

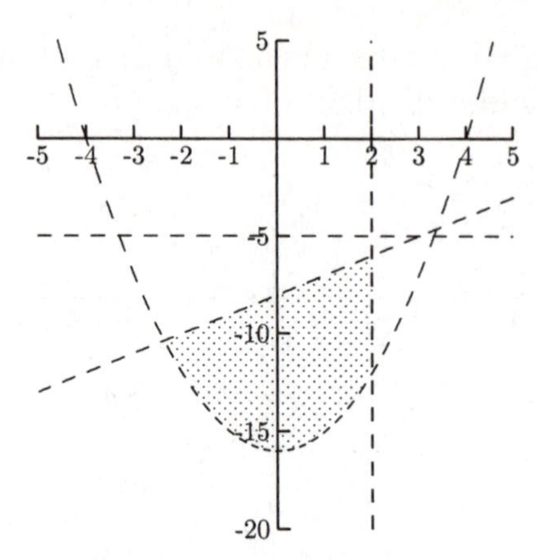

Fig. 10-45.

PRACTICE

Sketch the solutions to the inequalities.

1.

$$\begin{cases} 2x + y \geq 1 \\ -x + 2y \leq 4 \\ 5x - 3y \leq 15 \end{cases}$$

2.

$$\begin{cases} x + 2y \geq -6 \\ y \leq x \\ 5x + 2y \geq 10 \end{cases}$$

3.

$$\begin{cases} y \leq 9 - x^2 \\ x \leq 2 \\ y \leq -x \end{cases}$$

4.

$$\begin{cases} x + 2y < 10 \\ 2x + y < 8 \\ y < x \\ x \geq 0 \\ y \geq 0 \end{cases}$$

SOLUTIONS

1.

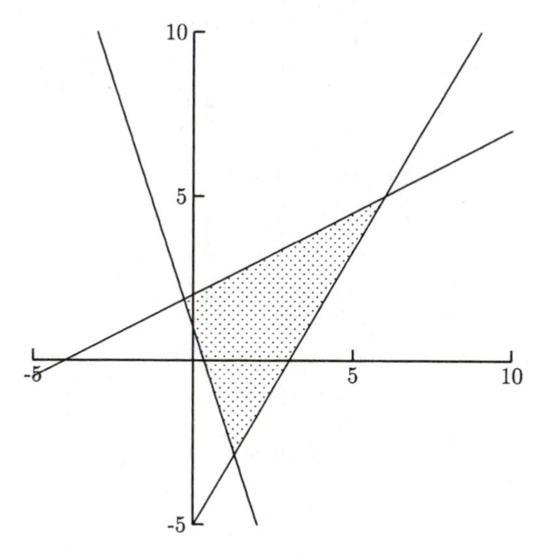

Fig. 10-46.

2.

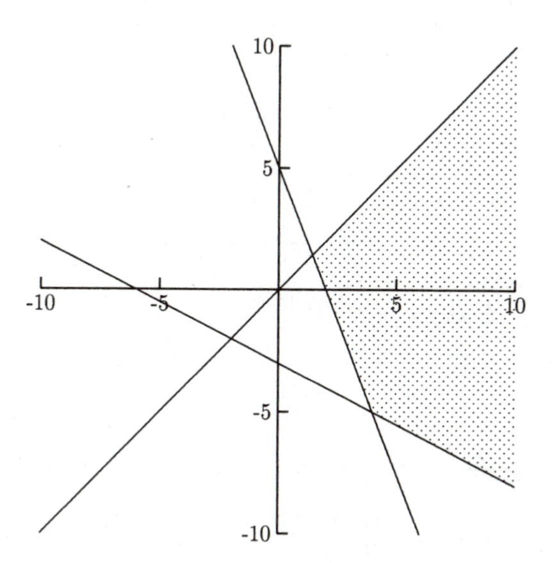

Fig. 10-47.

3.

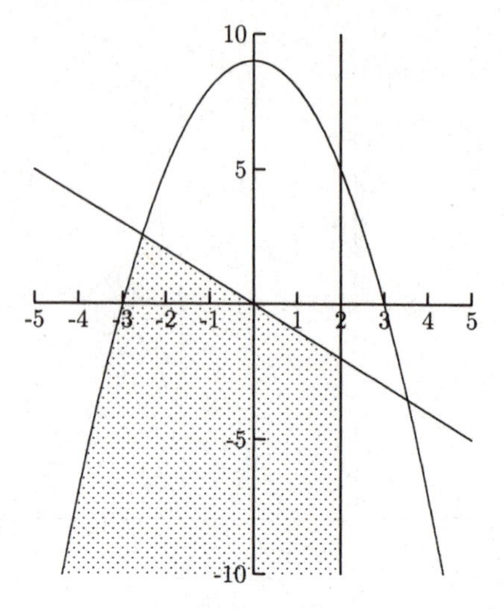

Fig. 10-48.

4.

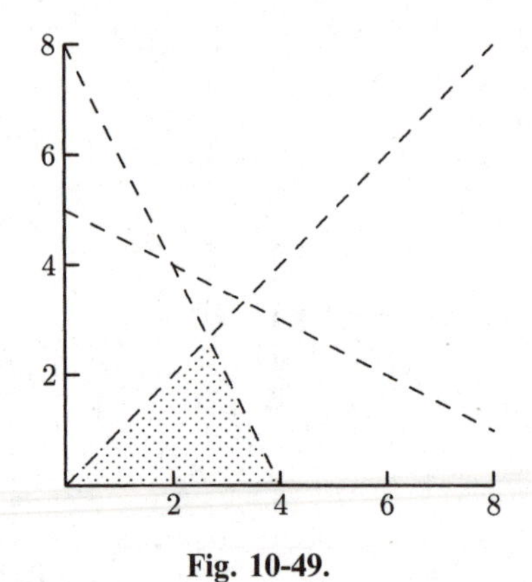

Fig. 10-49.

Chapter 10 Review

In some of the following problems, you will be asked to find such quantities as $x + 2y$ for a system of equations. Solve the system and put the solution in the formula. For example, if the solution is $x = 3$ and $y = 5$, $x + 2y$ becomes $3 + 2(5) = 13$.

1. What is $x + y$ for the system?

$$\begin{cases} x - 2y &= -1 \\ 4x + 2y &= 16 \end{cases}$$

a) 2 b) 3 c) 4 d) 5

2. What is $x + y$ for the system?

$$\begin{cases} y &= 4x - 10 \\ x + 2y &= -11 \end{cases}$$

a) −4 b) −5 c) −6 d) −7

3. What is $2x + y$ for the system?

$$\begin{cases} y &= -3x + 1 \\ y &= x + 9 \end{cases}$$

a) 3 b) 4 c) 5 d) 6

4. What is $x + 2y$ for the system?

$$\begin{cases} 4x - 3y &= -1 \\ 2x + 5y &= -7 \end{cases}$$

a) −3 b) −4 c) −5 d) −6

5. What is $x + y$ for the system?

$$\begin{cases} x^2 + y^2 &= 9 \\ x^2 - y^2 &= 9 \end{cases}$$

a) −2 and 5 b) −3 and 3 c) −4 and 4 d) −1 and 3

6. What is $x + y$ for the system?

$$\begin{cases} 2x - y &= -5 \\ y &= 4 - x^2 \end{cases}$$

a) 2 b) 3 c) 4 d) 5

7. The graph in Fig. 10-50 is the solution to which inequality?
a) $y > 2x + 2$ b) $y \geq 2x + 2$ c) $y < 2x + 2$ d) $y \leq 2x + 2$

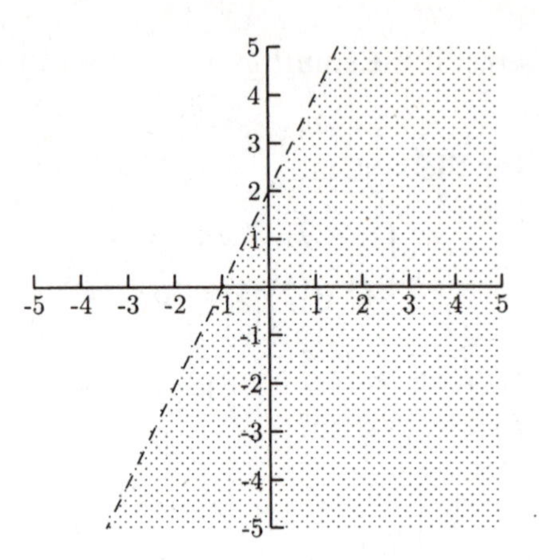

Fig. 10-50.

8. The graph in Fig. 10-51 is the solution to which inequality?
 a) $y > x^2 - 2x + 1$ b) $y \geq x^2 - 2x + 1$
 c) $y < x^2 - 2x + 1$ d) $y \leq x^2 - 2x + 1$

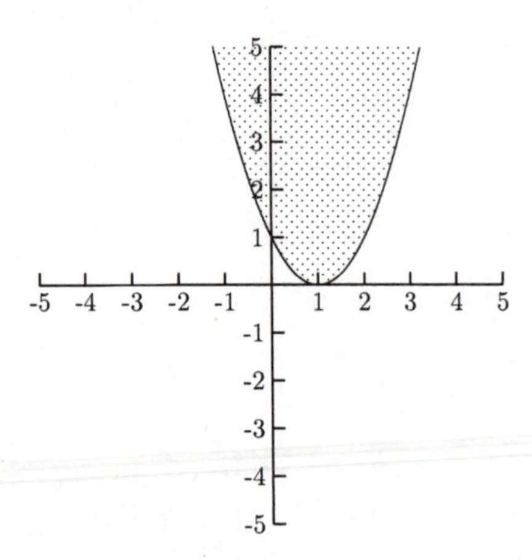

Fig. 10-51.

9. The graph in Fig. 10-52 is the solution to which system?
 a)

$$\begin{cases} y \leq -x^2 + 4x \\ y \geq x \end{cases}$$

b)

$$\begin{cases} y \le -x^2 + 4x \\ y \le x \end{cases}$$

c)

$$\begin{cases} y \ge -x^2 + 4x \\ y \ge x \end{cases}$$

d)

$$\begin{cases} y \ge -x^2 + 4x \\ y \le x \end{cases}$$

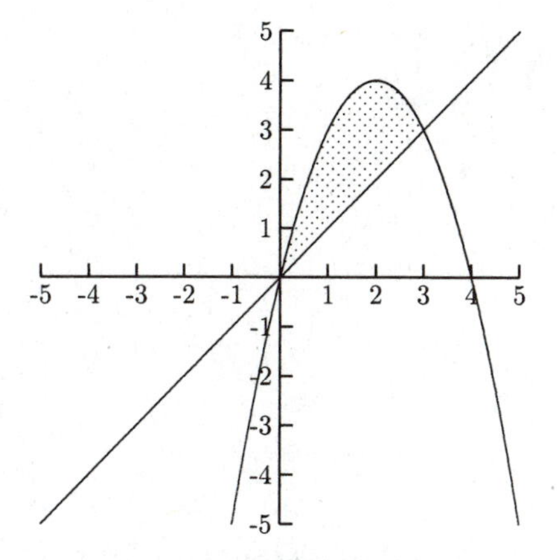

Fig. 10-52.

SOLUTIONS

1. d) 2. b) 3. a) 4. a) 5. b) 6. a) 7. c) 8. b) 9. a)

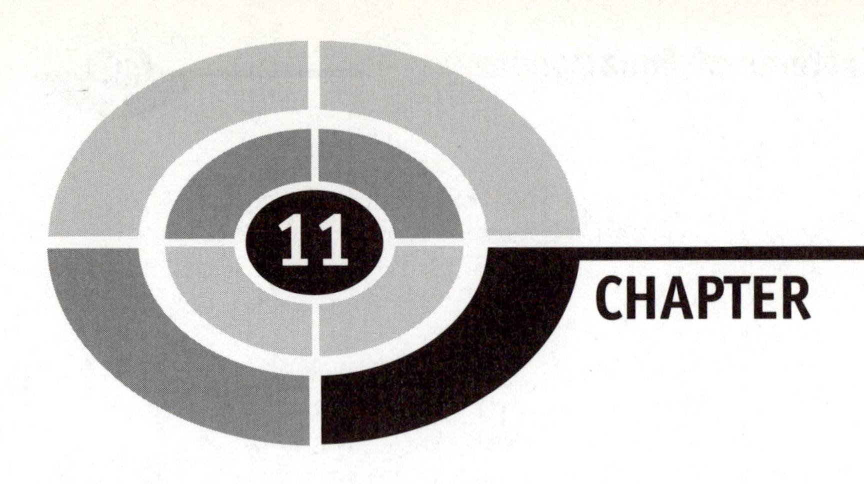

Exponents and Logarithms

Many things in nature and business grow (and decay) *exponentially*. To see how exponential growth and decay work, let us imagine a small country where no one moves away, no one moves in, and everyone marries and stays married. Suppose the population this year is 100,000. What will the population be in two generations if every couple has exactly two children? The population remains 100,000 in the next generation and 100,000 in the generation after that. If every couple has exactly one child, the population would decrease to 50,000 (a loss of 50,000) in the next generation and would decrease to 25,000 (a loss of 25,000) in the following generation. If every couple has four children, the population would increase to 200,000 (an increase of 100,000) in the next generation and 400,000 (an increase of 200,000) in the generation after that. The decreases get smaller with each generation, and the increases get larger. This effect is called *exponential* growth and decay.

An investment of money with compounded interest works the same way. For a fixed percentage increase per year, the exponential growth formula is $A = P(1 + r)^t$, where P is the initial amount, r is the percentage increase (as a decimal number), and t is the number of years. The formula for exponential decay is the same, except r is negative.

The following example will illustrate how compounding works. Suppose $100 is deposited in an account that earns 5% interest, compounded annually. *Compounded annually* means that interest is paid at the end of one year and that this interest earns interest in the next year. How much would be in this account after four years? After one year, the account has grown to $100 + 0.05(100) = 100 + 5 = \$105$. In the second year, the original $100 earns 5% interest plus the $5 earns 5% interest: $105 + (105)(0.05) = \$110.25$. Now this amount earns interest in the third year: $110.25 + (110.25)(0.05) = \$115.76$. Finally, this amount earns interest in the fourth year: $115.76 + (115.76)(0.05) = \$121.55$. If interest is not compounded, that is, the interest does not earn interest, the account would only be worth $120. The extra $1.55 is interest earned on interest.

Compound growth is not dramatic over the short term but it is over time. If $100 is left in an account earning 5% interest, compounded annually, for 20 years instead of four years, the difference between the compound growth and noncompound growth is a little more interesting. After 20 years, the compound amount is $265.33 compared to $200 for simple interest (noncompound growth). A graph of the growth of each type over 40 years is given in Fig. 11-1. The line is the growth for simple interest, and the curve is the growth for compound interest.

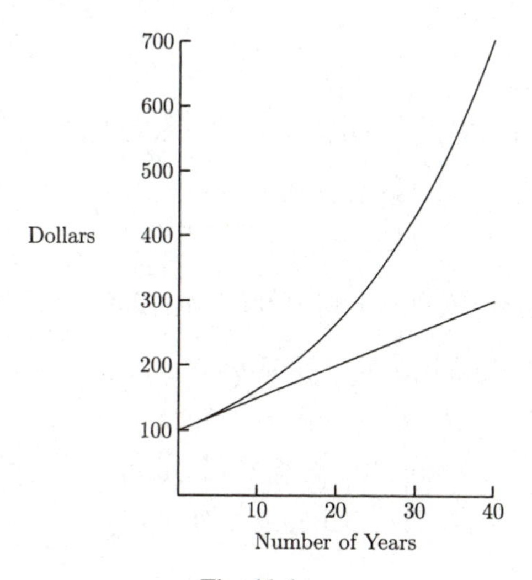

Fig. 11-1.

EXAMPLES

Find the compound amount.

- \$5000 after three years, earning 6% interest, compounded annually

  We will use the formula $A = P(1+r)^t$. $P = 5000$, $r = 0.06$, and $t = 3$. We want to know A, the compound amount.

  $$A = 5000(1+0.06)^3 = 5000(1.06)^3 = 5000(1.191016)$$
  $$= 5955.08$$

  The compound amount is \$5955.08.
- \$10,000 after eight years, $7\frac{1}{4}$% interest, compounded annually

  $$A = 10{,}000(1+0.0725)^8 = 10{,}000(1.0725)^8 \approx 10{,}000(1.7505656)$$
  $$\approx 17{,}505.66.$$

  The compound amount is \$17,505.66

PRACTICE

Find the compound amount.

1. \$800 after ten years, $6\frac{1}{2}$% interest, compounded annually
2. \$1200 after six years, $9\frac{1}{2}$% interest, compounded annually
3. A 20-year-old college student opens a retirement account with \$2000. If her account pays $8\frac{1}{4}$% interest, compounded annually, how much will be in the account when she reaches age 65?

SOLUTIONS

1.

$$A = 800(1+0.065)^{10} = 800(1.065)^{10} \approx 800(1.877137) \approx 1501.71$$

The compound amount is \$1501.71.

2.

$$A = 1200(1+0.095)^6 = 1200(1.095)^6 \approx 1200(1.72379) \approx 2068.55$$

The compound amount is \$2068.55.

3.

$$A = 2000(1+0.0825)^{45} = 2000(1.0825)^{45}$$
$$\approx 2000(35.420585) \approx 70{,}841.17$$

The account will be worth \$70,841.17.

Many investments pay more often than once a year, some paying interest daily. Instead of using the annual interest rate, we need to use the interest rate per period, and instead of using the number of years, we need to use the number of periods. If there are n compounding periods per year, then the interest rate per period is r/n and the total number of periods is nt. The compound amount formula becomes

$$A = P\left(1 + \frac{r}{n}\right)^{nt}.$$

EXAMPLES

Find the compound amount.

- \$5000 after three years, earning 6% annual interest
  - (a) compounded semiannually
  - (b) compounded monthly

  For (a), interest compounded semiannually means that it is compounded twice each year, so $n = 2$.

  $$A = 5000\left(1 + \frac{0.06}{2}\right)^{2(3)} = 5000(1.03)^6 \approx 5000(1.194052) \approx 5970.26$$

  The compound amount is \$5970.26.

  For (b), interest compounded monthly means that it is compounded 12 times each year, so $n = 12$.

  $$A = 5000\left(1 + \frac{0.06}{12}\right)^{12(3)} = 5000(1.005)^{36} \approx 5000(1.19668) \approx 5983.40$$

  The compound amount is \$5983.40.
- \$10,000 after eight years, earning $7\frac{1}{4}$% annual interest, compounded weekly.

  Interest that is paid weekly is paid 52 times each year, so $n = 52$.

  $$A = 10{,}000\left(1 + \frac{0.0725}{52}\right)^{52(8)}$$
  $$\approx 10{,}000(1.001394231)^{416} \approx 10{,}000(1.785317) \approx 17{,}853.17$$

  The compound amount is \$17,853.17.

PRACTICE

Find the compound amount.

1. \$800 after ten years, earning $6\frac{1}{4}\%$ annual interest
   (a) compounded quarterly
   (b) compounded weekly
2. \$9000 after five years, earning $6\frac{3}{4}\%$ annual interest, compounded daily (assume 365 days per year).
3. A 20-year-old college student opens a retirement account with \$2000. If she earns $8\frac{1}{4}\%$ annual interest, compounded daily, how much will be in the account when she is 65? (Assume 365 days per year.)

SOLUTIONS

1. (a) $n = 4$

$$A = 800\left(1 + \frac{0.0625}{4}\right)^{4(10)}$$
$$= 800(1.015625)^{40} \approx 800(1.85924) \approx 1487.39$$

   The compound amount is \$1487.39.

   (b) $n = 52$

$$A = 800\left(1 + \frac{0.0625}{52}\right)^{52(10)}$$
$$= 800(1.00120192)^{520} \approx 800(1.86754) \approx 1494.04$$

   The compound amount is \$1494.04.

2. $n = 365$

$$A = 9000\left(1 + \frac{0.0675}{365}\right)^{365(5)}$$
$$\approx 9000(1.000184932)^{1825} \approx 9000(1.4013959) \approx 12,612.56$$

   The compound amount is \$12,612.56.

3.

$$2000\left(1 + \frac{0.0825}{365}\right)^{365(45)} \approx 2000(1.000226027)^{16,425}$$
$$\approx 2000(40.93889) \approx 81,877.78$$

   The \$2000 investment will be worth \$81,877.78 when she is 65 years old.

The more often interest is compounded per year, the more interest is earned. An amount of \$1000 earning 8% annual interest, compounded annually, is worth \$1080 after one year. If interest is compounded quarterly, it is worth \$1082.43 after one year. And if interest is compounded daily, it is worth \$1083.28 after one year. What if interest is compounded each hour? Each second? It turns out that the most this investment could be worth (at 8% interest) is \$1083.29, when interest is compounded each and every instant of time. Each instant of time, a tiny amount of interest is earned. This is called *continuous* compounding. The formula for the compound amount for interest compounded continuously is $A = Pe^{rt}$, where A, P, r, and t are the same quantities as before. The letter e stands for a constant called Euler's number. It is approximately 2.718281828. You probably have an e or e^x key on your calculator. Although e is irrational, it can be approximated by rational numbers of the form

$$\left(1 + \frac{1}{m}\right)^m,$$

where m is a large rational number. The larger m is, the better the approximation for e. If we make the substitution $m = n/r$ and use some algebra, we can see how $(1 + r/n)^{nt}$ is very close to e^{rt}, for large values of n. If interest is compounded every minute, n would be 525,600, a rather large number!

EXAMPLE

- Find the compound amount of \$5000 after eight years, earning 12% annual interest, compounded continuously.

$$A = 5000e^{0.12(8)} = 5000e^{0.96} \approx 5000(2.611696) \approx 13{,}058.48$$

The compound amount is \$13,058.48.

PRACTICE

Find the compound amount.

1. \$800 after 10 years, earning $6\frac{1}{2}$% annual interest, compounded continuously.
2. \$9000, after 5 years, earning $6\frac{3}{4}$% annual interest, compounded continuously.
3. A 20-year-old college student opens a retirement account with \$2000. If she earns $8\frac{1}{4}$% annual interest, compounded continuously, how much will the account be worth by the time she is 65?

SOLUTIONS

1.

$$A = 800e^{0.065(10)} = 800e^{0.65} \approx 800(1.915540829) \approx 1532.43$$

The compound amount is \$1532.43.

2.

$$A = 9000e^{0.0675(5)} = 9000e^{0.3375} \approx 9000(1.401439608) \approx 12{,}612.96$$

The compound amount is =\$12,612.96.

3.

$$A = 2000e^{0.0825(45)} = 2000e^{3.7125} \approx 2000(40.95606882) \approx 81{,}912.14$$

The account will be worth \$81,912.14 by the time she is 65.

The compound growth formula for continuously compounded interest is used for other growth and decay problems. The general exponential growth model is $n(t) = n_0e^{rt}$, where $n(t)$ replaces A and n_0 replaces P. Their meanings are the same—$n(t)$ is still the compound growth, and n_0 is still the beginning amount. The variable t represents time in this formula; however, time will not always be measured in years. The growth rate and t need to have the same unit of measure. If the growth rate is in days, then t needs to be in days. If the growth rate is in hours, then t needs to be in hours, and so on. If the "population" is getting smaller, then the formula is $n(t) = n_0e^{-rt}$.

EXAMPLES

- The population of a city is estimated to be growing at the rate of 10% per year. In 2000, its population was 160,000. Estimate its population in the year 2005.

  The year 2000 corresponds to $t = 0$, so the year 2005 corresponds to $t = 5$; n_0, the population in year $t = 0$, is 160,000. The population is growing at the rate of 10% per year, so $r = 0.10$. The formula $n(t) = n_0e^{rt}$ becomes $n(t) = 160{,}000e^{0.10t}$. We want to find $n(t)$ for $t = 5$.

$$n(5) = 160{,}000e^{0.10(5)} \approx 263{,}795$$

  The city's population is expected to be 264,000 in the year 2005 (estimates and projections are normally rounded off).
- A county is losing population at the rate of 0.7% per year. If the population in 2001 is 1,000,000, what is it expected to be in the year 2008?

$n_0 = 1{,}000{,}000$, $t = 0$ is the year 2001, $t = 7$ is the year 2008, and $r = 0.007$. Because the county is losing population, we will use the decay model: $n(t) = n_0 e^{-rt}$. The model for this county's population is $n(t) = 1{,}000{,}000e^{-0.007t}$. We want to find $n(t)$ for $t = 7$.

$$n(7) = 1{,}000{,}000e^{-0.007(7)} \approx 952{,}181$$

The population is expected to be 952,000 in the year 2008.

- In an experiment, a culture of bacteria grew at the rate of 35% per hour. If 1000 bacteria were present at 10:00, how many were present at 10:45?

  $n_0 = 1000$, $r = 0.35$, t is the number of hours after 10:00. The growth model becomes $n(t) = 1000e^{0.35t}$. We want to find $n(t)$ for 45 minutes, or $t = 0.75$ hours.

$$n(0.75) = 1000e^{0.35(0.75)} = 1000e^{0.2625} \approx 1300$$

At 10:45, there were approximately 1300 bacteria present in the culture.

PRACTICE

1. The population of a city in the year 2002 is 2,000,000 and is expected to grow at 1.5% per year. Estimate the city's population for the year 2012.
2. A school is built for a capacity of 1500 students. The student population is growing at the rate of 6% per year. If 1000 students attend when it opens, will the school be at capacity in seven years?
3. A construction company estimates that a piece of equipment is worth \$150,000 when new. If it lost value continuously at an annual rate of 10%, what would its value be in 10 years?
4. Under certain conditions a culture of bacteria grows at the rate of about 200% per hour. If 8000 bacteria are present in a dish, how many will be in the dish after 30 minutes?

SOLUTIONS

1. $n_0 = 2{,}000{,}000$, $r = 0.015$ The growth formula is $n(t) = 2{,}000{,}000e^{0.015t}$ and we want to find $n(t)$ when $t = 10$.

$$n(10) = 2{,}000{,}000e^{0.015(10)} \approx 2{,}323{,}668$$

The population in the year 2012 is expected to be about 2.3 million.

2. $n_0 = 1000$, $r = 0.06$ The growth formula is $n(t) = 1000e^{0.06t}$. We want to find $n(t)$ when $t = 7$.

$$n(7) = 1000e^{0.06(7)} \approx 1522$$

Yes, the school will be at capacity in seven years.

3. $n_0 = 150{,}000$, $r = 0.10$ We will use the decay formula because value is being lost. The formula is $n(t) = 150{,}000e^{-0.10t}$. We want to find $n(t)$ when $t = 10$.

$$n(10) = 150{,}000e^{-0.10(10)} \approx 55{,}181.92$$

The equipment will be worth about \$55,000 after 10 years.

4. $n_0 = 8000$, $r = 2$ The growth formula is $n(t) = 8000e^{2t}$. We want to find $n(t)$ when $t = 0.5$.

$$n(0.5) = 8000e^{2(0.5)} \approx 21{,}746$$

About 21,700 bacteria will be present after 30 minutes.

A basic exponential function is of the form $f(x) = a^x$, where a is any positive number except 1. The graph of $f(x) = a^x$ comes in two shapes depending whether $0 < a < 1$ (a is positive but smaller than 1) or $a > 1$. Figure 11-2 shows the graph of $f(x) = (\frac{1}{2})^x$ and Fig. 11-3 is the graph of $f(x) = 2^x$.

We can sketch the graph of $f(x) = a^x$ by plotting points for $x = -3$, $x = -2, x = -1$, $x = 0$, $x = 1$, $x = 2$, and $x = 3$. If a is too large or too small, points for $x = -3$ and $x = 3$ might be too awkward to graph because

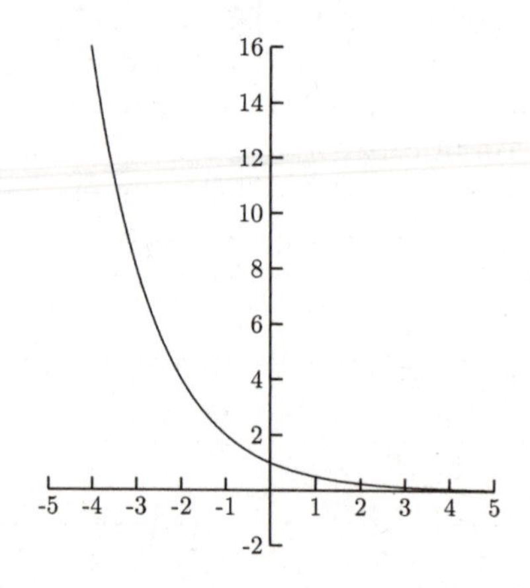

Fig. 11-2.

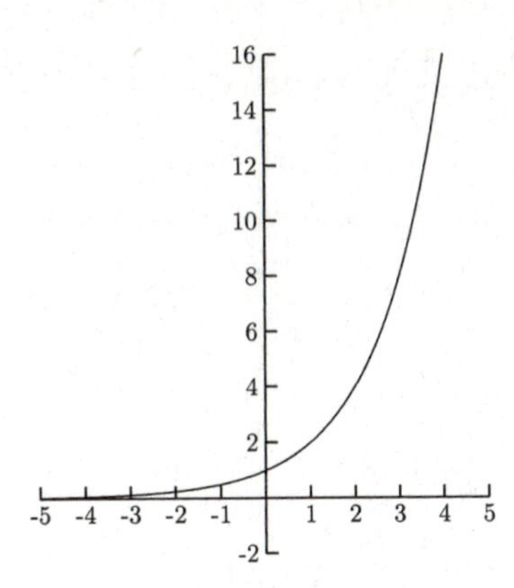

Fig. 11-3.

their y-values are too large or too close to 0. Before we begin sketching graphs, we will review the following exponent laws.

$$a^{-n} = \frac{1}{a^n} \qquad \left(\frac{1}{a}\right)^{-n} = a^n$$

EXAMPLE

Sketch the graphs.

- $f(x) = 2.5^x$

  We will begin with $x = -3, -2, -1, 0, 1, 2,$ and 3 in a table of values (Table 11-1).

Table 11-1

x	$f(x)$
-3	0.064 ($2.5^{-3} = \frac{1}{2.5^3}$) (Too hard to plot)
-2	0.16 ($2.5^{-2} = \frac{1}{2.5^2}$)
-1	0.40 ($2.5^{-1} = \frac{1}{2.5}$)
0	1
1	2.5
2	6.25
3	15.625

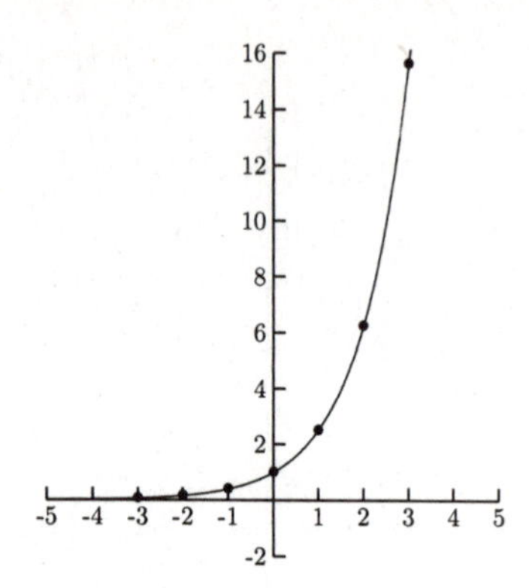

Fig. 11-4.

- $g(x) = (\frac{1}{3})^x$
  The values plotted are given in Table 11-2.

Table 11-2

x	$f(x)$
-3	$27\ ((\frac{1}{3})^{-3} = 3^3)$
-2	$9\ ((\frac{1}{3})^{-2} = 3^2)$
-1	$3\ ((\frac{1}{3})^{-1} = 3^1)$
0	1
1	0.33
2	0.11
3	0.037 (too hard to plot)

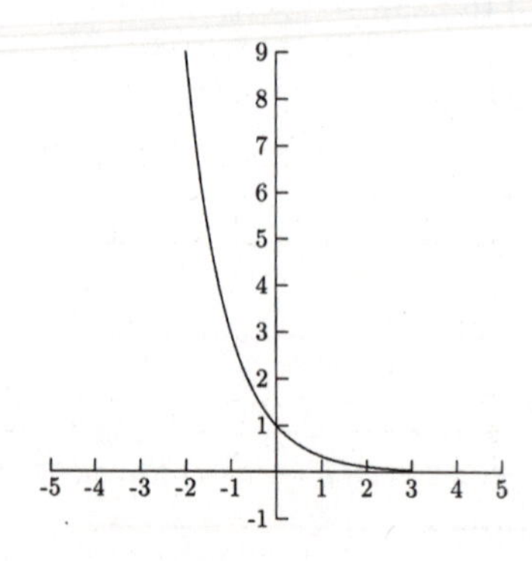

Fig. 11-5.

PRACTICE

Sketch the graphs.

1. $f(x) = (\frac{3}{2})^x$
2. $g(x) = (\frac{2}{3})^x$
3. $h(x) = e^x$ (Use the e or e^x key on your calculator.)

SOLUTIONS

1. The values plotted are given in Table 11-3

Table 11-3

x	$f(x)$
-3	$0.30\ ((\frac{3}{2})^{-3} = (\frac{2}{3})^3 = \frac{8}{27})$
-2	$0.44\ ((\frac{3}{2})^{-2} = (\frac{2}{3})^2 = \frac{4}{9})$
-1	$0.67\ ((\frac{3}{2})^{-1} = \frac{2}{3})$
0	1
1	1.5
2	2.25
3	3.375

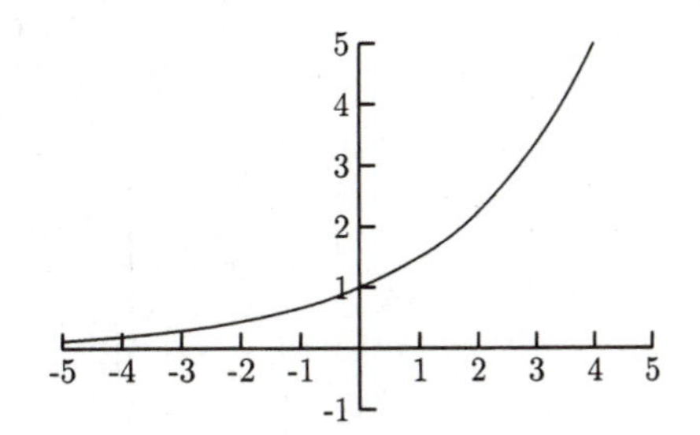

Fig. 11-6.

2. The values plotted are given in Table 11-4.

Table 11-4

x	$f(x)$
-3	$3.375\ ((\frac{2}{3})^{-3} = (\frac{3}{2})^3)$
-2	$2.25\ ((\frac{2}{3})^{-2} = (\frac{3}{2})^2)$
-1	$1.5\ ((\frac{2}{3})^{-1} = \frac{3}{2})$
0	1
1	0.67
2	0.44
3	0.30

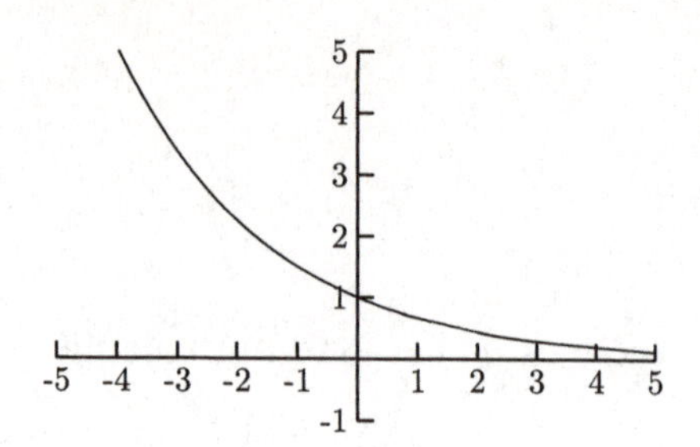

Fig. 11-7.

3. The values plotted are given in Table 11-5.

Table 11-5

x	$f(x)$
−3	0.05
−2	0.14
−1	0.37
0	1
1	2.72
2	7.39
3	20.09

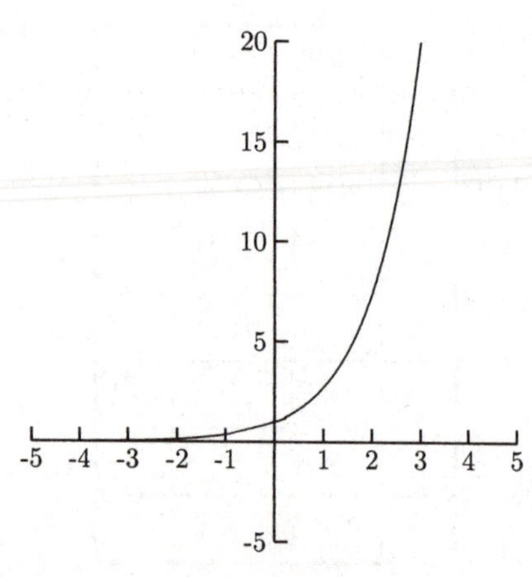

Fig. 11-8.

Transformations of the graphs of exponential functions behave in the same way as transformations of other functions.

- The graph of $f(x) = -2^x$ is the graph of $y = 2^x$ reflected about the x-axis (flipped upside down).
- The graph of $g(x) = 2^{-x}$ is the graph of $y = 2^x$ reflected about the y-axis (flipped sideways).
- The graph of $h(x) = 2^{x+1}$ is the graph of $y = 2^x$ shifted to the left 1 unit.
- The graph of $f(x) = -3 + 2^x$ is the graph of $y = 2^x$ shifted down 3 units.

Present Value

Suppose a couple wants to give their newborn grandson a gift of \$50,000 on his 20th birthday. They can earn $7\frac{1}{2}\%$ interest, compounded annually. How much should they deposit now so that it grows to \$50,000 in 20 years? To answer this question, we will use the formula $A = P(1 + r)^t$, where we know that $A = 50{,}000$ but are looking for P.

$$\begin{aligned} 50{,}000 &= P(1 + 0.075)^{20} \\ &= P(1.075)^{20} \\ \frac{50{,}000}{(1.075)^{20}} &= P = 11{,}770.66 \end{aligned}$$

The couple should deposit \$11,770.66 now so that the investment grows to \$50,000 in 20 years.

We say that \$11,770.66 is the *present value* of \$50,000 due in 20 years, earning $7\frac{1}{2}\%$ interest, compounded annually. The present value formula is $P = A(1 + r)^{-t}$ for interest compounded annually, and $P = A(1 + (r/n))^{-nt}$ for interest compounded n times per year.

EXAMPLE

- Find the present value of \$20,000 due in $8\frac{1}{2}$ years, earning 6% annual interest, compounded monthly.

$$P = 20{,}000\left(1 + \frac{0.06}{12}\right)^{-12(8.5)} = 20{,}000(1.005)^{-102} \approx 12{,}025.18$$

The present value is \$12,025.18.

PRACTICE

1. Find the present value of \$9000 due in five years, earning 7% annual interest, compounded annually.
2. Find the present value of \$50,000 due in 10 years, earning 4% annual interest, compounded quarterly.
3. Find the present value of \$125,000 due in $4\frac{1}{2}$ years, earning $6\frac{1}{2}\%$ annual interest, compounded weekly.

SOLUTIONS

1.

$$P = 9000(1.07)^{-5} \approx 6416.88$$

The present value is \$6416.88.

2.

$$P = 50{,}000\left(1 + \frac{0.04}{4}\right)^{-4(10)} = 50{,}000(1.01)^{-40} \approx 33{,}582.66$$

The present value is \$33,582.66.

3.

$$P = 125{,}000\left(1 + \frac{0.065}{52}\right)^{-52(4.5)}$$
$$= 125{,}000(1.00125)^{-234} \approx 93{,}316.45$$

The present value is \$93,316.45.

Logarithms

A common question for investors is, "How long will it take for my investment to double?" If \$1000 is invested so that it earns 8% interest, compounded annually, how long will it take to grow to \$2000? To answer the question using the compound growth formula, we need to solve for t in the equation $2000 = 1000(1.08)^t$. We will divide both sides of the equation by 1000 to get $2 = (1.08)^t$. Now what? It does not make sense to "take the tth root"of both sides. We need to use logarithms. Logarithms "cancel" exponentiation in the same way subtraction "cancels" addition and division "cancels" multiplication. Logarithms (or *logs*) are very useful in solving many science and business problems.

The logarithmic equation $\log_a x = y$ is another way of writing the exponential equation $a^y = x$. Verbally, we say, "log base a of x is (or equals) y." For $\log_a x$, we say, "(the) log base a of x."

EXAMPLES

Rewrite the logarithmic equation as an exponential equation.

- $\log_3 9 = 2$
  The base of the logarithm is the base of the exponent, so 3 will be raised to a power. The number that is equal to the log is the power, so the power on 3 is 2.

$$\log_3 9 = 2 \text{ rewritten as an exponent is } 3^2 = 9$$

- $\log_2 \frac{1}{8} = -3$
  The base is 2 and the power is -3.

$$2^{-3} = \frac{1}{8}$$

- $\log_9 3 = \frac{1}{2}$
  The base is 9 and the power is $\frac{1}{2}$.

$$9^{\frac{1}{2}} = 3$$

PRACTICE

Rewrite the logarithmic equation as an exponential equation.

1. $\log_4 16 = 2$
2. $\log_3 81 = 4$
3. $\log_{100} 10 = \frac{1}{2}$
4. $\log_a 4 = 3$
5. $\log_e 2 = 0.6931$
6. $\log_{(x+1)} 9 = 2$
7. $\log_7 \frac{1}{49} = -2$
8. $\log_8 4 = \frac{2}{3}$

SOLUTIONS

1. $\log_4 16 = 2$ rewritten as an exponential equation is $4^2 = 16$
2. $\log_3 81 = 4$ rewritten as an exponential equation is $3^4 = 81$
3. $\log_{100} 10 = \frac{1}{2}$ rewritten as an exponential equation is $100^{\frac{1}{2}} = 10$
4. $\log_a 4 = 3$ rewritten as an exponential equation is $a^3 = 4$

5. $\log_e 2 = 0.6931$ rewritten as an exponential equation is $e^{0.6931} = 2$
6. $\log_{(x+1)} 9 = 2$ rewritten as an exponential equation is $(x+1)^2 = 9$
7. $\log_7 \frac{1}{49} = -2$ rewritten as an exponential equation is $7^{-2} = \frac{1}{49}$
8. $\log_8 4 = \frac{2}{3}$ rewritten as an exponential equation is $8^{2/3} = 4$

Now we will work in the other direction, rewriting exponential equations as logarithmic equations. The equation $4^3 = 64$ written as a logarithmic equation is $\log_4 64 = 3$.

PRACTICE

Rewrite the exponential equation as a logarithmic equation.

1. $5^2 = 25$
2. $4^0 = 1$
3. $7^{-1} = \frac{1}{7}$
4. $125^{1/3} = 5$
5. $10^{-4} = 0.0001$
6. $e^{1/2} = 1.6487$
7. $8^x = 5$

SOLUTIONS

1. $5^2 = 25$ rewritten as a logarithmic equation is $\log_5 25 = 2$
2. $4^0 = 1$ rewritten as a logarithmic equation is $\log_4 1 = 0$
3. $7^{-1} = \frac{1}{7}$ rewritten as a logarithmic equation is $\log_7 \frac{1}{7} = -1$
4. $125^{1/3} = 5$ rewritten as a logarithmic equation is $\log_{125} 5 = \frac{1}{3}$
5. $10^{-4} = 0.0001$ rewritten as a logarithmic equation is $\log_{10} 0.0001 = -4$
6. $e^{1/2} = 1.6487$ rewritten as a logarithmic equation is $\log_e 1.6487 = \frac{1}{2}$
7. $8^x = 5$ rewritten as a logarithmic equation is $\log_8 5 = x$

Logarithm Properties

The first two logarithm properties we will learn are the cancelation properties. They come directly from rewriting one form of an equation in the other form.

$$\log_a a^x = x \qquad \text{and} \qquad a^{\log_a x} = x$$

When the bases of the exponent and logarithm are the same, they cancel. Let us see why these properties are true. What would the expression $\log_a a^x$

be? We will rewrite the equation $\log_a a^x = ?$ as an exponential equation: $a^? = a^x$. Now we can see that "?" is x. This is why $\log_a a^x = x$. What would $a^{\log_a x}$ be? Rewrite $a^{\log_a x} = ?$ as a logarithmic equation: $\log_a ? = \log_a x$, so "?" is x, and $a^{\log_a x} = x$.

EXAMPLE

- $5^{\log_5 2}$

  The bases of the logarithm and exponent are both 5, so $5^{\log_5 2}$ simplifies to 2.
- $10^{\log_{10} 8} = 8$
- $4^{\log_4 x} = x$
- $e^{\log_e 6} = 6$
- $29^{\log_{29} 1} = 1$
- $\log_m m^r = r$
- $\log_7 7^{ab} = ab$
- $\log_9 9^3 = 3$
- $\log_{16} 16^{-4} = -4$
- $\log_{10} 10^x = x$

PRACTICE

Use logarithm properties to simplify the expression.

1. $9^{\log_9 3}$
2. $10^{\log_{10} 14}$
3. $5^{\log_5 x}$
4. $\log_{15} 15^2$
5. $\log_{10} 10^{-8}$
6. $\log_e e^x$

SOLUTIONS

1. $9^{\log_9 3} = 3$
2. $10^{\log_{10} 14} = 14$
3. $5^{\log_5 x} = x$
4. $\log_{15} 15^2 = 2$
5. $\log_{10} 10^{-8} = -8$
6. $\log_e e^x = x$

Sometimes we will need to use exponent properties before using the property $\log_a a^x = x$.

$$\sqrt[n]{a^m} = a^{m/n} \text{ and } \frac{1}{a^m} = a^{-m}$$

EXAMPLES

- $\log_9 3 = \log_9 \sqrt{9} = \log_9 9^{1/2}$
  $= \frac{1}{2}$
- $\log_7 \frac{1}{49} = \log_7 \frac{1}{7^2}$
  $= \log_7 7^{-2} = -2$

- $\log_{10} \sqrt[4]{10} = \log_{10} 10^{1/4}$ $= \frac{1}{4}$
- $\log_{10} \sqrt[5]{100} = \log_{10} \sqrt[5]{10^2}$ $= \log_{10} 10^{2/5} = \frac{2}{5}$

PRACTICE

Use logarithm properties to simplify the expression.

1. $\log_7 \sqrt{7}$
2. $\log_5 \frac{1}{5}$
3. $\log_3 \frac{1}{\sqrt{3}}$
4. $\log_4 \frac{1}{16}$
5. $\log_{25} \frac{1}{5}$
6. $\log_8 \frac{1}{2}$
7. $\log_{10} \sqrt{1000}$

SOLUTIONS

1.

$$\log_7 \sqrt{7} = \log_7 7^{1/2} = \frac{1}{2}$$

2.

$$\log_5 \frac{1}{5} = \log_5 5^{-1} = -1$$

3.

$$\log_3 \frac{1}{\sqrt{3}} = \log_3 \frac{1}{3^{1/2}} = \log_3 3^{-1/2} = -\frac{1}{2}$$

4.

$$\log_4 \frac{1}{16} = \log_4 \frac{1}{4^2} = \log_4 4^{-2} = -2$$

5.

$$\log_{25} \frac{1}{5} = \log_{25} \frac{1}{\sqrt{25}} = \log_{25} \frac{1}{25^{\frac{1}{2}}} = \log_{25} 25^{-1/2} = -\frac{1}{2}$$

6. $2 = \sqrt[3]{8}$

$$\log_8 \frac{1}{2} = \log_8 \frac{1}{\sqrt[3]{8}} = \log_8 \frac{1}{8^{\frac{1}{3}}} = \log_8 8^{-1/3} = -\frac{1}{3}$$

7. $1000 = 10^3$

$$\log_{10}\sqrt{1000} = \log_{10}\sqrt{10^3} = \log_{10} 10^{3/2} = \frac{3}{2}$$

Two types of logarithms occur frequently enough to have their own notation. They are $\log_e$ and $\log_{10}$. The notation for $\log_e$ is "ln", pronounced "ell-in," and is called the *natural log*. The notation for $\log_{10}$ is "log" (no base is written) and is called the *common log*. The cancelation properties for these special logarithms are

$$\ln e^x = x \qquad e^{\ln x} = x$$

and

$$\log 10^x = x \qquad 10^{\log x} = x.$$

EXAMPLES

- $\ln e^{15} = 15$
- $e^{\ln 14} = 14$
- $\ln e^{-4} = -4$
- $10^{\log 5} = 5$
- $\log 10^{1/2} = \frac{1}{2}$
- $\log 10^{-4} = -4$

PRACTICE

Simplify.

1. $\ln e^5$
2. $\log 10^{\sqrt{x}}$
3. $10^{\log 9}$
4. $e^{\ln 6}$
5. $\log 10^{3x-1}$
6. $\ln e^{x+1}$

SOLUTIONS

1. $\ln e^5 = 5$
2. $\log 10^{\sqrt{x}} = \sqrt{x}$
3. $10^{\log 9} = 9$
4. $e^{\ln 6} = 6$
5. $\log 10^{3x-1} = 3x - 1$
6. $\ln e^{x+1} = x + 1$

Three More Important Logarithm Properties

The following three logarithm properties come directly from the exponent properties $a^m \cdot a^n = a^{m+n}$, $a^m/a^n = a^{m-n}$, and $a^{mn} = (a^m)^n$.

1. $\log_b mn = \log_b m + \log_b n$
2. $\log_b m/n = \log_b m - \log_b n$
3. $\log_b m^t = t \log_b m$

We will see why Property 1 works. Let $x = \log_b m$ and $y = \log_b n$. Rewriting these equations as exponential equations, we get $b^x = m$ and $b^y = n$. Multiplying m and n, we have $mn = b^x \cdot b^y = b^{x+y}$. Rewriting the equation $mn = b^{x+y}$ as a logarithmic equation, we get $\log_b mn = x + y$. Because $x = \log_b m$ and $y = \log_b n$, $\log_b mn = x + y$ becomes $\log_b mn = \log_b m + \log_b n$.

EXAMPLE

Use Property 1 to rewrite the logarithms.

- $\log_4 7x = \log_4 7 + \log_4 x$
- $\log_6 19t^2 = \log_6 19 + \log_6 t^2$
- $\log_9 3 + \log_9 27 = \log_9 3(27) = \log_9 81 = 2$
- $\ln 15t = \ln 15 + \ln t$
- $\log 100y^4 = \log 10^2 + \log y^4 = 2 + \log y^4$
- $\ln x + \ln \sqrt{y} = \ln x\sqrt{y}$

PRACTICE

Use Property 1 to rewrite the logarithms.

1. $\ln 59t$
2. $\log 0.10y$
3. $\log_{30} 148x^2$
4. $\log_6 3 + \log_6 12$
5. $\log_5 9 + \log_5 10$
6. $\log 5 + \log 20$

SOLUTIONS

1. $\ln 59t = \ln 59 + \ln t$
2. $\log 0.10y = \log 0.10 + \log y = \log 10^{-1} + \log y = -1 + \log y$
3. $\log_{30} 148x^2 = \log_{30} 148 + \log_{30} x^2$
4. $\log_6 3 + \log_6 12 = \log_6(3 \cdot 12) = \log_6 36 = \log_6 6^2 = 2$
5. $\log_5 9 + \log_5 10 = \log_5(9 \cdot 10) = \log_5 90$
6. $\log 5 + \log 20 = \log(5 \cdot 20) = \log 100 = \log 10^2 = 2$

EXAMPLES
Use Property 2 to rewrite the logarithms.

- $\log(x/4) = \log x - \log 4$
- $\log_{15} 3 - \log_{15} 2 = \log_{15}\left(\frac{3}{2}\right)$
- $\log_4\left(\frac{4}{3}\right) = \log_4 4 - \log_4 3 = 1 - \log_4 3$
- $\ln(5/x) = \ln 5 - \ln x$
- $\ln 16 - \ln t = \ln(16/t)$

PRACTICE
Use Property 2 to rewrite the logarithms.

1. $\log_4 10/9x$
2. $\log_2 \frac{7}{8}$
3. $\ln t/4$
4. $\log 100/x^2$
5. $\log_7 2 - \log_7 4$
6. $\log_8 x - \log_8 3$

SOLUTIONS

1.

$$\log_4 \frac{10}{9x} = \log_4 10 - \log_4 9x$$

2.

$$\log_2 \frac{7}{8} = \log_2 7 - \log_2 8 = \log_2 7 - \log_2 2^3 = (\log_2 7) - 3$$

3.

$$\ln \frac{t}{4} = \ln t - \ln 4$$

4.

$$\log \frac{100}{x^2} = \log 100 - \log x^2 = \log 10^2 - \log x^2 = 2 - \log x^2$$

5.

$$\log_7 2 - \log_7 4 = \log_7 \frac{2}{4} = \log_7 \frac{1}{2}$$

6.

$$\log_8 x - \log_8 3 = \log_8 \frac{x}{3}$$

The exponent property $\sqrt[n]{a^m} = a^{m/n}$ allows us to apply the third logarithm property to roots as well as to powers. The third logarithm property is especially useful in science and business applications.

EXAMPLE

Use Property 3 to rewrite the logarithms.

- $\log_4 3^x = x \log_4 3$
- $\frac{1}{3} \ln t = \ln t^{1/3}$
- $\log_6 \sqrt{2x} = \log_6 (2x)^{1/2} = \frac{1}{2} \log_6 2x$
- $\log x^2 = 2 \log x$
- $-3 \log 8 = \log 8^{-3}$
- $\ln \sqrt[4]{t^3} = \ln t^{3/4} = \frac{3}{4} \ln t$

PRACTICE

Use Property 3 to rewrite the logarithms.

1. $\ln 5^x$
2. $\log_{12} \sqrt{3}$
3. $\log \sqrt{16x}$
4. $\log_5 6^{-t}$
5. $2 \log_8 3$
6. $(x+6) \log_4 3$
7. $\log_{16} 10^{2x}$
8. $-2 \log_4 5$

SOLUTIONS

1. $\ln 5^x = x \ln 5$
2. $\log_{12} \sqrt{3} = \log_{12} 3^{1/2} = \frac{1}{2} \log_{12} 3$
3. $\log \sqrt{16x} = \log (16x)^{1/2} = \frac{1}{2} \log 16x$
4. $\log_5 6^{-t} = -t \log_5 6$
5. $2 \log_8 3 = \log_8 3^2 = \log_8 9$
6. $(x+6) \log_4 3 = \log_4 3^{x+6}$
7. $\log_{16} 10^{2x} = 2x \log_{16} 10$
8. $-2 \log_4 5 = \log_4 5^{-2} = \log_4 \frac{1}{5^2} = \log_4 \frac{1}{25}$

Sometimes we will need to use several logarithm properties to rewrite more complicated logarithms. The hardest part of this is to use the properties in the correct order. For example, which property should be used first on $\log (x/y^3)$? Do we first use the third property or the second property? We need to use the

second property first. For the expression $\log(x/y)^3$, we would use the third property first.

Going in the other direction, we need to use all three properties in the expression $\log_2 9 - \log_2 x + 3\log_2 y$. We need to use the second property to combine the first two terms.

$$\log_2 9 - \log_2 x + 3\log_2 y = \log_2 \frac{9}{x} + 3\log_2 y$$

We cannot use the first property on $\log_2 (9/x) + 3\log_2 y$ until we have used the third property to move the 3.

$$\log_2 \frac{9}{x} + 3\log_2 y = \log_2 \frac{9}{x} + \log_2 y^3 = \log_2 y^3 \frac{9}{x} = \log_2 \frac{9y^3}{x}$$

EXAMPLES

Rewrite as a single logarithm.

- $\log_2 3x - 4\log_2 y$

  We need use the third property to move the 4, then we can use the second property.

$$\log_2 3x - 4\log_2 y = \log_2 3x - \log_2 y^4 = \log_2 \frac{3x}{y^4}$$

- $3\log 4x + 2\log 3 - 2\log y$

$$\begin{aligned} 3\log 4x + 2\log 3 - 2\log y &= \log(4x)^3 + \log 3^2 - \log y^2 && \text{Property 3} \\ &= \log 4^3x^3 \cdot 3^2 - \log y^2 && \text{Property 1} \\ &= \log 576x^3 - \log y^2 = \log \frac{576x^3}{y^2} && \text{Property 2} \end{aligned}$$

- $t\ln 4 + \ln 5$

$$t\ln 4 + \ln 5 = \ln 4^t + \ln 5 = \ln(5 \cdot 4^t) \qquad (\text{not } \ln 20^t)$$

Expand each logarithm.

- $\ln 3\sqrt{x}/y^2$

$$\begin{aligned} \ln \frac{3\sqrt{x}}{y^2} &= \ln 3(x^{1/2}) - \ln y^2 = \ln 3 + \ln x^{1/2} - \ln y^2 \\ &= \ln 3 + \frac{1}{2}\ln x - 2\ln y \end{aligned}$$

- $\log_7 4/10xy^2$

$$\log_7 \frac{4}{10xy^2} = \log_7 4 - \log_7 10xy^2$$
$$= \log_7 4 - (\log_7 10 + \log_7 x + \log_7 y^2)$$
$$= \log_7 4 - (\log_7 10 + \log_7 x + 2\log_7 y)$$
$$\text{or} \quad \log_7 4 - \log_7 10 - \log_7 x - 2\log_7 y$$

PRACTICE

For 1–5, rewrite each as a single logarithm. For 6–10, expand each logarithm.

1. $2\log x + 3\log y$
2. $\log_6 2x - 2\log_6 3$
3. $3\ln t - \ln 4 + 2\ln 5$
4. $t\ln 6 + 2\ln 5$
5. $\frac{1}{2}\log x - 2\log 2y + 3\log z$
6. $\log 4x/y$
7. $\ln 6/\sqrt{y}$
8. $\log_4 10x/\sqrt[3]{z}$
9. $\ln \sqrt{4x}/5y^2$
10. $\log \sqrt{2y^3/x}$

SOLUTIONS

1.

$$2\log x + 3\log y = \log x^2 + \log y^3 = \log x^2y^3$$

2.

$$\log_6 2x - 2\log_6 3 = \log_6 2x - \log_6 3^2 = \log_6 2x - \log_6 9 = \log_6 \frac{2x}{9}$$

3.

$$3\ln t - \ln 4 + 2\ln 5 = \ln t^3 - \ln 4 + \ln 5^2$$
$$= \ln \frac{t^3}{4} + \ln 25$$
$$= \ln 25\frac{t^3}{4} = \ln \frac{25t^3}{4}$$

4.

$$t \ln 6 + 2 \ln 5 = \ln 6^t + \ln 5^2 = \ln[25(6^t)]$$

5.

$$\begin{aligned}\frac{1}{2}\log x - 2\log 2y + 3\log z &= \log x^{1/2} - \log(2y)^2 + \log z^3 \\ &= \log x^{1/2} - \log 2^2 y^2 + \log z^3 \\ &= \log x^{1/2} - \log 4y^2 + \log z^3 \\ &= \log \frac{x^{1/2}}{4y^2} + \log z^3 = \log z^3 \frac{x^{1/2}}{4y^2} \\ &= \log \frac{z^3 x^{1/2}}{4y^2} \text{ or } \log \frac{z^3\sqrt{x}}{4y^2}\end{aligned}$$

6.

$$\log \frac{4x}{y} = \log 4x - \log y = \log 4 + \log x - \log y$$

7.

$$\ln \frac{6}{\sqrt{y}} = \ln 6 - \ln \sqrt{y} = \ln 6 - \ln y^{1/2} = \ln 6 - \frac{1}{2}\ln y$$

8.

$$\begin{aligned}\log_4 \frac{10x}{\sqrt[3]{z}} &= \log_4 10x - \log_4 \sqrt[3]{z} = \log_4 10x - \log_4 z^{1/3} \\ &= \log_4 10 + \log_4 x - \frac{1}{3}\log_4 z\end{aligned}$$

9.

$$\begin{aligned}\ln \frac{\sqrt{4x}}{5y^2} &= \ln \sqrt{4x} - \ln 5y^2 = \ln(4x)^{1/2} - \ln 5y^2 \\ &= \frac{1}{2}\ln 4x - (\ln 5 + \ln y^2) = \frac{1}{2}(\ln 4 + \ln x) - (\ln 5 + 2\ln y) \\ &\text{or } \frac{1}{2}\ln 4 + \frac{1}{2}\ln x - \ln 5 - 2\ln y\end{aligned}$$

10.

$$\log\sqrt{\frac{2y^3}{x}} = \log\left(\frac{2y^3}{x}\right)^{1/2} = \frac{1}{2}\log\frac{2y^3}{x}$$
$$= \frac{1}{2}(\log 2y^3 - \log x) = \frac{1}{2}(\log 2 + \log y^3 - \log x)$$
$$= \frac{1}{2}(\log 2 + 3\log y - \log x) \text{ or } \frac{1}{2}\log 2 + \frac{3}{2}\log y - \frac{1}{2}\log x$$

With these logarithm properties we can solve some logarithm and exponent equations. We can use these properties to rewrite equations either in the form "log = log" or "log = number." When the equation is in the form "log = log," the logs cancel. When the equation is in the form "log = number," we can rewrite the equation as an exponential equation.

The Change of Base Formula

There are countless bases for logarithms but calculators usually have only two logarithms—log and ln. How can we use our calculators to approximate $\log_2 5$? We can use the change of base formula; but first, let us use logarithm properties to find this number. Let $x = \log_2 5$. Then $2^x = 5$. Take the common log of each side.

$\log 2^x = \log 5$ Now use the third logarithm property.

$x\log 2 = \log 5$ Divide both sides by the number log 2.

$$x = \frac{\log 5}{\log 2} \approx \frac{0.698970004}{0.301029996} \approx 2.321928095$$

This means that $2^{2.321928095}$ is very close to 5.

We just proved that $\log_2 5 = \log_{10} 5/\log_{10} 2$. Replace 2 with b, 5 with x, and 10 with a and we have the change of base formula.

$$\log_b x = \frac{\log_a x}{\log_a b}$$

This formula converts a logarithm with old base b to new base a. Usually, the new base is either e or 10.

EXAMPLE

- Evaluate $\log_7 15$. Give your solution accurate to four decimal places.

$$\log_7 15 = \frac{\log 15}{\log 7} \approx \frac{1.176091259}{0.84509804} \approx 1.3917$$
$$= \frac{\ln 15}{\ln 7} \approx \frac{2.708050201}{1.945910149} \approx 1.3917$$

PRACTICE

Evaluate the logarithms. Give your solution accurate to four decimal places.

1. $\log_6 25$
2. $\log_{20} 5$

SOLUTIONS

1.

$$\log_6 25 = \frac{\ln 25}{\ln 6} \approx \frac{3.218875825}{1.791759469} \approx 1.7965$$
$$= \frac{\log 25}{\log 6} \approx \frac{1.397940009}{0.7781525} \approx 1.7965$$

2.

$$\log_{20} 5 = \frac{\ln 5}{\ln 20} \approx \frac{1.609437912}{2.995732274} \approx 0.5372$$
$$= \frac{\log 5}{\log 20} \approx \frac{0.698970004}{1.301029996} \approx 0.5372$$

The change of base formula can be used to solve equations like $4^{2x+1} = 8$ by rewriting the equation in logarithmic form and using the change of base formula. The equation becomes $\log_4 8 = 2x + 1$. Because $\log_4 8 = \ln 8 / \ln 4$, the equation can be written as $2x + 1 = \ln 8 / \ln 4$.

$$2x + 1 = \frac{\ln 8}{\ln 4}$$
$$2x = -1 + \frac{\ln 8}{\ln 4}$$
$$x = \frac{1}{2}\left(-1 + \frac{\ln 8}{\ln 4}\right) = \frac{1}{4}$$

Chapter 11 Review

1. $1200 is deposited into an account that earns $7\frac{1}{2}\%$ annual interest, compound monthly. How much will be in the account in ten years?
   a) $2534.48 b) $2473.24 c) $2941.63 d) $1277.14

2. How much is an investment worth if $10,000 is invested for fifteen years, earning 5% annual interest, compounded continuously?
   a) $20,789.28 b) $21,170.00 c) $21,137.04 d) $21,162.37

3. A couple wants to present a $25,000 gift to their newborn grandson on his 20th birthday. If they can earn 8% annual interest, compounded quarterly, how much do they need to invest now?
   a) $121,886 b) $5047 c) $18,165 d) $5128

4. The graph in Figure 11.9 is the graph for what function?
   a) $y = (\frac{2}{3})^{x-1}$ b) $y = (\frac{2}{3})^{x+1}$ c) $y = (\frac{3}{2})^{x-1}$ d) $y = (\frac{3}{2})^{x+1}$

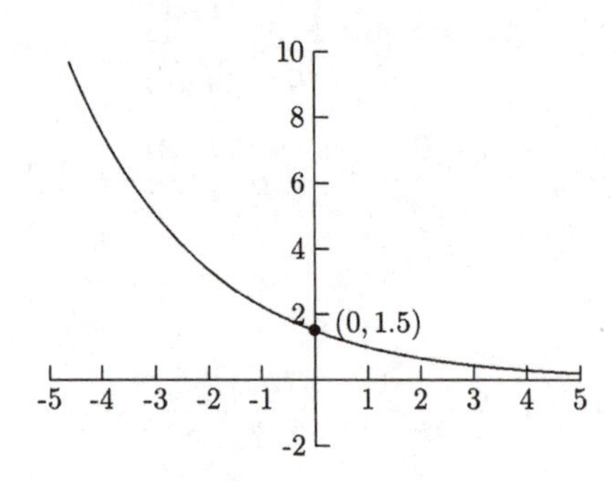

Fig. 11-9.

5. The population of a country is growing at an annual rate of 1.5%. If the population in 2004 is 4 million, estimate the population for the year 2015.
   a) 3.5 million b) 4.1 million c) 4.7 million d) 5.2 million

6. The population of a certain species of bird in a region is approximated by the function $n(t) = 115e^{-0.01t}$ where t is the number of years after 1998 and $n(t)$ is the population in thousands. Estimate the bird population in the region for the year 2005.
   a) About 107,000 b) About 111,000 c) About 117,000
   d) About 121,000

7. Rewrite $t = \log_b m$ as an exponential equation
   a) $t = m^b$ b) $b^m = t$ c) $m = b^t$ d) $m^t = b$
8. $\log_9 \frac{1}{81} =$
   a) -1 b) -2 c) 2 d) Does not exist
9 $\log \sqrt{10} =$
   a) -1 b) $-\frac{1}{2}$ c) $\frac{1}{2}$ d) 2
10. $\ln e^{2t} =$
   a) 2 b) t c) $2t$ d) Cannot be determined without a calculator
11. $\ln x - 2\ln y =$
   a) $\ln \frac{x}{2y}$ b) $2\ln \frac{2}{y}$ c) $\ln \frac{x}{y^2}$ d) $\frac{\ln x}{2\ln y}$
12. $\log_5 x\sqrt{y} =$
   a) $\log_5 x + \frac{1}{2}\log y$ b) $\log_5 x + \sqrt{\log_5 y}$ c) $\sqrt{\log_5 x + \log_5 y}$
   d) $(\log_5 x)\sqrt{\log_5 y}$
13. $\log_8 10 =$
   a) $\frac{\ln 10}{\ln 8}$ b) $\frac{\ln 8}{\ln 10}$ c) $\frac{\ln 8}{\ln 10}$ d) $\frac{\log 8}{\log 10}$

SOLUTIONS

1. a) 2. b) 3. d) 4. a) 5. c) 6. a) 7. c) 8. b)
9. c) 10. c) 11. c) 12. a) 13. a)

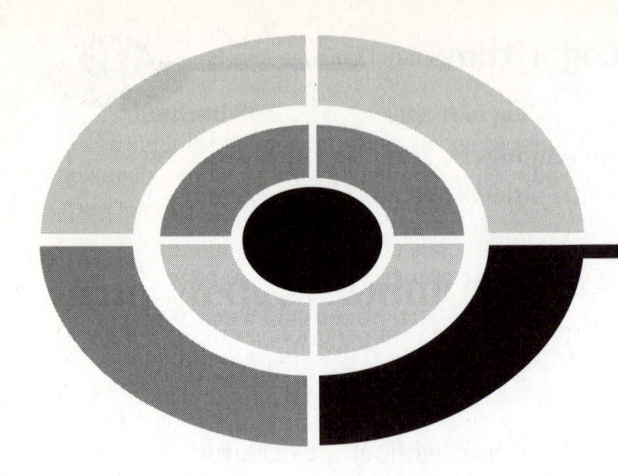

Final Exam

1. What is the slope of the line $y = 4$?
   a) 4 b) 0 c) -4 d) The slope does not exist.

2. What is the range for the function $f(x) = x^2 + 2x - 3$?
   a) $[-4, \infty)$ b) $(-\infty, -4]$ c) $[-1, \infty)$ d) $(-\infty, -1]$

3. What kind of triangle do the points $(-3, -1)$, $(1, -3)$, and $(-2, 0)$ form?
   a) Isosceles triangle (exactly two sides are equal) b) Right triangle
   c) Equilateral triangle (all three sides are equal)
   d) None of the above

4. What is the solution for $|6x + 1| = 5$?
   a) $x = -\frac{2}{3}$ b) $x = \frac{2}{3}$
   c) $x = \frac{2}{3}$ and $x = -\frac{2}{3}$ d) $x = \frac{2}{3}$ and $x = -1$

5. \$5000 is deposited into a retirement account. If it earns 8% annual interest, compounded annually, what will it be worth in 25 years?
   a) \$36,223.23 b) \$11,098.20 c) \$34,242.38 d) \$36,947.28

6. What is $x+y$ for the solution to the system?

$$\begin{cases} 5x-2y & =-2 \\ x+2y & =-10 \end{cases}$$

a) –3 b) –4 c) –5 d) –6

7. What is the solution for the inequality $x^2-2x-3 \leq 0$?
a) $[-1, 3]$ b) $(-\infty, -1] \cup [3, \infty)$
c) $(-\infty, -1) \cup (3, \infty)$ d) $(-\infty, -1] \cup [-\infty, 3]$

8. Which of the following correctly completes the square for $y=x^2+6x-4$?
a) $y=x^2+6x+9-4$ b) $y=(x^2+6x+9)-4+9$
c) $y=x^2+6x+9$ d) $y=(x^2+6x+9)-4-9$

9. Is $f(x)=3/x^2+2$ an even function, odd function, or neither?
a) Even b) Odd c) Neither d) Cannot be determined

10. What is the midpoint for the points (2, 5) and (–1, 6)?
a) $(\frac{1}{2}, \frac{11}{2})$ b) $(-\frac{3}{2}, \frac{1}{2})$ c) $(\frac{3}{2}, -\frac{1}{2})$ d) $(\frac{7}{2}, \frac{5}{2})$

11. What is the present value of \$100,000 due in ten years, earning 6% annual interest, compounded monthly?
a) \$54,963 b) \$54,881 c) \$55,840 d) \$48,780

12. The graph of a polynomial function whose leading term is $-6x^5$
(a) goes up on the left and up on the right.
(b) goes up on the left and down on the right.
(c) goes down on the left and up on the right.
(d) goes down on the left and down on the right.

13. The graph shown in Fig. A-1 is the graph of what function?
a) $y=(\frac{3}{2})^{x+1}$ b) $y=(\frac{3}{2})^{x-1}$ c) $y=(\frac{2}{3})^{x+1}$ d) $y=(\frac{2}{3})^{x-1}$

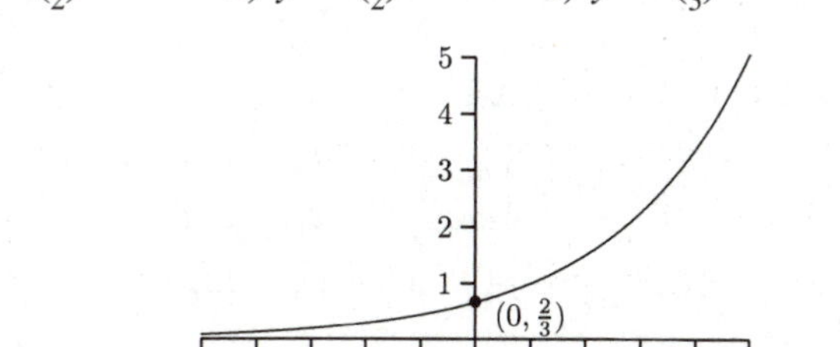

Fig. A-1.

The next three problems refer to the function $f(x)$ whose graph is shown in Fig. A-2.

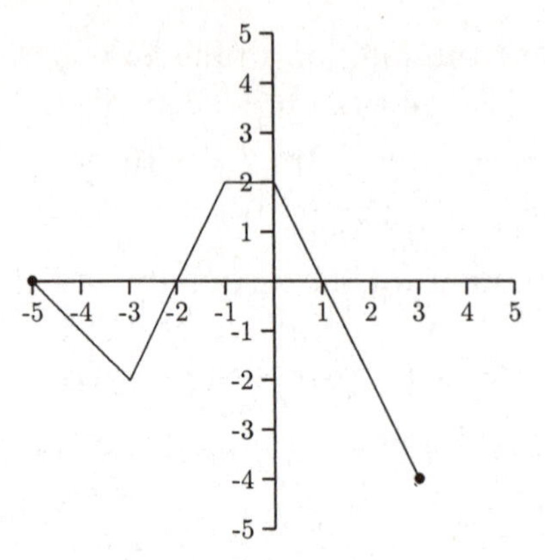

Fig. A-2.

14. What is the domain for $f(x)$?
    a) $[-5, 3]$ b) $[-5, -4]$ c) $[-4, 2]$ d) $[0, -4]$

15. What is $f(-3)$?
    a) 2 b) -2 c) 0 d) -4

16. For what interval(s) of x is this function decreasing?
    a) $(-2, 2)$ b) $(-5, -2) \cup (0, -4)$
    c) $(-3, 1)$ d) $(-5, -3) \cup (0, 3)$

17. A purchasing agent rents a car during a business trip. Her bill for Wednesday was \$33 for driving 45 miles. Her bill for Thursday was \$39 for driving 60 miles. Find an equation that gives the daily cost in terms of the number of miles driven.
    a) $y = 0.40x + 31.80$ b) $y = 2.50x - 79.50$
    c) $y = 0.40x + 44.40$ d) $y = 0.40x + 15$

18. Evaluate $\log_7 \frac{1}{\sqrt{7}}$.
    a) -2 b) $\frac{1}{2}$ c) -1 d) $-\frac{1}{2}$

19. What is the domain for $f(x) = \sqrt{x^2 - 9}$?
    a) $[3, \infty)$ b) $(-\infty, -3) \cup (-3, 3) \cup (3, \infty)$
    c) $(-\infty, -3] \cup [3, \infty)$ d) $[-3, \infty) \cup [3, \infty)$

20. What are the x-intercepts of the polynomial function $f(x) = (x + 4)(x + 1)(x - 3)^2(x - 5)$?
    a) $-4, -1, 3, 5$ b) $-4, -1, 9, 5$ c) $4, 1, 3, 5$ d) $4, 1, 9, 5$

21. What is the distance between $(2, 3)$ and $(-5, 6)$?
    a) $\sqrt{40}$ b) $\sqrt{122}$ c) $\sqrt{58}$ d) $\sqrt{2}$

22. The cost per unit of a product is given by the function $C(x) = 0.05x^2 - 40x + 8002.5$, where x is the number of units produced and C is in dollars. How many units should be produced to minimize the cost per unit?
a) 400 b) 450 c) 500 d) 550

23. Expand the logarithm $\ln(xy^2/z)$.
a) $\ln x + 2\ln y - \ln z$ b) $\dfrac{\ln x - 2\ln y}{\ln z}$
c) $\ln x + \ln y - \ln z$ d) $\ln x + (\ln y)^2 - \ln z$

24. Are the lines $6x - 2y = 5$ and $2x + 6y = 9$ parallel, perpendicular, or neither?
a) Parallel b) Perpendicular
c) Neither d) Cannot be determined

25. What is $x + y$ for the solution to the system?

$$\begin{cases} y = 2x - 5 \\ y = 3x - 7 \end{cases}$$

a) 2 b) -1 c) 0 d) 1

26. Find the x- and y-intercepts for $y = (x+1)/(x-3)$.
(a) The x-intercepts are -1 and 3, and the y-intercept is $-\frac{1}{3}$.
(b) The x-intercepts are -1 and 3, and there is no y-intercept.
(c) The x-intercept is -1, and the y-intercept is $-\frac{1}{3}$.
(d) There is no x-intercept, the y-intercept is $-\frac{1}{3}$.

27. Which of the following lines is perpendicular to the line $y = -\frac{2}{3}$?
a) $y = \frac{3}{2}$ b) $y = -\frac{3}{2}x$ c) $x = 2$ d) None

28. Evaluate $g(u+v)$ for $g(x) = 12x + 10$.
a) $12u + 12v + 10$ b) $12u + v + 10$
c) $(u+v)(12x+10)$ d) $12x + 10 + u + v$

29. What is the solution for $|x+3| < 4$?
a) $(-\infty, -7) \cup (1, \infty)$ b) $(-7, 1)$ c) $(-\infty, 1)$ d) $(-1, 1)$

30. What is the domain for $f \circ g(x)$ when $f(x) = x^2$ and $g(x) = \sqrt{x}$?
a) $(-\infty, \infty)$ b) $[0, \infty)$ c) $(-\infty, 0) \cup (0, \infty)$ d) $(0, \infty)$

31. What are the zeros for the polynomial function $f(x) = x^4 - 16$?
a) ± 4 b) ± 2 c) $\pm 2, -4$ d) $\pm 2, \pm 2i$

32. Find the equation of the line containing the points $(-1, 0)$ and $(0, 1)$.
a) $y = -x + 1$ b) $y = x + 1$ c) $y = -x - 1$ d) $y = x - 1$

33. What is the center and radius for the circle whose equation is $(x+5)^2+(y-6)^2=9$?
(a) The center is $(5,-6)$, and the radius is 81.
(b) The center is $(5,-6)$, and the radius is 3.
(c) The center is $(-5,6)$, and the radius is 81.
(d) The center is $(-5,6)$, and the radius is 3.

34. What is the domain for $f(x)=\frac{x+1}{x^2-4}$?
a) $(2,\infty)$ b) $(-\infty,-2)\cup(-2,2)\cup(2,\infty)$
c) $(-\infty,2)\cup(2,\infty)$ d) $(-\infty,-1)\cup(-1,2)\cup(2,\infty)$

35. Is $a=-3$ a lower bound for the real zeros of the polynomial function $f(x)=x^4-x^3+x^2+x-4$?
a) Yes b) No c) Cannot be determined

36. $x^2+x-1=0$ is equivalent to
a) $(x+\frac{1}{4})^2=\frac{5}{4}$ b) $(x+\frac{1}{2})^2=\frac{3}{2}$
c) $(x+\frac{1}{2})^2=\frac{5}{4}$ d) $(x+\frac{1}{2})^2=\frac{3}{4}$

37. Rewrite $\log_5 3x$ in base 8.
a) $\dfrac{\log_8 3x}{\log_8 5}$ b) $\dfrac{\log_8 5}{\log_8 3x}$ c) $\dfrac{\log_8 3x}{\ln 5}$ d) $\dfrac{\log_8 5}{\log 3x}$

38. What is the solution for the inequality $x^2>1$?
a) $(1,\infty)$ b) $(-\infty,-1)\cup(1,\infty)$
c) $(1,\infty)\cup(-1,\infty)$ d) $(-1,1)$

39. The graph of $-\frac{1}{2}f(x)$ is the graph of $f(x)$
(a) reflected about the x-axis and vertically stretched.
(b) reflected about the x-axis and vertically flattened.
(c) reflected about the y-axis and vertically stretched.
(d) reflected about the y-axis and vertically flattened.

40. Evaluate $f(-2)$ for $f(x)=6$.
a) -2 b) -12 c) 6 d) Cannot be determined

41. Find an equation of the line whose slope is $\frac{5}{3}$ and contains the point $(6,8)$.
a) $5x-3y=6$ b) $5x-3y=22$
c) $3x-5y=-22$ d) $3x-5y=-6$

42. What are the zeros for the polynomial function $f(x)=6x^3-11x^2+6x-1$?

a) $\frac{1}{2}$, 3, 1 b) $\frac{1}{2}$, $\frac{1}{3}$, 1 c) $-\frac{1}{2}$, $\frac{1}{3}$, 1 d) Cannot be determined.

43. The population of a certain type of fish in a lake is approximated by the function $n(t) = 25e^{0.024t}$, where t is the number of years after 2000 and $n(t)$ is the size of the population in hundreds. Estimate the size of the fish population in the lake for the year 2006.
a) About 2300 b) About 2500
c) About 2700 d) About 2900

44. If $f(x) = x^3$ and $g(x) = 1/(x+1)$, find $f \circ g(-2)$.
a) -1 b) $-\frac{1}{7}$ c) $-\frac{1}{8}$ d) 8

45. What is $x + y$ for the solution for the system?

$$\begin{cases} y = 2x^2 - x + 3 \\ 3x - y = -1 \end{cases}$$

a) 2 b) 3 c) 5 d) 6

46. What is the vertex for $y = \frac{1}{2}x^2 + 3x - 4$?
a) $(6, 32)$ b) $(-3, -\frac{17}{2})$ c) $(-\frac{3}{2}, -\frac{25}{4})$ d) $(3, \frac{19}{2})$

47. What is the solution for $|7 - x| > 2$?
a) $(-\infty, 5) \cup (9, \infty)$ b) $(-\infty, 5) \cup (-\infty, 9)$
c) $(9, 5)$ d) $(5, 9)$

48. The solid graph in Fig. A-3 is the graph of $f(x) = x^3$. The dashed graph is the graph of which function?
a) $y = (x+1)^3 + 1$ b) $y = (x+1)^3 - 1$ c) $y = (x-1)^3 + 1$
d) $y = (x-1)^3 - 1$

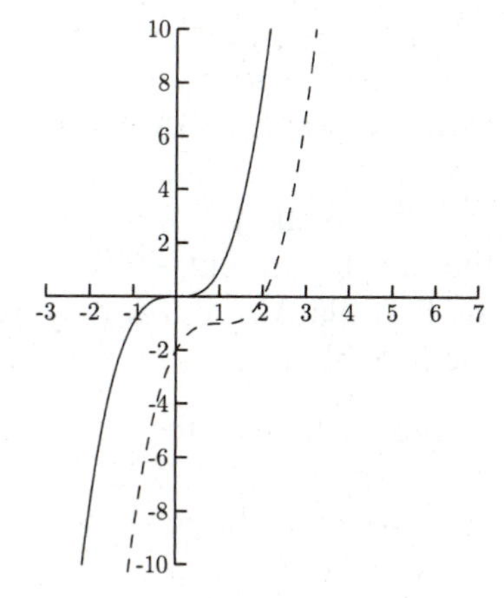

Fig. A-3.

49. According to the Rational Zero Theorem, which of the following is *not* a possible zero for the polynomial function $f(x) = 12x^4 - x^2 + 9$?
a) 4 b) $\frac{1}{3}$ c) $\frac{9}{2}$ d) All are possible rational zeros

50. Evaluate $g(3)$ for

$$g(x) = \begin{cases} 8 & \text{if } x \leq -1 \\ x+7 & \text{if } x > -1 \end{cases}$$

a) 8 b) 24 c) 10 d) 8 and 10

51. Rewrite $m^t = u$ as a logarithm equation.
a) $\log_m t = u$ b) $\log_m u = t$ c) $\log_u m = t$ d) $\log_u t = m$

52. For the function $f(x) = -\frac{1}{2}x^2 + 3x + 4$
(a) the maximum functional value is $\frac{17}{2}$.
(b) the minimum functional value is $\frac{17}{2}$.
(c) the maximum functional value is 3.
(d) the minimum functional value is 3.

53. What are the zeros for the function $P(x) = x^3 + 3x^2 - 2x - 8$? (Hint: $x = -2$ is a zero.)
a) $-2, \dfrac{-1 \pm \sqrt{17}}{2}$ b) $-2, \dfrac{-1 \pm \sqrt{17i}}{2}$
c) $-2, -2 \pm \sqrt{17}\,i$ d) $-2, 2 \pm 3i$

54. What is the center and radius for the circle whose equation is $x^2 + y^2 + 8x + 6y = 11$?
(a) The center is $(-4, -3)$, and the radius is 6.
(b) The center is $(-4, -3)$, and the radius is $\sqrt{11}$.
(c) The center is $(4, 3)$, and the radius is 6.
(d) The center is $(4, 3)$, and the radius is $\sqrt{11}$.

55. To complete the square, what numbers should be used to fill in the blanks for $y = -\frac{2}{3}(x^2 - 6x + \_) + 2 + \_$?
(a) Use 9 for the first blank and 6 for the second blank.
(b) Use 9 for the first blank and -6 for the second blank.
(c) Use 9 for the first blank and 9 for the second blank.
(d) Use 9 for the first blank and -9 for the second blank.

56. Find the quotient and remainder for $(4x^3 - x + 2) \div (x^2 + 1)$.
(a) The quotient is $4x^2 + 3x + 3$, and the remainder is 5.
(b) The quotient is $4x$, and the remainder is $-5x + 2$.
(c) The quotient is $4x$, and the remainder is $-3x + 2$.
(d) The quotient is $4x$, and the remainder is $3x + 2$.

57. Is $f(x) = x^3 - 4$ an even function, odd function or neither?
    a) Even b) Odd c) Neither d) Cannot be determined.

58. $e^{\ln 2} =$
    a) e^2 b) $2e^{\ln}$ c) 2 d) ln 2

59. Find an equation of the circle with center (8, 5) containing the point (5,9).
    a) $(x-8)^2 + (y-5)^2 = 5$ b) $(x-8)^2 + (y-5)^2 = 25$
    c) $(x+8)^2 + (y+5)^2 = 5$ d) $(x+8)^2 + (y+5)^2 = 25$

60. Solve for x: $\dfrac{x+2}{x-2} > 0$.
    a) $(-2, \infty)$ b) $(-2, 2)$
    c) $(-2, \infty) \cup (2, \infty)$ d) $(-\infty, -2) \cup (2, \infty)$

61. What is the vertex for $y = x^2 - 8x + 1$?
    a) $(-4, 49)$ b) $(8, 1)$ c) $(4, -15)$ d) $(-8, 129)$

62. The graph of $f(x) = (x+4) - 5$ is the graph of $f(x)$
    (a) shifted to the left 4 units and down 5 units.
    (b) shifted to the left 4 units and up 5 units.
    (c) shifted to the right 4 units and down 5 units.
    (d) shifted to the right 4 units and up 5 units.

63. The graph in Fig. A-4 is the solution to which system?
    (a)

$$\begin{cases} y < x - 1 \\ y < -3x + 3 \end{cases}$$

    (b)

$$\begin{cases} y < x - 1 \\ y > -3x + 3 \end{cases}$$

    (c)

$$\begin{cases} y > x - 1 \\ y < -3x + 3 \end{cases}$$

    (d)

$$\begin{cases} y > x - 1 \\ y > -3x + 3 \end{cases}$$

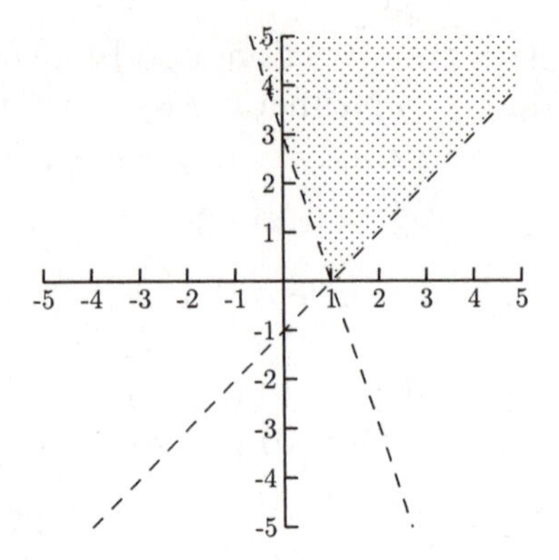

Fig. A-4.

64. Solve for x: $\dfrac{2x+5}{x-6} \leq 0$.
    a) $[-\frac{5}{2}, 6]$ b) $[-\frac{5}{2}, 6)$ c) $(-\frac{5}{2}, 6)$ d) $(-\frac{5}{2}, 6]$

65. In the equation $x^2 + y^2 = 25$, is y a function of x?
    a) Yes b) No c) Cannot be determined

66. What is the vertex for $y = \frac{1}{2}(x-4)^2 - 3$?
    a) $(4, -3)$ b) $(-4, -3)$ c) $(2, -3)$ d) $(-2, -3)$

67. Find the quotient and remainder for $(3x^4 - 5x + 2) \div (x - 4)$.
    (a) The quotient is $3x^3 + 12x^2 - 48x + 187$, and the remainder is -746.
    (b) The quotient is $3x + 7$, and the remainder is 30.
    (c) The quotient is $3x^3 - 12x^2 - 48x - 197$, and the remainder is 790.
    (d) The quotient is $3x^3 + 12x^2 + 48x + 187$, and the remainder is 750.

68. What are the intercepts for $y = x^2 + 2x - 24$?
    (a) The x-intercepts are -4 and 6, and the y-intercept is -24.
    (b) The x-intercepts are 8 and 9, and the y-intercept is -24.
    (c) The x-intercepts are -8 and -9, and the y-intercept is -24.
    (d) The x-intercepts are 4 and -6, and the y-intercept is -24.

69. A property manager wants to fence the back of an office building for storage. The side against the building will not be fenced. If 100 feet of fencing is available and if the area to be fenced is rectangular, what is the maximum area?
    a) 1000 square feet b) 1250 square feet
    c) 1500 square feet d) Cannot be determined.

70. The solid graph in Fig. A-5 is the graph of $f(x) = |x|$. The dashed graph is the graph of which function?
a) $y = 3|x|$ b) $y = -3|x|$ c) $y = \frac{1}{3}|x|$ d) $y = -\frac{1}{3}|x|$

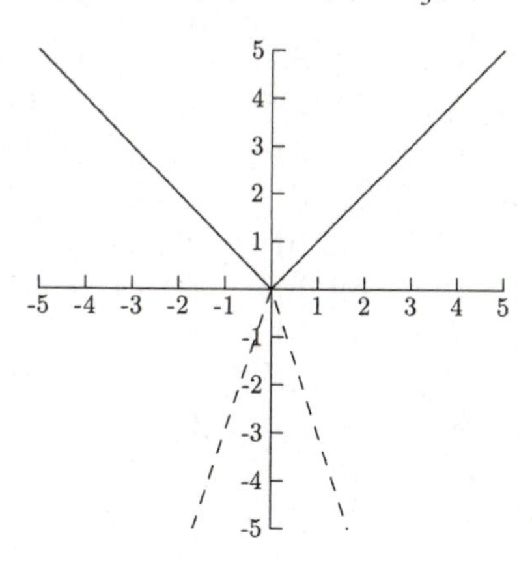

Fig. A-5.

71. State the zeros and their multiplicity for the polynomial function $f(x) = x^2(x+2)^3(x-6)^5$.
(a) The zeros are 2 (multiplicity 1), 3 (multiplicity 2), and 5 (multiplicity 6).
(b) The zeros are 0 (multiplicity 2), 2 (multiplicity 3), and –6 (multiplicity 5).
(c) The zeros are 0 (multiplicity 2), –2 (multiplicity 3), and 6 (multiplicity 5).
(d) The zeros are 2 (multiplicity 3), and –6 (multiplicity 5).

72. Solve for x: $\dfrac{3-x}{x+3} > 5$.
a) $(-3, \infty) \cup (-2, \infty)$ b) $(-3, 3)$
c) $(-\infty, -3) \cup (3, \infty)$ d) $(-3, -2)$

73. Put the quotient $(7-4i)/(1+3i)$ in the form $a+bi$, where a and b are real numbers.
a) $\dfrac{-1}{2} + \dfrac{17}{2}i$ b) $-\dfrac{1}{2} - \dfrac{5}{2}i$ c) $7 - \dfrac{4}{3}i$ d) $\dfrac{19}{10} + \dfrac{17}{10}i$

74. What numbers should be used to fill in the blank for $y = (x^2 - 10x + \_) + 6 + \_$?
(a) Put 25 in the first blank and 25 in the second blank.
(b) Put –25 in the first blank and 25 in the second blank.
(c) Put 25 in the first blank and –25 in the second blank.
(d) Put –25 in the first blank and –25 in the second blank.

75. Evaluate $(f(a+h)-f(a))/h$ for $f(x)=4x-1$.

a) h b) 1 c) $4h+8$ d) 4

SOLUTION

1. b)	2. a)	3. b)	4. d)	5. c)	6. d)	7. a)
8. d)	9. a)	10. a)	11. a)	12. b)	13. b)	14. a)
15. b)	16. d)	17. d)	18. d)	19. c)	20. a)	21. c)
22. a)	23. a)	24. b)	25. d)	26. c)	27. c)	28. a)
29. b)	30. b)	31. d)	32. b)	33. d)	34. b)	35. a)
36. c)	37. a)	38. b)	39. b)	40. c)	41. a)	42. b)
43. d)	44. a)	45. c)	46. b)	47. a)	48. d)	49. a)
50. c)	51. b)	52. a)	53. a)	54. a)	55. a)	56. b)
57. c)	58. c)	59. b)	60. d)	61. c)	62. a)	63. d)
64. b)	65. b)	66. a)	67. d)	68. d)	69. b)	70. b)
71. c)	72. d)	73. b)	74. c)	75. d)		

INDEX

ABOUT THE AUTHOR

Rhonda Huettenmueller has taught mathematics at the college level for over 14 years. Popular with students for her ability to make higher math understandable and even enjoyable, she incorporates many of her teaching techniques in this book. She received her Ph.D. in mathematics from the University of North Texas.